Governing the Air

Governing the Air

The Dynamics of Science, Policy, and Citizen Interaction

edited by Rolf Lidskog and Göran Sundqvist

The MIT Press
Cambridge, Massachusetts
London, England

For information about special quantity discounts, please email special_sales@mitpress .mit.edu.

This book was set in Sabon by Toppan Best-set Premedia Limited. Printed and bound in the United States of America.

Library of Congress Cataloging-in-Publication Data

Governing the air : the dynamics of science, policy, and citizen interaction / edited by Rolf Lidskog and Göran Sundqvist.
p. cm. — (Politics, science, and the environment)
Includes bibliographical references and index.
ISBN 978-0-262-01650-6 (hardcover : alk. paper) — ISBN 978-0-262-51642-6 (pbk. : alk. paper)
1. Air quality management—Government policy—Europe. 2. Air—Pollution—Europe—Prevention—International cooperation. 3. Air—Pollution—Europe—Prevention—Citizen participation. 4. Air quality—Political aspects—Europe. 5. International relations—Europe. I. Lidskog, Rolf. II. Sundqvist, Göran.
TD883.7.E85G68 2012
363.739'2561094—dc22

2011010006

10 9 8 7 6 5 4 3 2 1

Contents

Series Foreword

As our understanding of environmental threats deepens and broadens, it is increasingly clear that many environmental issues cannot be understood, analyzed, or acted upon in a simple way. The multifaceted relationships between human beings, social and political institutions, and the physical environment in which they are situated extend across disciplinary as well as geopolitical confines and cannot be analyzed or resolved in isolation.

The purpose of this series is to address the increasingly complex questions of how societies come to understand, confront, and cope with both the sources and the manifestations of present and potential environmental threats. Works in the series may focus on matters political, scientific, technical, social, or economic. What they share is attention to the intertwined roles of politics, science, and technology in the recognition, framing, analysis, and management of environmentally related contemporary issues and a manifest relevance to the increasingly difficult problems of identifying and forging environmentally sound public policy.

Peter M. Haas
Sheila Jasanoff

Preface

More than 20 years ago, we started to investigate the role of expertise in environmental regulation and policymaking, and we have been involved in researching this topic ever since. With backgrounds as sociologists, we have worked in multidisciplinary settings in which linkages among science, technology, and the environment are studied. This experience gradually led us to bring various streams of research closer together in order to gain more nuanced and deeper insight into the role of science, policy, and citizens in environmental governance.

A decade ago we were involved in the multidisciplinary program International and National Abatement Strategies for Transboundary Air Pollution (ASTA). This program aimed to produce scientific knowledge in support of international measures to control transboundary air pollution in Europe. An important task of the program was to develop policy-relevant science and to increase policymakers' understanding of this environmental problem. Through this program, we had the opportunity to study the interface between science and policymaking in greater detail.

As part of the ASTA program, in 2005 we arranged the international workshop "Toward Robust European Air Pollution Policies: Constraints and Prospects for a Wider Dialogue between Scientists, Experts, Decision Makers, and Citizens." The lively discussions among policymakers, negotiators, natural scientists, and social scientists became the starting point for this book, considering how better to understand the shaping of international environmental governance. The workshop inspired us to continue this discussion from a social science perspective. To that end, we invited scholars from international relations and science and technology studies to contribute their perspectives to forging a better understanding of science–policy–citizen dynamics in international environmental governance. This book is the result of that invitation.

We thank in particular the ASTA program director Professor Peringe Grennfelt at the IVL Swedish Environmental Research Institute for his excellent guidance of the research program and firm commitment to the role of social science in developing knowledge of transboundary air pollution. We also thank Martin Letell—our former Ph.D. student and member of the ASTA social science team, now a researcher at the University of Gothenburg—for stimulating discussions of how to make transboundary air pollution governable.

We presented a draft of the final chapter at the International Sociological Association's 2010 World Congress in Gothenburg, and we thank participants in that session for their constructive comments. We are also grateful to three anonymous reviewers for providing constructive comments on an earlier version of the manuscript. In addition, we extend our thanks to the Foundation for Strategic Environmental Research, the Swedish Research Council for Environment, Agricultural Sciences, and Spatial Planning, and the Swedish Research Council for grants that made our work on this book possible.

Finally, we thank the chapter authors for their productive cooperation and strong commitment to our project and to the Centre for Urban and Regional Studies, Örebro University, and the Centre for Technology, Innovation and Culture, University of Oslo, for providing stimulating and supportive research environments during the completion of this book.

Örebro and Oslo, June 2011
Rolf Lidskog
Göran Sundqvist

Contributors

Leendert van Bree was trained in medicinal chemistry and received a Ph.D. in environmental toxicology. He has more than 25 years of experience in health risk assessment of environmental pollutants, has coordinated a large number of (international) studies, and has completed several advisory tasks for national and international bodies and agencies. His professional interests include environmental quality, health impact and benefit assessments, sustainable and climate-proof cities, and environmental health governance. He is currently senior researcher on health, well-being, and quality of life at the Netherlands Environmental Assessment Agency. He has published widely in the academic literature, particularly in toxicology.

Michael J. Chadwick now retired, has held several senior posts including director of the Stockholm Environment Institute. Professor Chadwick's research interests have included air pollution, climate change, and, in particular, policy and institutional response. His extensive publications include many on the assessment and implementation of air quality policy.

Lisa Emberson is director of the Stockholm Environment Institute's "Managing Environmental Systems" research theme. Her extensive international research focuses on the effects of tropospheric ozone on agricultural yields, forest productivity, and the functioning of terrestrial seminatural ecosystems and aims to provide information relevant for policy formulation.

John Forrester is a research fellow at the Stockholm Environment Institute, University of York, and an affiliate of the York Centre for Complex Systems Analysis, where he works on social systems. His work focuses on the theory and practice of multistakeholder engagement with science policy issues, usually around environmental matters. He currently works on projects using agent-based models and Bayesian belief networks to understand these relationships. He is coauthor of "Facilitating the Local Governance of Air Pollution Using GIS for Participation" in *Applied Geography* (2005) and "Science and Public Participation: The Democratization of Science" in the *Blackwell Encyclopedia of Sociology* (2006).

Peter M. Haas is professor of political science at the University of Massachusetts–Amherst. He has published many books and articles on issues relating to global governance, international relations theory, and the role of science and institutions in environmental governance.

Kevin Hicks is research associate at the Stockholm Environment Institute. His research concentrates on air pollution impacts to terrestrial ecosystems and the transfer of scientific knowledge to policy. He is coauthor of several journal articles on regional policy process and on the need for a greater global perspective in assessment.

Atsushi Ishii is associate professor at the Center for Northeast Asian Studies, Tohoku University. His research interests include global environmental and resource politics, science and technology studies, and Japanese environmental politics. His recent publications include "The Social and Political Complexities of Learning in CCS Demonstration Projects" in *Global Environmental Change* 21 (2011) and "An Alternative Explanation of Japan's Whaling Diplomacy in the Post-Moratorium Era" in *Journal of International Wildlife Law and Policy* 10 (2007).Other articles have appeared in, among other English-language journals, *Climate Policy*, and *Review of European Community & International Environmental Law* as well as in numerous Japanese books and journals.

Johan Kuylenstierna is director of the Stockholm Environment Institute at the University of York. His research concerns climate change as well as regional and global air pollution impacts. He is also concerned with effective communication of scientific information to policymakers to promote effective decision making. He coordinates the bulletin on climate and development, *Tiempo*.

Rolf Lidskog with Ph.D.s in sociology and ethics, serves as professor of sociology at the Centre for Urban and Regional Studies, Örebro University. His research interests include environmental policy and politics at the international and national levels, especially the role of expertise in environmental governance. He is coauthor of the book *Transboundary Risk Governance (*2009) and coeditor of *Consuming Cities: The Urban Environment in the Global Economy after Rio* (1999).

Arthur Petersen is chief scientist in the PBL Netherlands Environmental Assessment Agency and professor of science and environmental public policy at the VU University Amsterdam. He holds doctoral degrees in atmospheric science and philosophy of science and is a visiting professor at the London School of Economics and Political Science and a research affiliate at the Massachusetts Institute of Technology.

Håkan Pleijel is professor in applied environmental science at the University of Gothenburg. He is affiliated with the Department of Plant and Environmental Sciences as a scientist and teacher. His research is focused on the variation in pollution exposure in time and space and its effects on plants. He has been active in expert groups of the Convention on Long-Range Transboundary Air Pollution and is the editor of the book *Air Pollution & Climate Change: Two Sides of the Same Coin?* (2009).

Dieter Schwela is research fellow at the Stockholm Environment Institute. His research is on human exposure and risk assessment, risk-management procedures, and strategies for air quality management, particularly in the countries of Asia and Africa. He is coauthor of the book *Urban Air Pollution in Asian Cities* (2008).

Henrik Selin is associate professor in the Department of International Relations at Boston University. He is the author of *Global Governance of Hazardous Chemicals: Challenges of Multilevel Management* (MIT Press, 2010) and the coeditor of *Changing Climates in North American Politics: Institutions, Policy-making, and Multilevel Governance* (MIT Press, 2009) and *Transatlantic Environment and Energy Politics: Comparative and International Perspectives* (2009). In addition, Selin is the author and coauthor of more than two dozen peer-reviewed journal articles and book chapters as well as numerous reports, reviews, and commentaries.

Bernd Siebenhüner is professor of ecological economics at Carl von Ossietzky University of Oldenburg. He is deputy leader of the Global Governance Project. His research interests are, among others, international organizations, global environmental governance, social learning, corporate sustainability strategies, climate, and biodiversity governance. Key publications include "Learning in International Organisations in Global Environmental Governance" in *Global Environmental Politics* 8(4) (2008) and "How Do Scientific Assessments Learn?" in *Environmental Science & Policy* 5 (2002). He is also coeditor of a special issue of *Ecological Economics* 53 (2005) on biodiversity conservation, access and benefit sharing, and traditional knowledge.

Julie Simon is a part-time researcher at the University of York's Environment Department, where she is studying science–policy networks and completing her Ph.D., which uses social network analysis to investigate the interface between science and policy and focuses on the parallel areas of European air pollution and biodiversity policy.

Carolyn Snell is lecturer at the Department of Social Policy and Social Work at the University of York, where her research interests include environmental policy, local sustainable development, public participation, and environmental justice. She is coauthor of several recent publications on local air pollution and policy networks.

Casey Stevens is a Ph.D. candidate in political science at the University of Massachusetts–Amherst. He is currently working on a dissertation that focuses on the idea of ecological resilience and its political and social causes and consequences.

Göran Sundqvist is professor of science, technology, and culture at the Centre for Technology, Innovation, and Culture at the University of Oslo and senior research fellow of science and technology studies at the University of Gothenburg. His research interests include the sociology of expertise, the science–policy interface, and environmental regulation. He is author of *The Bedrock of Opinion: Science, Technology, and Society in the Siting of High-Level Nuclear Waste* (2002) and coeditor of the special issue *Radioactive Waste Management: Technocratic Dominance in an Age of Participation* of the *Journal of Integrative Environmental Sciences* (2010).

Willemijn Tuinstra is assistant professor of environmental policy at the Open University in the Netherlands. She is also advisor on science–policy interactions for the Netherlands Environmental Assessment Agency. She was previously involved in work for the European Environment Agency, Wageningen University, and the International Institute for Applied Systems Analysis, focusing on linkages between air pollution and climate change, uncertainty management, stakeholder involvement, and other aspects of integrated assessment. She holds a Ph.D. in environmental sciences from Wageningen University.

Stacy D. VanDeveer is associate professor of political science at the University of New Hampshire. His research interests include international environmental policymaking and its domestic impacts, the connections between environmental and security issues, and the role of expertise in policymaking. In addition to authoring and coauthoring more than 50 articles, book chapters, working papers, and reports, he has coedited several books, including *Comparative Environmental Politics* (MIT Press, forthcoming); *Changing Climates in North American Politics* (MIT Press, 2009); *Transatlantic Environment and Energy Politics* (2009); *EU Enlargement and the Environment* (2005); and *Saving the Seas* (1997).

Jørgen Wettestad is senior research fellow at the Fridtjof Nansen Institute in Oslo. His main research interests are energy in the European Union; environmental and resource politics, specifically climate politics at the domestic, European, and global levels; the design and functioning of emissions trading systems; and regime theory, specifically questions related to the effectiveness and design of international environmental institutions, including the science–politics relationship. His most recent book is *EU Emissions Trading: Initiation, Decision-Making, and Implementation* (coauthored with Jon B. Skjærseth, 2008).

Steven Yearley is professor of the sociology of scientific knowledge at the University of Edinburgh and is currently director of the Economic and Social Research Council Genomics Forum there. He works on the sociology of science and environmental sociology and is well known for his studies (with colleagues at the Stockholm Environment Institute) of public participation in urban air quality modeling.

1

Transboundary Air Pollution Policy in Transition

Rolf Lidskog and Göran Sundqvist

Climate change, *transboundary air pollution*, *urban air quality*, *greenhouse gases*, *sulfur dioxides*, and *fine particles*—a cluster of related words is circulating in news reports, political discussions, and public debates. These words describe the importance of clean air for human health, a livable society, and a sustainable environment. But what is clean air, or, rather, what makes air polluted? And when pollutants are traced and explored, what kinds of actions should then be taken? How can the air be rendered governable? To what extent and in what way is it possible to steer society, regulate activities, abate pollution, and control emissions? And what authorities have the capacities to exert control in a globalized and fragmented world?

Simple expressions such as *transboundary air pollution*, *ozone depletion*, and *climate change* raise a range of issues that concern the understanding of air pollution and society. This book's point of departure is that governing the air entails a process in which boundaries between society, science, and nature are intermingled and constantly renegotiated. Earlier demarcations are transgressed, and new ones configured. Changes in scientific understandings and political organizations influence each other and often do so in unforeseeable ways.

The book's empirical focus is transboundary air pollution. This environmental problem has a fairly long regulatory history, not least in Europe, where a United Nations (UN) convention, the Convention on Long-Range Transboundary Air Pollution (CLRTAP), was established in the late 1970s. Many commentators consider this convention to have been very successful. As such, it has been of great interest to social scientists, international relations (IR) scholars in particular, seeking to understand and explain what has been achieved and why. However, this success story of air pollution regulation currently faces a number of challenges. The organizational context has recently changed due to the

European Union's (EU) strengthened political role and its development of an air policy for all its member states. Scientific advances have led to a more complex and dynamic view of ecosystems. There has also been a gradual shift in focus from aquatic and terrestrial ecosystems to human health; this shift means that epistemic networks, established concepts, and developed models may become obsolete—or at least less relevant.

The 1980s were the heyday of transboundary air pollution as a public issue, and the term *acid rain* was on everyone's lips. Since the mid-1990s, however, this environmental problem has gradually been eclipsed by climate change. Nevertheless, because both transboundary air pollution and climate change concern atmospheric processes and substances emitted by the burning of fossil fuels, activities are under way to connect the two regulatory frameworks to achieve synergetic effects (Pleijel and Grennfelt 2007). Underlying these activities is the belief that the regulation of transboundary air pollution can gain a new mission by transferring its experience of success to the context of a less successful international regime.

Air pollution policy is in transition and can develop in various directions. It is therefore useful to assess critically how various actors have developed strategies to make the governing of the air more socially robust, politically viable, and environmentally effective. This book's focus is not confined to the limited area of air pollution. Current debate in this field provides ample opportunity to analyze the relationships among science, policy, and the public in international environmental governance in general, making a theoretical contribution to the academic field of IR. As in many other environmental areas—not least that of climate change—science and expertise are given a pivotal role in developing effective abatement strategies.

This book has two objectives: first, to evaluate international air pollution policy critically and, second, to improve the theoretical understanding of the dynamics among science, policy, and citizens. To achieve the first objective, it is necessary to place the regulation of air pollution in its political and scientific contexts, explain its historical development, and relate it to other relevant environmental issues. The empirical focus is primarily air pollution in Europe, but regulatory efforts in other parts of the world are also considered. The second objective aims to deepen our understanding of the links among science, policy, and citizens in international environmental governance. By discussing and combining approaches and findings from the academic fields of IR and science and technology studies (STS), we deepen our understanding of science–

policy–citizen relationships and consequently of what makes an environmental regime effective. This book is based on the belief that the new political complexity, new scientific understandings, and recent theoretical developments in social science concerning international environmental governance need to be exploited. This book aims to cross-fertilize the fields of IR and STS in light of the new complexity in European air policy, based on 40 years of practical regulatory work and 30 years of social scientific studies of the experience gained.

The chapters in this volume approach these objectives from various angles: reviewing earlier findings, analyzing the current situation, and discussing future development paths. Instead of subordinating all the chapters to a single theoretical approach, this book uses various theoretical perspectives with the explicit aims of more deeply analyzing the underlying drivers of the development of international air pollution policies and improving our theoretical understanding.

The book is structured around three related themes: policy and institutions, expertise and learning, and citizens and involvement. As demonstrated in the chapters, it is misleading to regard these themes as discrete categories. Institutions obviously have to do with knowledge and learning as well as with public participation, and the use of expertise and public participation is related to policy. In fact, a guiding theme here is that processes in disparate fields greatly influence each other. At the same time, these three themes highlight some crucial aspects of the current understanding of environmental problems in general and of transboundary problems in particular.

Before turning to the individual chapters, we elaborate on the current context of air pollution policy—in particular on changes in scientific understanding, policy development, and social science insights.

Transboundary Pollution and Spatial Effects

Today, climate change is at the top of the environmental agenda. Media reports attribute current weather conditions to climate change; scientists develop tools for projecting future consequences; and political representatives ponder how this problem can be made politically manageable—that is, how space can be created for concerted political action. A fundamental belief of those involved in combating climate change is that the biosphere does not recognize political boundaries, so international cooperation is indispensible. This understanding is in no way restricted to the climate-change issue. Most environmental problems rest on the understanding that pollutants cross boundaries and that previous ways of distinguishing between spatial scales—local, national, regional, and global—have been superseded. Local emissions may have global consequences; global activities may have local consequences; and emissions from a nation-state may mainly affect other nation-states.

What now seems a commonplace notion—that many environmental problems have a transboundary nature—is not given by nature but something that has been historically established. However, the paradigmatic example is neither today's discussion of climate change nor the long-range transport of sulfur dioxides discovered in the late 1960s. It was earlier, in the wake of the Second World War, that the transboundary nature of pollutants was first placed at the center of international political concern.

The atomic bombing of Hiroshima and Nagasaki in 1945 and the atmospheric nuclear bomb testing in the 1950s made radiation a key matter of international concern and regulation. Radioactivity was eventually found to have genetic effects that transcend the temporal aspects of a single human life. More important, however, the effects of nuclear bombs were not restricted to the targets in that radioactive fallout was spread far and wide by winds. The notion of the "mist of death" or "death dust" encompassed the understanding that a nuclear bomb and even bomb testing can kill at a distance, far from the immediate circle of destruction, creating a worldwide "A-bomb disease" (Weart 1988, 199; Sundqvist 2002, 54). Since that time, the atmosphere has been understood as both a global commons and a carrier of air pollution; a forceful reminder was given with the Chernobyl disaster, which spread radioactivity over most of Europe in 1986.

Since the Second World War, the fear of radioactivity has been a key driver of the development of our understanding of the long-term transportation of pollutants. Meteorology and atmospheric chemistry became important scientific disciplines, and researchers specialized in measuring small amounts of invisible pollutants, tracing them from their sources to their eventual disposition via atmospheric transportation over long distances. Today, this understanding of pollutant patterns is widespread in society, and it is generally accepted that environmental degradation and human health effects are caused at least in part by emissions from sources far away. Pollutants travel in land, water, and air; however, these flows do not have uniform effects, and pollutants are deposited at various places around the globe. The consequences of this deposition are also differentiated: ecosystems differ in their sensitivity to pollutant uptake, and societies differ in how they are affected and have different capacities for resilience.

Many environmental problems have both transboundary and spatial characteristics, and the consequences of long-range air pollution are spatially distributed in an uneven way. Some nation-states are net exporters of airborne pollutants, whereas others are net importers, and some will be more severely hurt by a changing regional or global atmosphere. The discovery of long-range air pollution did not result in the dissolution of spatial boundaries, though. Instead, it meant the transgression of earlier demarcations between local, national, and international environmental problems and the establishment of new ones. A new political geography of winners and losers was created, and new and different incentives for international cooperation were established.

The Discovery of Transboundary Pollution

In the late 1960s, the long-distance travel of sulfur dioxides from local sources, together with its transnational effects, was discovered. Acidification of freshwater and its effects on fish stocks were connected to airborne pollutants that traveled from the United Kingdom and continental Europe to Scandinavian rivers, lakes, and forests (Lundgren 1998). Europe was suffering not only from a Cold War, but from tacit, creeping, and unintentional chemical warfare with a new political geography—not between the North Atlantic Treaty Organization (NATO) and the Warsaw Pact, but between net exporters and net importers of air pollutants.

Although the emergent understanding of radioactive dust had already paved the way, new insights into the transboundary character of airborne pollutants resulted in a shift in the understanding of the environment. This shift became visible in science, environmental consciousness, and political cooperation. Scientific investigations began to focus on detecting airborne pollutants, following them from source to effect, and on learning about chemical transformations in the atmosphere.

There were initially some decades of naive hope that pollutants would be diluted to harmlessness in the atmosphere after release through tall smokestacks. Environmental consciousness slowly became established, however, encompassing an understanding that the effects of local sources have a new temporal–spatial pattern: delayed in time and extensively spread out in space. Due to the spatial characteristics of long-range air pollution, initiatives for multilateral political cooperation were undertaken. In Europe, it was discovered that transboundary air pollutants constituted a threat to a common European atmosphere (Letell 2002, 25–27; VanDeveer 2004, 313; Sundqvist, chap. 7 in this volume). The transboundary nature of air pollution supported cooperation, but it also engendered competition and conflict between nation-states.

At the first international political meeting on environmental issues, the UN Conference on the Human Environment in Stockholm in 1972, a declaration was agreed upon. One of its best-known principles says, "States have, in accordance with the Charter of the United Nations and the principles of international law, the sovereign right to exploit their own resources pursuant to their own environmental policies, and the responsibility to ensure that activities within their jurisdiction or control do not cause damage to the environment of other States or of areas beyond the limits of national jurisdiction" (UN 1973, Principle 21).

The transboundary nature of many emissions into water and air implies the violation of this principle and indicates that no nation-state has complete sovereign rule over its territory. The adoption of this principle placed the transboundary nature of air pollution at center stage, and the task it indicated was how to regulate this kind of pollution. But this task is not easy to complete. Whereas the victim countries—those that are ecologically vulnerable to specific problems—are obviously motivated to foster international agreements to reduce the emissions at their sources, the main producers of the pollutants are less motivated. There seems to be a paradox that the stronger the need for concerted action, the harder it appears to be to achieve it; when mutual action is most needed, the mutual benefits are not obvious. It is important to note, however, that this apparent paradox seems not to be generally applicable. In the case of efforts to deal with transboundary air pollution beginning in the 1970s, it was possible to see beyond short-term winners and losers and to develop concerted actions and effective abatement strategies. But how was this cooperation accomplished?

Regulating Transboundary Air Pollution

In the late 1960s, Scandinavian researchers argued, contrary to earlier beliefs, that air pollutants can travel several thousand kilometers before deposition and damage occurred. This assertion made the burning of fossil fuels for transportation, heating, and electricity production not only a local problem, but also an international one. Even though there was little doubt as to the acid output from power plants based on fossil fuels, the idea of long-range air pollution was highly controversial among both researchers and politicians in Europe in the early 1970s. However, this situation changed, and in a process involving scientific development and political initiatives an international UN convention was adopted in 1979: the CLRTAP. This convention's aim was and is to reduce emissions of airborne cross-boundary pollutants within the UN Economic Commission for Europe (UNECE) region. Thirty-two parties—29 European nation-states, the United States, Canada, and the European Community—signed the convention, which was the first multilateral instrument addressing atmospheric environmental issues (Munton, Sooros, Nikitina, et al. 1999, 167). The CLRTAP was later followed by protocols regulating specific substances (see table 1.1). Today, 51 parties have ratified the convention.

CLRTAP is one of the oldest international environmental conventions, and several researchers have analyzed its formation and development. Their descriptions differ somewhat in emphasis but agree in positively assessing the convention, which, compared with other environmental regimes, is considered a success. The historical account given in these studies unanimously emphasizes the role of science in developing the convention (Sooros 1993; Hajer 1995a; Munton, Sooros, Nikitina, et al.

Table 1.1
CLRTAP, Its Protocols, and Where They Were Signed (Year of Entering into Force)

1979	Convention on Long-Range Transboundary Air Pollution (1988)
1984	Protocol on Long-Term Financing of the Cooperative Program for Monitoring and Evaluation of the Long-Range Transmission of Air Pollutants in Europe (EMEP Protocol), Geneva (1988)
1985	Protocol on the Reduction of Sulfur Emissions or Their Transboundary Fluxes by at Least 30 Percent (Sulfur Protocol), Helsinki (1987)
1988	Protocol Concerning the Control of Nitrogen Oxides or Their Transboundary Fluxes (NO_x Protocol), Sofia (1991)
1991	Protocol Concerning the Control of Emissions of Volatile Organic Compounds or Their Transboundary Fluxes (VOCs Protocol), Geneva (1997)
1994	Protocol on Further Reduction of Sulfur Emissions (Second Sulfur Protocol), Oslo (1998)
1998	Protocol on Heavy Metals, Århus (1998)
1998	Protocol on Persistent Organic Pollutants (POPs Protocol), Århus (2003)
1999	Protocol to Abate Acidification, Eutrophication, and Ground-Level Ozone, Gothenburg (2005)

1999; Wettestad 1999; Underdal and Hanf 2000; Tuinstra, Hordijk, and Kroeze 2006).

The CLRTAP has become more complicated over time. Abatement strategies were initially negotiated based on the flat-rate reduction of emissions of pollutants (i.e., sulfur dioxides and nitrogen oxides), meaning the same emission cuts for all countries. These strategies were later called the "first-generation protocols." In the 1990s, the development of a second generation of protocols started, resulting in the Protocol on Further Reduction of Sulfur Emissions (Second Sulfur Protocol, 1994) and the Protocol to Abate Acidification, Eutrophication, and Ground-Level Ozone (multipollutant/multieffect protocol, 1999). These protocols focused on varying national reduction rates based on the approach of critical loads—that is, effects in relation to what nature can withstand—and cost effectiveness. They also covered more substances. The multipollutant/multieffect protocol was signed in Gothenburg in November 1999 and regulates four types of compounds (i.e., sulfur dioxide, nitrogen oxides, ammonia, and volatile organic compounds) that affect human health, natural ecosystems, materials, and crops through acidification, eutrophication, and ground-level ozone (UNECE 1999, Article 2). The signatories have called it "a smart protocol," one of the most sophisticated and most scientifically based protocols ever signed (Thompson 1999).

However, the negotiation processes connected with the protocols have not run completely smoothly. Several diverging standpoints and conflicting interests became apparent, not least in the former Federal Republic of Germany and the United Kingdom (Boehmer-Christiansen 2000; Sprinz and Wahl 2000). Of note, in describing this process, researchers have regarded the overcoming of these conflicts as a token of CLRTAP's success. Consensual science is often singled out as the key factor in reaching agreements; for example, international scientific work is said to have been crucial in persuading British decision makers of the need for further international control measures (Hajer 1995b; Hanf 2000). Thus, from the late 1960s to the early 1980s acid rain was transformed from an esoteric branch of scientific research in certain specialized fields of atmospheric chemistry and ecology into a household word. Science played a decisive role in both developing the regulatory regime and making it feasible.

However, not only science but also East–West diplomacy was important for the establishment of CLRTAP. In fact, diplomacy accounts for the important role given to science within the regime (Lidskog and Sundqvist 2002). Politicians tried to mitigate the East–West conflict by proposing cooperation in the field of environmental protection. The context of CLRTAP's creation was the stalemate between the NATO states and the Warsaw Pact in Europe. The Soviet Union, despite its earlier boycott of the UN Conference on the Human Environment in Stockholm in 1972, initiated discussions of international cooperation on environmental issues at the Helsinki Conference on Security and Cooperation in Europe in 1975. Air pollution was soon chosen as the specific area of cooperation, and CLRTAP became an important common social project for European states across the Iron Curtain, acknowledging long-range transport of air pollutants as a serious environmental problem. The politicians' search for politically less controversial issues on which to cooperate was an important factor in explaining the scientific nature of the regime. This emphasis has meant the scientization of the European air pollution problem. Policymakers demanded scientific cooperation and saw it as a key factor in the establishment of the CLRTAP regime, and scientists accepted an expanded role for science in environmental governance, which gave them considerable room for technocratic maneuvering.

As a concrete result of the political support for scientific cooperation, several monitoring stations were set up in European countries in the 1970s, and emission data were exchanged. In 1977, the European Monitoring and Evaluation Programme (EMEP, officially the Cooperative Programme for Monitoring and Evaluation of the Long-Range Transmission of Air Pollutants in Europe) was established. EMEP enabled the development, support, and coordination of national monitoring programs, thereby facilitating the evaluation of the emission-reduction protocols.

EMEP was integrated into CLRTAP and became a mechanism for producing, distributing, and exchanging information on pollution flows, effects, and mitigation options. EMEP started in the 1970s, and the first protocol under the CLRTAP, signed in 1984, concerned maintaining financial support for EMEP. Today, EMEP encompasses some 100 stations in Europe, monitoring sulfur dioxide, nitrogen dioxides, ground-level ozone, and other substances. A channel for exchanging standardized scientific information and empirical data was established, enabling the growth and spread of a common knowledge base concerning both the seriousness of the acid rain issue and ecosystem mechanisms.

Current Situation: Growing Political and Scientific Complexity

Since the mid-1990s, the political organization of air regulation has changed in Europe, largely due to initiatives taken by an expanded EU via the European Council and the European Parliament. The EU has adopted the Air Quality Framework Directive (1996); presented an acidification strategy (1997); issued the National Emission Ceilings Directive (2001); launched a process for developing a thematic strategy for air quality, Clean Air for Europe (CAFE) (2001); and adopted a new thematic strategy on air pollution (2005).

The EU's thematic strategy on air pollution and the UN's CLRTAP share a number of similarities in how they conceptualize the problem. They also differ in a number of ways, not least concerning organizational structure, geographical scope, and ways of organizing the science–policy relationship (Selin and VanDeveer, chap. 3 in this volume). This means that various political organizations are managing the same problems; linkages and overlaps exist but are not often made explicit or critically discussed. Furthermore, air quality and climate change are becoming increasingly interlinked, and policies developed for abating one problem will influence the other.

As a political body, the EU is much stronger than CLRTAP. The EU's established directives on air quality standards for several pollutants are legally binding on all its member states. In contrast, CLRTAP relies on initiatives from participating countries and has no authority to determine binding regulations or to use sanctions should any party fail to honor the commitments. In this situation, science has served as an important policy support in the development of agreed-on abatement strategies. Scientific networks were formed and gradually became formally adopted as integral parts of the CLRTAP organization.

The scientific understanding has also changed. It is not ecological damage but human health that is today considered the most urgent problem connected with transboundary air pollution (Sliggers 2004; Pleijel and Grennfelt 2007). Studies indicate that air pollution contributes significantly to a wide range of acute and chronic health problems (Kunzli and Tager 1999). The most severe effects relate to an increased risk of premature death, expressed by the estimated 3.6 million life years lost through some 348,000 annual deaths attributable to current particulate matter exposure in the EU (WHO 2006, 95). Air pollution also results in a number of less severe health consequences. Exposure to particulate matter, for example, results in 100,000 hospital admissions per year in the EU due to respiratory and cardiac emergencies (WHO 2006, 95).

When people's own health is at stake, they become more motivated to take action and support further abatement strategies. The EU Air Quality Framework Directive explicitly focuses on air quality in big cities, and "problem cities" such as Athens and London are singled out (Nikolaou 2003; EC 2008). Dying fish stocks and acid lakes, located in rural areas in peripheral Scandinavian countries, are no longer a strong driving force of citizen and politician engagement in continental Europe to combat air pollution.

The European Commission's focus on air quality can also be understood in conjunction with the project of European integration (Sundqvist and Letell 2005). What is more of a shared resource than the air we breathe? Air quality can be identified not only as a local/urban problem but also as a pan-European problem, which makes it an ideal integrative project. From being an issue located on the boundary between East and West during the Cold War era and politically supported as a possible communicator over that boundary, air pollution is today once again part of the macropolitical project to support European integration. This time, however, the political context is not two political blocs locked in a cold war, but nation-states and citizens reluctant to embrace the EU's integrative ambitions. European regulatory efforts to control air pollution are in a formative phase, and the interactions among science, policy, and citizens seem pivotal in determining how they will develop.

Science, Policy, and Citizens

The new political geography caused by transboundary air pollution is not inherent in nature but has been constructed through social processes (Beck 1995; Yearley 1996; Lidskog 2004). Some actors may try to advance claims about the transboundary nature of the problem, thus changing the opportunity structure for political action. Others may try to prevent an issue from being perceived as global or regional, or they may try to weaken the resolve to address an issue that is already on the international agenda. These attempts are made because the regulation of certain environmental problems goes right to the heart of domestic policies concerning, for example, energy, transport, and employment. There may be reasons other than environmental or scientific ones for wishing to see certain environmental issues treated either as matters of international priority or as matters solely within the scope of domestic policymaking.

As in all political areas, it is not the object per se, but those who speak for it or claim to represent it who are the driving causal agents in the political process. Hence, there is a need for a social scientific explanation of why biodiversity conservation and climate change are framed as global issues, acidification and marine pollution as regional issues, and freshwater quality and household wastes as local issues (cf. Martello and Jasanoff 2004).

By negotiating and constructing boundaries among the global, regional, national, and local levels, an identity is shaped between a specific environmental problem and a specific spatial level (Lidskog, Soneryd, and Uggla 2009). An image of a geographical location at risk is constructed and becomes the object of action. An important task for social science is to analyze the configuration of a problem as a socially and politically shaped process of image building. But what actors have the power to identify environmental issues, construct their spatiality, and thereby propose the kinds of measures and cooperation that are important to agree on? And what institutions limit and enable these actors' maneuvering room?

Three Approaches: Institutions, Networks, and Co-production

The most common way to understand international environmental governance from a social science perspective is the *institutional approach*. This approach is based on rationalistic assumptions rooted in exchange theory, which says that when interest-based nation-states see opportunities for mutual benefit, international cooperation will occur. International institutions are therefore created and shaped by the tension between expectations of political benefit and the costs of achieving them (Keohane 1988). International cooperation is a possible result of nation-states' negotiations of benefits and costs. When costs are low and benefits high, cooperation can be expected to happen.

In general, an institution is understood as a "set of rules, decision-making procedures, and programs that define social practices, assign roles to the participants in these practices, and guide interactions among the occupants of individual roles. . . . Institutions that deal explicitly with environmental or resource issues are commonly known as environmental or resource regimes" (Young 2002, 5). An international regime refers to "sets of implicit and explicit principles, norms, rules, and decision-making procedures around which actor expectations converge in a given area of international relations" (Krasner 1983, 2). Such institutions or regimes result from negotiations between interest-driven parties and create consensus and cooperation that go beyond the involved parties' interests.

This means that the institutional approach studies regimes, which in this context are certain kinds of institutions that deal with environmental issues. According to Oran Young (2002), a leading proponent of the institutional approach, a regime can be studied as "thin" or "thick"—that is, as rules on paper or as rules in application. The distinction is between what is formal and what is informal, explicit and implicit. In the institutional approach, the three most crucial questions are, according to Young (2002, 11–12): (1) How important is a specific regime for the condition of a specific natural resource, and what difference does it make? (2) How effective is the regime in relation to determined criteria, such as sustainability, efficiency, or equity? and (3) How can regimes be designed or redesigned to improve the environment? Young argues that the approach is still very heterogeneous, that results from studies of specific environmental issues are difficult to compare, and that the cumulative aspect of the tradition—its progress—could be much better.

Whereas the institutional approach has traditionally emphasized the role of the interest-driven nation-state, assuming that its preferences and interests are stable and given, the *epistemic-community approach*, which provides a foundation for understanding international negotiations and cooperation, places a greater emphasis on science. According to Peter M. Haas, the leading proponent of the latter approach, environmental regimes are driven not only "by state power, but by the application of scientific understanding about ecological systems to the management of environmental policy issues with which decision makers are unfamiliar" (1997, 200). Therefore, knowledge—consensual and trusted—is a requirement for successful environmental cooperation.

The concept of the "epistemic community" has been introduced as an analytical tool for investigating the role of science in the formation and development of regimes. An epistemic community is a knowledge-based transnational network of professionals holding political power through cognitive authority (Haas 1992, 3). Members of an epistemic community can come from various disciplines but share a body of basic knowledge according to which they interpret phenomena in a similar way. They also share values and policies—that is, convictions concerning how to enhance human welfare.

The epistemic-community approach is also based on a rational and interest-based foundation, but, instead of exchanging and competing nation-states, we find epistemic communities that work on a consensual basis. According to this approach, epistemic communities are strong vehicles for achieving international cooperation and establishing institutions that go beyond nation-states' interests. As in the institutional approach, the results from this approach are international institutions designed for governing specific environmental problems.

Experts' extensive involvement has contributed to the notion that regimes serve as institutions for learning that in turn produce convergent state policies (Haas 1989, 377; cf. Social Learning Group 2001). This expert influence is strengthened by the transnational nature of epistemic communities: scientists, through the power of their network, are able to resist political temptations to subordinate their advice to existing national concerns. What explains international cooperation is not political benefits but adaptation to scientific understanding.

Whereas the epistemic-community approach focuses strongly on international networks of scientific experts and on scientific consensus as a necessary condition for political and practical achievements, the STS approach directs attention to the preliminary character of knowledge and the fact that knowledge is always locally embedded.

The *STS approach* forcefully argues against a linear model of the relationship between science and policy, in which the role of science is that of "speaking truth to power" (Haas and Stevens, chap. 5 in this volume). Instead of assuming a strong separation between science and policy, STS scholars advocate a view in which scientific knowledge and political order are seen as two sides of the same coin: science and policy are co-produced (Wynne 1996; Jasanoff 1996a, 2004b; Jasanoff and Wynne 1998).

In contrast to the first two approaches, the STS approach was not developed in the IR research field. However, it has gradually gained a foothold in this field, not least due to its sophisticated way of understanding expert knowledge and its connections to political power. Some IR studies are influenced by findings from STS, and some STS studies focus on the international regulation of environmental problems (e.g., Jasanoff 1996b; Social Learning Group 2001; Lidskog and Sundqvist 2002). The STS approach's potential is still underexploited in IR research, in partly because researchers in this field are unaware of it and in part because STS researchers have not in any deeper sense taken advantage of most knowledge developed in the IR field.

The STS approach is more relativistic, sociological, and actor oriented, questioning some basic assumptions of the institutional and epistemic-community approaches—for example, that interests and preferences are inherent, that international cooperation is the result of interest-based negotiations among nation-states, and that scientific knowledge is consensual. On the contrary, STS argues that preferences, interests, and knowledge are unstable and changing. Science is often uncertain, controversial, and interpreted differently by different actors (Sundqvist, chap. 7 in this volume). Knowledge and interests are not factors that explain social institutions, but phenomena that need to be explained. Moreover, actors' various strategies and learning processes lead to changes in preferences, interests, and knowledge; therefore, these latter phenomena cannot provide a strong foundation for explaining international cooperation.

The STS approach focuses on knowledge as something contingent and on science and policy as co-produced. However, compared with both the institutional and the epistemic-community approaches, it provides a less coherent framework and is therefore less capable of evaluating the development of international cooperation. The co-production thesis, however, might be a starting point for developing a stronger STS-inspired theoretical position in the field of IR research. Co-production means that policy influences the production and stabilization of knowledge and that knowledge simultaneously supports and justifies policy; actors may therefore accept a particular knowledge claim because it supports their policy strategies. A specific understanding of nature can be used to solve social problems, and new social orders can create new understandings of nature. In this way, causes and effects become functionally interrelated—that is, knowledge production is also policy production, and vice versa. Co-production means that uncertain or contested science can grow stronger if the policy context is "right," or a weak policy context can become stronger through the support of science (Jasanoff 2004b).

The STS approach emphasizes the role social conditions play in enabling science to acquire the power to influence and shape policy. Science and policy are always interdependent, which implies that science can shape regimes if social conditions are favorable (cf. Haas and Stevens, chap. 5 in this volume). Because of the interdependence of science and policy, the role of science in environmental policy can never be prescribed but is instead open to investigation. There is no other way to understand how transboundary air pollution functions in society than to investigate various actors' articulations, strategies, and practices.

Some important conclusions can be drawn from the STS approach; for example, scientific knowledge is not an independent, fixed entity entering political negotiations from the outside but should be understood as dependent on political processes. As Sheila Jasanoff puts it, "Scientific development is tied to society at all points" (2004a, 16). Nor is the boundary between science and policy fixed: it changes and is negotiated among actors. However, in everyday understanding the boundary between science and policy is often considered stable. Therefore, an important objective for STS scholars is to make visible what is invisible—the fluid and changing pattern of the interdependence of science and policy. To do this, they have to identify the political connections of scientific knowledge.

Converging Approaches?

All three approaches have drawn criticism. The institutional approach has been criticized for its strong focus on interest-oriented nation-states and for not giving science a distinct role in international negotiations; the epistemic-community approach has been criticized for ascribing too much power to science, giving a pivotal role to consensual knowledge; and the STS approach has been criticized for focusing primarily on the microlevel of the scientific enterprise and neglecting everything but actors. The existence of all three approaches, all trying to understand the same object, creates good opportunities for theoretical development, and both controversies and cross-fertilization can be noted between them (Social Learning Group 2001; Lidskog and Sundqvist 2002; Haas 2004; Young 2004). This debate obviously has modified the positions of proponents of the three approaches, leading to convergence in certain respects or at least to overlapping emphases.

Today, all three approaches suggest that improving our understanding of international environmental governance calls for closer scrutiny of the roles of expertise and learning, citizens and stakeholders, and institutions. All three stress that international regimes can stimulate processes of social learning, affecting social identities and changing discourses. This means that regimes not only change a nation-state's external environment but also facilitate learning and define roles for actors. A specific regime may foster new ways of perceiving a problem and its effects and of successfully coping with it (Young 1999; 2008). This is a more dynamic view of how to understand the internal mechanisms of international regimes than was first developed by the institutional approach, which argued that international regimes' performance can be reduced to the interests of nation-states. The dynamic perspective is an important distinctive mark of STS but is now taken into account in all three approaches.

Furthermore, the three approaches focus on knowledge and on the interplay between science and policy, which is considered crucial in international environmental governance. Their emphasis is on the interaction between science and policy, with causal powers not being attached exclusively and a priori to either of them, even though the institutional approach tends to focus on the importance of policy and the epistemic-community approach on science. All three approaches are interested in assessing factors that make transboundary environmental problems governable and want to investigate such questions as: How are scientific integrity and political involvement balanced, and how does this balance affect the credibility of science and the legitimacy of political decisions? What configurations of science–policy relationships have developed and with what results? Do differences between international organizations in terms of their political and expert cultures significantly influence abatement strategies' effectiveness and legitimacy? We firmly believe that, when exploring these questions, more pertinent answers can be achieved if perspectives and findings from all three approaches are considered.

The Missing Link: Citizen Involvement

Alongside the focus on the relationship between science and policy in international environmental governance, there is growing emphasis on the relationships between science and the public. This emphasis is most energetically cultivated by the STS approach (Irwin and Michael 2003; Jasanoff and Martello 2004; Jasanoff 2005; Yearley 2006), though it is still underdeveloped. A broader, critical focus on citizen involvement in science-based issues is a strong trend more generally in today's social science, supported by the work of scholars such as Ulrich Beck, the governance approach, and neo-Foucauldian work on governmentality. This trend has exerted an important influence, leading to new understandings of environmental policymaking—for example, as governance (Hempel 1996; Haas 2008), "subpolitics" (Beck 1992, chap. 8), or advanced liberal steering mechanisms (Barry, Osborne, and Rose 1996; Rose 1999).

Both policy and science have recently claimed that public inclusion is a requirement for creating policies that are socially robust, politically legitimate, and trustworthy in the eyes of the public (see, e.g., EC 2001; Nowotny, Scott, and Gibbons 2001; Renn 2008). Although citizens' role in environmental governance has been described for some years now, the meaning and implications of public inclusion remain scantily researched and poorly understood (Gouldson, Lidskog, and Wester-Herber 2007), in part because public involvement in regulating complex issues such as biotechnology, climate change, and transboundary air pollution has not yet been widely adopted (Yearley, chap. 9 in this volume). Many decision makers still ponder how to reconcile the growing complexity of an issue with the need for public awareness, inclusion, and engagement.

So how should citizen involvement be integrated into international environmental governance? Some see citizens as contributing agents, whereas others see them as passive accepters of reached agreements who provide the agreements with posteriori legitimacy. Many argue that citizens interpret issues based on preferences, interests, and knowledge that differ from those of scientific experts and political representatives (Jasanoff and Martello 2004; Corell and Sundqvist 2005; Leach, Scoones, and Wynne 2005).

At the same time that the complexity of many environmental issues is increasing, there is a growing need to garner public support for proposed courses of action. The European Commission views this situation as indicating a crisis in the credibility of expertise and public unease with science. The commission's white paper on governance argues that the credibility of expertise must be high; otherwise, regulation will be in serious trouble (EC 2001). Because regulatory policymaking is considered an important feature of European integration as such, the whole EU project might lose its legitimacy if the credibility of expertise is considered too low (Sundqvist 2003). The commission has set up an expert group of STS scholars on science and governance to provide advice on the process of increasing transparency and expert credibility (Wynne and Felt 2007). The solution, according to the commission, is to increase transparency in expert work, which it hopes will lead to increased public trust. By means of "democratized expertise," including increased transparency, the quality and credibility of expertise will be restored. The commission is aware of the need to restore public trust in science, but its concern still amounts more to words than deeds, and few concrete initiatives have been taken to bring citizens and science closer to each other (Irwin 2006).

A starting point for this urgent task is to reconsider the old one-way communication process, from experts to laypeople, and to focus on more symmetrical processes of mutual learning. IR scholars usually take learning to mean the recognition and acceptance of new information that can change involved actors' motives and behavior (Young 1999, 25). Whereas some studies restrict their focus to learning mechanisms operative between scientific experts and political representatives, others pay attention to the wider community of stakeholders (Fischer 2004; Held and Koenig-Archibugi 2005). The regulation of transboundary air pollution has implied largely an emphasis on the former case, in which networks between scientists and policymakers have developed and become extensively institutionalized and formalized through the development of organizational structures of interaction. However, the current focus on the need for public support implies an additional emphasis on the relationship between science and society and on the development of a broader understanding of the air pollution problem. Researchers in science communication and risk management have strongly emphasized that public involvement cannot be fostered simply by means of one-way information campaigns, going from science and policy to the public. On the contrary, communication is needed in which citizens are welcome to take part as active subjects having experience, knowledge, and perspectives that are important for making regulation more robust, effective, and trusted (Wynne 2001; Löfstedt 2005; Renn 2008).

In general, bringing science closer to society entails cultivating both scientific citizenship and socially sensitive expertise (Elam and Bertilsson 2003; Lidskog 2008). Creating socially robust regulation implies merging science with democracy, which means that citizens and scientists together are mobilized in shaping policy. The public understanding of an issue must be included when seeking that public's support for a policy; we must ask not only what is true regarding a policy, but also what that policy is about and what its social purpose is (Wynne 2003, 405).

A Successful Regime?

CLRTAP is one of the most studied international environmental regimes, and most policy evaluators have deemed it a success. As a comparative study of regime effectiveness states, "Indeed, CLRTAP's success in enhancing cooperation, bestowing authority, and facilitating learning is perhaps unique among international environmental regimes" (Munton, Sooros, Nikitina, et al. 1999, 235). The close interaction between scientific experts and policymakers in CLRTAP has been emphasized as the key explanatory factor for this success.

The shift from uniform emission cuts (the same for all parties) to effect-oriented strategies based on the critical-load concept in the early 1990s has been described as "a revolution in the management of transborder pollutants" (Levy 1993, 100; Ishii, chap. 6, and Sundqvist, chap. 7, in this volume). Combining the critical-load concept and the Regional Acidification Information System (RAINS) model allowed negotiators to justify nonuniform sulfur emission cuts as politically neutral, scientifically sound, and a smart alternative to arbitrary fixed-percentage reductions (Patt 1998, 10). Adopting the critical-load concept changed the cooperative atmosphere, shifting the dynamic from singling out "bad guys and dirtbags" to determining specific regions' vulnerability. The debate became more "sophisticated and relaxed" in both the scientific and the political contexts (Wettestad 1999, 101). Decisive for science-based regulation was the multilateral trust fund for the long-term financing of EMEP, considered the backbone of efforts to reach further agreements on emission reductions (Wettestad 1997, 245).

Scientists initially complained about uncertainties and oversimplification in relation to both critical loads and the RAINS model. However, policymakers, bureaucrats, and scientific experts eventually succeeded in defending the concept and the model as a productive way forward, considering the uncertainties as gaps in knowledge that could be bridged by further research (Tuinstra, Hordijk, and Amann 1999, 38). Critical loads and RAINS became tools of communication between scientific expertise, on one hand, and negotiations and political decisions, on the other. A "buffer body" was thus established between science and policy (Wettestad 1999, 33), enabling coordination of work and stabilizing the roles and boundaries in the CLRTAP process (Tuinstra, Hordijk, and Amann 1999, 35). However, this effort involved many questions: Would the scientific community in the long run accept this way of doing science? And would policymakers within the CLRTAP regime be ready to accept such a complicated foundation for abatement solutions?

After the success of the multipollutant/multieffect protocol signed in Gothenburg in 1999, some experts began worrying about the increased complexity of abatement strategies based on the critical-load concept, which regulates several compounds and effects in a single protocol. They argued that protocols should not be so "smart" that only a few experts can understand their mechanisms for distributing the signatories' commitments (Sundqvist 2003, 66). Some argued that the protocol might mark the end of the road for abatement strategies based on the critical-load concept (Tuinstra, Hordijk, and Amann 1999). Since 1999, no more protocols have been determined under CLRTAP; instead, a revision of the multipollutant/multieffect protocol has been started (Selin and VanDeveer, chap. 3 in this volume). The protocol's great complexity has not only made air pollution work more complicated, but also made it much harder to communicate the reasoning underlying the determined abatement strategies to those not involved in the work (Pleijel and Grennfelt 2007, 51).

The CLRTAP regime's success, which politicians, involved scientific experts, and social science analysts often cite, has also been questioned. The problems that CLRTAP deals with are far from being solved, and some of the most important emission cuts, primarily those of sulfur dioxides, have been caused by factors exogenous to the international regulation (e.g., economic restructuring, as in the case of the British electricity and coal industries, and technical developments, such as the introduction of low-sulfur fossil fuels and desulfurization equipment for coal-burning plants). The extent to which CLRTAP should be seen as a success—in terms of being the main cause of emissions cuts—is not uncontested, as some of the chapters in this volume discuss (chaps. 2 and 3). The development of parallel and partly overlapping regulatory regimes in Europe also merits investigation (chapters 3 and 4). Furthermore, the shift in the problem formulation—from a perspective emphasizing ecosystem effects to one emphasizing urban air quality and human health—implies that new groups of experts are being mobilized and enrolled in the regulatory work (chap. 8). Established relationships between policymakers and scientific experts within the CLRTAP organization are thereby being challenged, and opportunities for new types of coproduction between science and policy may arise that are different from those that have previously characterized CLRTAP (chap. 7). Earlier ways of mobilizing knowledge may change, giving rise to new opportunities for learning (chaps. 5 and 6).

Taken together, the current situation—marked by changes in political organizations, expert involvement, and science–policy dynamics—suggests an uncertain future direction for international cooperation on transboundary air pollution. Moreover, the issue of public trust has become a central concern of regulatory work. Without public legitimacy, decision makers and experts may not succeed in implementing further abatement strategies, and today there are serious concerns that the earlier one-way information flow will remain in effect. Various forms of public inclusion and citizen involvement are being proposed as a general strategy for creating public acceptance and support. Thus far, the issue of citizen involvement has to some extent been a blind spot in international studies of air policy work. This book helps remedy this lack, including both practical examples of citizen involvement in air pollution work and theoretical contributions regarding how to understand the role of citizens in developing the regulation of transboundary air pollution (chaps. 9,

10, and 11). As emphasized in the last chapter, two important tasks are to develop research into citizen involvement and to foster a broader and deeper understanding of science–policy–citizen dynamics in international environmental governance.

This Book: Science, Policy, and Citizens

The regulation of transboundary air pollution results from a combination of social order, political negotiations, and scientific practices. This book aims to contribute to new understandings of international environmental governance that take into consideration changes in political organization, scientific understanding, and civil society. Policy and institutions, expertise and learning, and citizens and involvement are accordingly three key areas to take into account when considering how the regulation of transboundary air pollution can be configured to be trustworthy, effective, and relevant.

In the volume, prominent social scientists specializing in international environmental governance make substantial contributions to the study of air pollution governance. They all share an emphasis on the need to assess previous experience if we are to understand the current situation, and they all suggest possible future directions for research, concerning both theoretical understandings and practical work.

Part I, "Policy and Institutions," focuses on the institutional dimensions of international efforts to combat air pollution. The three chapters in this section employ mainly an institutional approach, focusing on regime effectiveness, institutional linkages, and institutional design of the interface between science and policy. Chapter 2 focuses on the effectiveness of CLRTAP, and chapters 3 and 4 are devoted to the new political complexity, in which two organizations—the CLRTAP and the EU—compete and cooperate regarding how to regulate European air pollution.

In "The Improving Effectiveness of CLRTAP: Due to a Clever Design?" Jørgen Wettestad questions the regime's effectiveness and asks whether CLRTAP is in fact a cleverly designed regime. After discussing two ways to define effectiveness, as "relative improvement" or as "distance to optimum," and compiling his own top-five list of obstacles to effectiveness, he gives the answer that CLRTAP is a moderately effective regime. He views the regime in this light because the problem it addresses is malign and not easy to deal with and therefore has not yet been solved. At the same time, it should be noted that the regime's problem-solving capacity has improved over time. One of Wettestad's top-five obstacles to regime effectiveness is "differences regarding the perception of the problem," and he suggests that the way past this obstacle is via international knowledge-improvement efforts. Though Wettestad does not explicitly state it, he comes close to the epistemic-community approach when arguing that the CLRTAP regime is a good example of the establishment of this kind of knowledge capacity. He also focuses on the flexible organization of the science–policy relationship, citing, for example, how the CLRTAP Working Group on Strategies has functioned as a flexible arena for negotiation between science and policy, building consensus on both scientific and political matters. Wettestad's chapter is a good example of bridge building between theoretical approaches because it is grounded in the institutional approach but simultaneously includes perspectives from other approaches.

In "Institutional Linkages and European Air Pollution Politics," Henrik Selin and Stacy VanDeveer focus on political complexity in studying cooperation and conflict between the CLRTAP and the EU in European air pollution regulation. They aim to present a conceptual framework for studying institutional linkages in international environmental policy formation and implementation that comprises multiple theoretical approaches but is still grounded in a focus on institutions. The background for their discussion is the growing number of international institutions that deal with environmental governance, the overlaps and linkages between which are not yet well understood or studied. Selin and VanDeveer argue that a recent development in one institution cannot be fully understood if the linkages to other institutions are not taken into account. After making basic distinctions between vertical and horizontal linkages as well as between governance (institutional) and actor linkages, they state that linkages can have either synergetic or disruptive effects. Of great interest in their study of the CLRTAP and EU linkages are the horizontal linkages between two institutions acting on the same level, located in the same geographical area, and regulating the same kinds of problems. The CLRTAP–EU horizontal linkages are numerous, and there is reason to ask whether there really is a need for two European institutions for air regulation. Selin and VanDeveer conclude, however, that the similarities and overlaps have been more synergetic than disruptive, meaning that cooperation has led to more ambitious targets for both institutions.

The first part ends with the chapter "Transboundary Science for Transnational Air Pollution Policies in Europe." Bernd Siebenhüner follows up on the comparison of the CLRTAP and the EU by closely examining the interface between scientific knowledge and political decision making in these two institutional settings. He discerns three factors characterizing the design of the interplay—interaction, participation, and reflexive mechanisms—and questions addressing how, who, and what in science–policy interactions. The last factor—reflexive mechanisms—is more concerned with quality in that it enables reflection on current and past experience and how to feed them into policy processes. What Siebenhüner finds in his analysis is that the many workshops organized by CLRTAP, which are an essential feature of the convention's work, have strongly promoted close collaboration between scientists and policymakers and made reflection possible. This collaboration has also led to a science-driven policy process. In comparison, the EU's CAFE program is characterized as a policy-driven process in which external reviews are commonly used reflexive mechanisms. Siebenhüner concludes by arguing that the EU uses scientific expertise from outside in the form of reviews, whereas CLRTAP's work, done through workshops, leads to direct and ongoing contacts with scientific experts. He does not play the processes off against each other but considers both of them successful ways to organize science–policy interaction.

In part II, "Expertise and Learning," attention is turned to the roles of science, expertise, and learning. This part starts by giving a broad overview of international environmental regimes and how they have designed scientific bodies to influence policy and then presents more detailed case studies of expert strategies and learning processes in the area of European air pollution policy.

In chapter 5, "Organized Science, Usable Knowledge, and Multilateral Environmental Governance," Peter M. Haas and Casey Stevens provide an overview of how and when usable knowledge—that is, knowledge with both a substantive core and a mechanism for being transferred from the world of science to the world of policy—has contributed to effective multilateral environmental regimes. The authors assess more than 30 existing regimes that have involved scientific bodies to evaluate "when knowledge speaks to power" and to identify the conditions under which epistemic communities can be established. More specifically, they argue that science and policy must be kept apart: expertise and knowledge claims should be developed independently of the policy process, and only after consensus has been reached should scientific experts interact more closely with policymaking. In such cases, they argue, epistemic communities may serve to link knowledge and decision making. The two basic failures that often occur, according to Haas and Stevens, are that scientific bodies are not insulated from political control and that science tries to influence policy before consensus has been reached. When issues are highly politicized and involve high stakes—such as in the case of climate change—it is difficult for science to inform policy. The CLRTAP case has been more successful in this respect, and other regimes have something to learn from its accomplishments; it has enabled the creation of autonomous scientific bodies that are given the opportunity to develop consensual knowledge that helps foster consensus in the policy field. The authors conclude that knowledge speaks to power only under certain conditions and that social scientists have a mission to investigate these conditions in greater depth.

In chapter 6, "Scientists Learn Not Only Science but Also Diplomacy: Learning Processes in the European Transboundary Air Pollution Regime," Atsushi Ishii addresses the question of whether learning has had beneficial effects on the CLRTAP regime. By critically discussing ideas from the institutional, epistemic-community, and STS approaches, Ishii develops a theoretical framework that includes three learning modes in relation to policy change: one adaptive (change of policy tools), one reformative (change of policy tools and policy goals), and one paradigmatic (a fundamental change in the understanding of policy tools and goals). He then applies this framework in a detailed case study of the process that led to CLRTAP's Second Sulfur Protocol, signed in Oslo in 1994, which is based on the critical-load concept and the RAINS model. He concludes that paradigmatic learning took place when the critical-load concept was applied in the preparation of this protocol. Applying the critical-load approach entailed the creation of "diplomatory science," a new kind of science developed and used in diplomatic settings. This chapter is also a good example of cross-fertilization between theoretical approaches.

In chapter 7, "Fewer Boundaries and Less Certainty: The Role of Experts in European Air Policy," Göran Sundqvist starts from some key STS findings suggesting that science is Janus faced. According to this view, science is on one side uncertain, which is well known to scientists, but on the other a source of certainty for the public because uncertainties are concealed in public presentations of scientific results. This duality leads to separation between groups and false expectations of what science can provide to policy. When these expectations are not met, public trust in science disappears, and a rift emerges between science and society. Sundqvist tests these STS findings according to a historical analysis of four formative phases in the development of the CLRTAP. He confronts the IR focus on the separation between science, policy, and the public with the STS aim of blurring the boundaries between these three domains. He finds that CLRTAP activities have usually not been characterized by strong divides between these domains and their understandings. Thus, STS seems to oversell its findings regarding how scientific experts present their results to policymakers and the public. In addition, he considers IR wrong to argue that separation between experts and policymakers is necessary for success. Nevertheless, his conclusion should not detract from the fact that STS provides useful tools for studying how scientists present their findings in public and how communication among scientists, policymakers, and the public can be improved. In doing so, STS challenges both IR and current trends within CLRTAP that support stronger boundaries between scientists and policymakers regarding health issues.

In chapter 8, "Co-producing Policy-Relevant Science and Science-Based Policy: The Case of Regulating Ground-Level Ozone," Rolf Lidskog and Håkan Pleijel present a study of a topic rarely considered by STS and IR: the establishment and development of ground-level ozone regulation. By investigating this new object of policy and regulation, the chapter sheds light on how science-based policy and policy-relevant science are co-produced. By discussing various phases in the problem-formulation trajectory for ground-level ozone, the authors demonstrate the importance both of science in environmental regulation and of political considerations in the expert community. Political actors designed a setting for science–policy interaction in which science is both politically and publicly trusted. Scientific activities were formally separated from political negotiations, but behind this separation lay a close interaction and integration of science and policy. Measurement is developed in a social context in which scientists make political calculations regarding how to design policy-relevant knowledge and in which powers other than scientific ones also influence how environmental problems are represented. In this sense, CLRTAP can be seen as a learning environment in which expertise and stakeholders interact, resulting in the co-production of policy-relevant science and science-based policy. Lidskog and Pleijel demonstrate that instruments such as monitoring technologies and mapping activities not only present environment problems, but also play an important role in shaping an understanding of these problems. Scientific measures and political considerations co-produce a dynamic and changing governable space.

Part III, "Citizens and Involvement," focuses on "the missing link" in air pollution policies: the lack of public involvement. Whereas in the 1980s the acidification issue was strongly debated by nongovernmental organizations, the media, and citizens, today's concerns are more scientifically oriented and directed largely toward global warming. The three chapters in this section start from the perspective of public involvement and put forward practical examples and theoretical considerations of citizen engagement in air pollution policymaking. They also address critical issues, not least the decisive question of why it is important for decision makers to involve citizens. The chapters make new contributions to the social scientific study of air policy issues, long characterized by a focus on international institutions and the interplay between science and policy, leaving citizens aside.

In chapter 9, "Citizen Engagement with the Politics of Air Quality: Lessons for Social Theory, Science Studies, and Environmental Sociology," Steven Yearley presents air pollution as an exemplar of environmental problems. Air pollution is an externality; it is complex and variable (with several problems being given a single name); it is international; and its effects are difficult to measure. In addition, air pollution's social dimensions are obvious with regard to the uneven distribution of environmental ills. Yearley's primary concern is to investigate and reflect on what social theory can contribute to the understanding of air pollution and CLRTAP's achievements. In particular, he investigates three important approaches currently being discussed in social theory dealing with environmental problems: STS, ecological modernization, and Beck's theory of the risk society. The STS approach provides tools for studying the politics of public engagement but has little to say about CLRTAP: the convention seems to have been designed to make it difficult for citizens to exert influence. The theory of ecological modernization is based on the assumption that conflicts between economic growth and environmental improvement have been transcended, resulting in a convergent path for all to follow. Yearley finds that the development of CLRTAP is hard to interpret from the ecological modernization perspective, however, because it seems to be a policy-driven process. Last, the air pollution problem, compared with other environmental problems, has been poorly understood and little discussed as a risk issue. Thus, none of the dominant approaches widely employed in understanding other environmental issues has been used in analyzing CLRTAP. Yearley's answer to this gap can be interpreted as a challenge to social scientists. New ground can be broken to the benefit of both air pollution work and social theory—for example, when trying to develop a new rationale and novel methods for involving citizens in issues that are not on a local scale.

In chapter 10, "Framing Air Pollution and Health Problems: How to Include Stakeholder Perspectives," Arthur Petersen, Leendert van Bree, and Willemijn Tuinstra analyze the European AIRNET Thematic Network. This multistakeholder project's aim was to interpret the results of research supported by EU Framework Programs in the field of air pollution and public health and to make policy recommendations. The author analyze AIRNET from a science–policy–society interface view that aims to support the democratization of expertise and to suggest guidelines for dealing with framing and uncertainties. They argue that the concept of framing is extremely important but is largely overlooked in discussions of and strategies for stakeholder communication. Framing processes determine the scope of a problem and the knowledge and actions that are considered relevant. They also determine what is at stake and who is to be involved. All multistakeholder projects entail tension between the various frames that should be made explicit and discussed if communication is to be strengthened. The authors' analysis of the AIRNET project concludes that framing issues were not initially discussed; the project was established in line with a technocratic perspective in which boundaries between stakeholders as well as the why and how of stakeholder participation were not considered changeable—which they must be for improvement to occur. These topics were raised only toward the end of the project. The chapter's major conclusion is that to get a better start in multistakeholder projects, framing issues should be consciously reflected on and widely discussed from the outset.

In chapter 11, "Governance of Air Quality and Stakeholder Engagement: Lessons and Experience from International Cases," John Forrester and his colleagues present an international outlook on public participation in air policy work. The chapter provides insight into how the Eurocentric CLRTAP experience has spread to other parts of the world, not least South Asia and southern Africa. It examines stakeholder involvement in four cases: CLRTAP, the Malé Declaration (South Asia), the Air Pollution Information Network for Africa, and the United Kingdom's urban Air Quality Management Areas. The authors handle these very different cases by performing a two-step analysis, focusing first on the science–policy interface and then on how citizen engagement is taking place. They start with the assumptions that all parties have the right to be seen as stakeholders and that all stakeholders should be engaged early on and take part at all levels and stages of the process. These assumptions are based on the ideal of democratic rights and on research results suggesting that robust social knowledge is decisive in linking science and policy ("speaking truth to power"), which can be realized only through multistakeholder engagement. A prerequisite is that citizens should be brought into the process at an early stage and not—as mostly happens—late in the implementation stage. The authors conclude from the four cases that it is more difficult for citizens to engage with regional problems than with more local problems. However, for citizens to engage with local problems, such as urban air quality, it is crucial to understand how the problem became localized—that is, to investigate its regional connections.

The book ends with the chapter "Science–Policy–Citizen Dynamics in International Environmental Governance," in which Rolf Lidskog and Göran Sundqvist refer to general questions on the future of international air pollution work and how to study science–policy–citizen interactions in this field and more generally. Their discussion is based on the insights gained from the book's individual chapters and provides an overview of the conditions and possibilities for the future regulation of transboundary environmental issues. It is important to apply the lessons from CLRTAP when pondering future abatement strategies in this field and when approaching other environmental problems such as climate change. Furthermore, there is a need for ongoing theoretical development concerning the analysis of science–policy–citizen dynamics. Lidskog and Sundqvist argue that STS insights can nurture and improve the understanding of the roles of science and citizens in international regimes; at the same time, STS needs to make better use of findings from the institutional and epistemic-community approaches regarding science–policy–citizen dynamics in international environmental regimes.

European air pollution policy is in transition. Changing scientific understandings, growing political complexity, new political handling demands, and a need for citizen engagement together pose a great challenge to air pollution policies. We need to understand what is happening and to take the current situation into account when developing new strategies for abating transboundary air pollution and other transboundary environmental problems. In this demanding work, improved analyses from social scientists will make important contributions.

References

Barry, Andrew, Thomas Osborne, and Nicholas Rose, eds. 1996. *Foucault and Political Reason: Liberalism, Neo-Liberalism, and Rationalities of Government.* Chicago: University of Chicago Press.

Beck, Ulrich. 1992. *The Risk Society: Towards a New Modernity*. London: Sage Publications.

Beck, Ulrich. 1995. *Ecological Politics in an Age of Risk*. Cambridge, UK: Polity Press.

Boehmer-Christiansen, Sonja. 2000. The British Case: Overcompliance by Luck or Policy? In *International Environmental Agreements and Domestic Politics: The Case of Acid Rain*, ed. Arild Underdal and Kenneth Hanf, 279–312. Aldershot: Ashgate.

Corell, Elisabeth, and Göran Sundqvist. 2005. Knowledge Processes in Decision-Making on Sustainability: Challenges for the Future. In *Global Challenges: Furthering the Multilateral Process for Sustainable Development*, ed. Angela Churie Kallhauge, Gunnar Sjöstedt, and Elisabeth Corell, 232–246. Sheffield, UK: Greenleaf.

EC (Commission of the European Communities). 2001. *White Paper on Governance: Report of the Working Group "Democratising Expertise and Establishing Scientific Reference Systems."* Brussels: Commission of the European Communities.

EC (Commission of the European Communities). 2008. *Directive 2008/50/EC of the European Parliament and of the Council of 21 May 2008 on Ambient Air Quality and Cleaner Air for Europe.*

Elam, Mark, and Margareta Bertilsson. 2003. Consuming, Engaging, and Confronting Science: The Emerging Dimensions of Scientific Citizenship. *European Journal of Social Theory* 6 (2): 233–251.

Fischer, Frank. 2004. Professional Expertise in a Deliberative Democracy: Facilitating Participatory Inquiry. *Good Society* 13 (1): 21–27.

Gouldson, Andrew, Rolf Lidskog, and Misse Wester-Herber. 2007. The Battle for Hearts and Minds: Evolutions in Organisational Approaches to Environmental Risk Communication. *Environment and Planning C* 25 (1): 56–72.

Haas, Peter M. 1989. Do Regimes Matters? Epistemic Communities and Mediterranean Pollution Control. *International Organization* 43 (3): 377–403.

Haas, Peter M. 1992. Introduction: Epistemic Communities and International Policy Coordination. *International Organization* 46 (1): 1–35.

Haas, Peter M. 1997. Scientific Communities and Multiple Paths to Environmental Management. In *Saving the Seas: Values, Scientists, and International Governance*, ed. L. Anathea Brooks and Stacy D. VanDeveer, 193–228. College Park: Maryland Sea Grant College.

Haas, Peter M. 2004. When Does Power Listen to Truth? A Constructivist Approach to the Policy Process. *Journal of Environmental Public Policy* 11 (4): 569–592.

Haas, Peter M., ed. 2008. *International Environmental Governance*. Aldershot, UK: Ashgate.

Hajer, Marten. 1995a. Acid Rain in Great Britain: Environmental Discourse and the Hidden Politics of Institutional Practice. In *Greening Environmental Policy: The Politics of Sustainable Future*, ed. Frank Fisher and Michael Black, 145–164. London: Paul Shapman.

Hajer, Marten. 1995b. *The Politics of Environmental Discourse: Ecological Modernization and the Policy Process*. Oxford, UK: Oxford University Press.

Hanf, Kenneth. 2000. The Problem of Long-Range Transport of Air Pollution and the Acidification Regime. In *International Environmental Agreements and Domestic Politics: The Case of Acid Rain*, ed. Arild Underdal and Kenneth Hanf, 21–48. Aldershot, UK: Ashgate.

Held, David, and Mathias Koenig-Archibugi, eds. 2005. *Global Governance and Public Accountability*. Malden, MA: Blackwell.

Hempel, Lamont C. 1996. *Environmental Governance: The Global Challenge*. Washington, DC: Island Press.

Irwin, Alan. 2006. The Politics of Talk: Coming to Terms with the "New" Scientific Governance. *Social Studies of Science* 36 (2): 299–320.

Irwin, Alan, and Mike Michael. 2003. *Science, Social Theory, and Public Knowledge*. Maidenhead, UK: Open University Press.

Jasanoff, Sheila. 1996a. Beyond Epistemology: Relativism and Engagement in the Politics of Science. *Social Studies of Science* 26: 393–418.

Jasanoff, Sheila. 1996b. Science and Norms in Global Environmental Regimes. In *Earthly Goods: Environmental Change and Social Justice*, ed. Fen Oslev Hampson and Judith Reppy, 173–197. Ithaca, NY: Cornell University Press.

Jasanoff, Sheila. 2004a. Ordering Knowledge, Ordering Society. In *States of Knowledge: The Co-Production of Science and Social Order*, ed. Sheila Jasanoff, 13–45. London: Routledge.

Jasanoff, Sheila, ed. 2004b. *States of Knowledge: The Co-Production of Science and Social Order*. London: Routledge.

Jasanoff, Sheila. 2005. *Designs of Nature: Science and Democracy in Europe and the United States*. Princeton, NJ: Princeton University Press.

Jasanoff, Sheila, and Marybeth Long Martello, eds. 2004. *Earthly Politics: Local and Global in Environmental Governance*. Cambridge, MA: MIT Press.

Jasanoff, Sheila, and Brian Wynne. 1998. Science and Decisionmaking. In *Human Choice and Climate Change*, vol. 1: *The Societal Framework*, ed. Steve Rayner and Elisabeth L. Malone, 1–87. Columbus, OH: Battelle Press.

Keohane, Robert O. 1988. International Institutions: Two Approaches. *International Studies Quarterly* 32 (4): 379–396.

Krasner, Stephen D. 1983. *International Regimes*. Ithaca, NY: Cornell University Press.

Kunzli, Nino, and Ira B. Tager. 1999. Comments on PM2.5 and Mortality in Europe in Long Term Prospective Cohort Studies: Cause–Effect or Statistical Association? *Environmental Health Perspectives* 107: A234–A235.

Leach, Melissa, Ian Scoones, and Brian Wynne. 2005. *Science and Citizens: Globalization and the Challenge of Engagement*. London: Zed Books.

Letell, Martin. 2002. *Acid Narratives: National Security and the Configuration of SO_2*. STS Research Reports 5. Section for Science and Technology Studies. Gothenburg: University of Gothenburg.

Levy, Marc A. 1993. European Acid Rain: The Power of Tote-Board Diplomacy. In *Institutions for the Earth: Sources of Effective International Environmental Protection*, ed. Peter M. Haas, Robert O. Keohane, and Marc A. Levy, 75–132. Cambridge, MA: MIT Press.

Lidskog, Rolf. 2004. Making Science Matter: New Conditions for Policy-Relevant Science in Environmental Policy-Making. *European Spatial Research and Policy* 11 (2): 5–25.

Lidskog, Rolf. 2008. Scientised Citizens and Democratised Science: Re-Assessing the Expert–Lay Divide. *Journal of Risk Research* 11 (1–2): 69–86.

Lidskog, Rolf, Linda Soneryd, and Ylva Uggla. 2009. *Transboundary Risk Governance*. London: Earthscan.

Lidskog, Rolf, and Göran Sundqvist. 2002. The Role of Science in Environmental Regimes: The Case of LRTAP. *European Journal of International Relations* 8 (1): 77–101.

Löfstedt, Ragnar. 2005. *Risk Management in Post-Trust Societies*. Basingstoke, UK: Palgrave Macmillan.

Lundgren, Lars J. 1998. *Acid Rain on the Agenda: A Picture of a Chain of Events in Sweden, 1966–1968*. Lund: Lund University Press.

Martello, Marybeth Long, and Sheila Jasanoff. 2004. Introduction: Globalization and Environmental Governance. In *Earthly Politics: Local and Global in Environmental Governance*, ed. Sheila Jasanoff and Marybeth Long Martello, 1–29. Cambridge, MA: MIT Press.

Munton, Don, Marvin Sooros, Elena Nikitina, and Marc A. Levy. 1999. Acid Rain in Europe and North America. In *The Effectiveness of International Environmental Regimes: Causal Connections and Behavioural Mechanisms*, ed. Oran R. Young, 155–248. Cambridge, MA: MIT Press.

Nikolaou, K. 2003. Air Quality in European Urban Areas and the New EC Directives. *Journal of Environmental Protection and Ecology* 4 (2): 477–482.

Nowotny, Helga, Peter Scott, and Michael Gibbons. 2001. *Re-Thinking Science: Knowledge and the Public in an Age of Uncertainty*. Cambridge: Polity Press.

Patt, Anthony. 1998. *Analytic Frameworks and Politics: The Case of Acid Rain in Europe*. ENRP Discussion Paper E-98-20. Cambridge, MA: Kennedy School of Government, Harvard University.

Pleijel, Håkan, and Peringe Grennfelt. 2007. European Air Pollution—the Historic Legacy. In *Transboundary Air Pollution: Scientific Understanding and Environmental Policy in Europe*, ed. Håkan Pleijel, 27–53. Lund, Sweden: Studentlitteratur.

Renn, Ortwin. 2008. *Risk Governance: Coping with Uncertainty in a Complex World*. London: Earthscan.

Rose, Nikolas. 1999. *Powers of Freedom: Reframing Political Thought*. Cambridge, UK: Cambridge University Press.

Sliggers, Johan. 2004. Blue Skies for Ever. In *Clearing the Air: 25 Years of the Convention on Long-Range Transboundary Air Pollution*, ed. Johan Sliggers and Willem Kakebeeke, 149–167. New York: United Nations.

Social Learning Group. 2001. *A Comparative History of Social Responses to Climate Change, Ozone Depletion, and Acid Rain*. Vol. 1 of *Learning to Manage Global Environmental Risks*. Politics, Science, and the Environment. Cambridge, MA: MIT Press.

Sooros, Marvin S. 1993. Arctic Haze and Transboundary Air Pollution. In *Polar Politics*, ed. Oran R. Young and Gail Osherenko, 186–222. Ithaca, NY: Cornell University Press.

Sprinz, Detlef F., and Andreas Wahl. 2000. Reversing (Inter)national Policy—Germany's Response to Transboundary air Pollution. In *International Environmental Agreements and Domestic Politics: The Case of Acid Rain*, ed. Arild Underdal and Kenneth Hanf, 139–166. Aldershot, UK: Ashgate.

Sundqvist, Göran. 2002. *The Bedrock of Opinion: Science, Technology, and Society in the Siting of High-Level Nuclear Waste*. Dordrecht, Netherlands: Kluwer Academic.

Sundqvist, Göran. 2003. Recovery in the Acid Rain Story: Transparency and Credibility in Science-Based Environmental Regulation. *Journal of Environmental Policy and Planning* 5 (1): 57–79.

Sundqvist, Göran, and Martin Letell. 2005. Configuring the Urban Traveler: Local Adaptation of European Air Policy. *Acta Sociologica* 48 (3): 205–219.

Thompson, Jan. 1999. Introduction to the Ministerial Declaration on Long-Range Transboundary Air Pollution. Mimeo. Gothenburg, 30 November 1999.

Tuinstra, Willemijn, Leen Hordijk, and Markus Amann. 1999. Using Computer Models in International Negotiations: The Case of Acidification in Europe. *Environment* 41 (9): 33–42.

Tuinstra, Willemijn, Leen Hordijk, and Carolien Kroeze. 2006. Moving Boundaries in Transboundary Air Pollution: Co-Production of Science and Policy under the Convention on Long-Range Transboundary Air Pollution. *Global Environmental Change* 16 (4): 349–363.

UN (United Nations). 1973. *Report of the United Nations Conference on the Human Environment, Stockholm, 5–16 June 1972*. A/Conf.48/14/Rev.1. New York: United Nations Publications.

Underdal, Arild, and Kenneth Hanf, eds. 2000. *International Environmental Agreements and Domestic Politics: The Case of Acid Rain.* Aldershot, UK: Ashgate.

UNECE (United Nations Economic Commission for Europe). 1999. *Protocol to the 1979 Convention on Long-Range Transboundary Air Pollution to Abate Acidification, Eutrophication, and Ground-Level Ozone.* Geneva: UNECE.

VanDeveer, Stacy D. 2004. Ordering Environments: Regions in European International Environmental Cooperation. In *Earthly Politics: Local and Global in Environmental Governance*, ed. Sheila Jasanoff and Marybeth Long Martello, 309–334. Cambridge, MA: MIT Press.

Weart, Spencer R. 1988. *Nuclear Fear: A History of Images.* Cambridge, MA: Harvard University Press.

Wettestad, Jørgen. 1997. Acid Lessons? LRTAP Implementation and Effectiveness. *Global Environmental Change* 7 (3): 235–249.

Wettestad, Jørgen. 1999. *Designing Effective Environmental Regimes: The Key Conditions.* Cheltenham, UK: Edward Elgar.

WHO (World Health Organization). 2006. *Health Risks of Particulate Matter from Long-Range Transboundary Air Pollution.* Copenhagen: WHO.

Wynne, Brian. 1996. SSK's Identity Parade: Signing Up, Off-and-On. *Social Studies of Science* 26 (2): 357–391.

Wynne, Brian. 2001. Creating Public Alienation: Expert Cultures of Risk and Ethics on GMOs. *Science as Culture* 10 (4): 445–481.

Wynne, Brian. 2003. Seasick on the Third Wave? Subverting the Hegemony of Propositionalism. *Social Studies of Science* 33 (3): 401–417.

Wynne, Brian, and Ulrike Felt. 2007. *Taking European Knowledge Society Seriously. Report of the Expert Group on Science and Governance to the Science, Economy, and Society Directorate, Directorate-General for Research, European Commission.* Brussels: European Communities.

Yearley, Steven. 1996. *Sociology, Environmentalism, Globalization: Reinventing the Globe.* London: Sage.

Yearley, Steven. 2006. Bridging the Science–Policy Divide in Urban Air-Quality Management: Evaluating Ways to Make Models More Robust Through Public Engagement. *Environment and Planning C: Government & Policy* 24 (5): 701–714.

Young, Oran R., ed. 1999. *The Effectiveness of International Environmental Regimes: Causal Connections and Behavioral Mechanisms.* Cambridge, MA: MIT Press.

Young, Oran R. 2002. *The Institutional Dimensions of Environmental Change.* Cambridge, MA: MIT Press.

Young, Oran R. 2004. Institutions and the Growth of Knowledge: Evidence from International Environmental Regimes. *International Environmental Agreement: Politics, Law, and Economics* 4 (2): 215–228.

Young, Oran R. 2008. Institutions and Environmental Change: The Scientific Legacy of a Decade of IDGEC Research. In *Institutions and Environmental Change: Principal Findings, Applications, and Research Frontiers*, ed. Oran R. Young, Leslie A. King, and Heike Schroeder, 3–45. Cambridge, MA: MIT Press.

Young, Oran R., and Marc A. Levy, with the assistance of Gail Osherenko. 1999. The Effectiveness of International Environmental Regimes. In *The Effectiveness of International Environmental Regimes: Causal Connections and Behavioral Mechanisms*, ed. Oran R. Young, 1–32. Cambridge, MA: MIT Press.

I

Policy and Institutions

2

The Improving Effectiveness of CLRTAP: Due to a Clever Design?

Jørgen Wettestad

This chapter identifies and discusses a "top-five" list of central obstacles to the improvement of regime effectiveness and related important institutional cures and techniques and subsequently applies this framework in an analysis of the Convention on Long-Range Transboundary Air Pollution (CLRTAP) regime (building on and updating analyses carried out in Wettestad 1999, 2001, 2002a, 2002b, and 2005a).[1] Because effectiveness has seemingly steadily improved, and the basic problem addressed by CLRTAP must be characterized as malign, should we (re)assess CLRTAP as a truly cleverly designed regime? The chapter addresses this key question.

The second section briefly introduces the understanding of effectiveness as utilized in this chapter. Drawing on the work of Arild Underdal, which also underpins the analyses of regime effectiveness carried out in the work by Edward Miles and his colleagues (2002), it introduces and discusses the two central dimensions of "relative improvement" and "distance to collective optimum." The third section then explains effectiveness and identifies promising institutional techniques to enhance regime effectiveness. In this venture, I take the major perspectives from Miles and his coauthors (2002) as a main point of departure: the distinction between "institutional and problem-solving capacity" and "characteristics of the problem(s)." Drawing on this framework, I put forward and discuss my own suggested "top-five" list of central problematic characteristics or obstacles to the improvement of effectiveness and then offer some related central institutional cures and techniques. The fourth section applies these concepts and approaches in a discussion of the CLRTAP regime, and the fifth section winds up the chapter with some concluding reflections.

The Concept of Effectiveness: "Distance to Optimum" versus "Relative Improvement"

Although this chapter's aim is not to discuss the concept of effectiveness per se, it is necessary to introduce a basic understanding of this complex and central issue. A very important idea first put forward by Arild Underdal in the beginning of the 1990s (Underdal 1990, 1992) was that there are at least two crucial and quite different ways to conceptualize and measure the effectiveness of international regimes: *distance to collective optimum* and *relative improvement.* In my view, this perspective still holds up.

The "distance to collective optimum" perspective is "the appropriate perspective if we want to determine to what extent a collective problem is in fact 'solved' under present arrangements" (Underdal 1992, 231). But what, then, is a "problem solution"? What constitutes the collective optimum and the maximum that can be achieved? Whenever we are dealing with collective decisions that can be made only through agreement, this situation would be the Pareto frontier, where "no further increase in benefits to . . . any party can be obtained without thereby leaving one or more prospective partner(s) worse off" (Underdal 1992, 233). However, applying this criterion to specific cases is very difficult. As Underdal notes, "The Pareto frontier can be determined only for a given negotiation setting, including a given set of actors and a certain set of issues and issue linkages.. A change in any of these elements may affect the range of politically feasible solutions" (1992, 234). In dynamic regimes that operate over decades, the negotiation setting undergoes numerous changes and becomes a moving target. Hence, Underdal suggests instead focusing on more easily measurable proxies. One sensible rule of thumb, he explains, would be to look for independent expert advice that would indicate to decision makers what the (technically) "perfect" solution would be. When expert advice cannot be found (or experts heavily disagree!), the fall-back strategy would be to look for some official declaration of a joint goal or purpose (e.g., limit and reduce transboundary air pollution) to serve as a point of reference.

The "relative improvement" perspective is "clearly the notion we have in mind when considering whether and to what extent 'regimes matter'" (Underdal 1992, 231). This perspective is basically counterfactual. What is the hypothetical "state of affairs" that would have obtained if, instead of the present regime, we were left in a "no-regime" condition? However, similar to the challenge of measuring the distance to the collective

		Distance to collective optimum	
		GREAT	SMALL
Relative improvement	HIGH	Important, but still imperfect	Important and (almost) perfect
	LOW	Insignificant and suboptimal	Unimportant, yet (almost) optimal

Figure 2.1
Dimensions of regime effectiveness.

optimum, carrying out a counterfactual regime assessment is a tall order because a number of factors at the international, national, and subnational levels must be taken into consideration. Underdal advises us to look for whatever predictions we can find in negotiation documents on what will happen if the instrument in question is or is not adopted, preferably documents that can be seen as "nonpartisan" inputs. Such predictions can then give us a clue about what well-informed actors see as the main causal forces to watch and "control for" in the setting at hand. Other than that, the task of determining what would otherwise have happened simply calls for the best judgment that the analyst can produce on the basis of available sources.

Seen together, these perspectives are clearly complementary. As a key insight, Underdal states that "even a regime leading to substantial improvement may fall short of being 'perfect'" (1992, 231). Figure 2.1 shows an elaboration of the relationship between the perspectives.

Explaining Effectiveness and the Focused Issue of Institutional Design: Five Key Features

With regard to effectiveness and the particularly intriguing question of institutional techniques for improving effectiveness, I find the two fundamental perspectives *problem characteristics* and *problem-solving capacity* helpful. Problem characteristics are fundamental aspects of the environmental problems addressed by the regimes. For instance, is the underlying collaborative problem one of complicated transboundary effects and competition over collective goods, implying the need for painful and often costly behavioral changes, or is it more a simple coordination problem where modest behavioral adjustments will do? Moreover, to what extent is knowledge about the problem uncertain and disputed? Problem-solving capacity is then a combination of the institutional efforts established and the entrepreneurial efforts made to address and solve the environmental or resource problems.[2] A core idea is that some regimes are more effective than others either because the problems they deal with are more benign or because they are addressed by more effective problem-solving instruments and efforts.

Hence, institutional design and "techniques" form a central part of problem-solving capacity; such techniques and smart ways of designing international institutions and commitments can obviously be seen as ways to overcome cooperative obstacles flowing from malign problem characteristics. The next step is to identify the central obstacles to the improvement of effectiveness and some related important institutional cures and techniques.

The first obstacle is often marked *differences in capabilities among the actors to establish/implement abatement policies*. Hence, an important institutional "cure" and technique is the establishment of a funding mechanism, which can help states with weak financial or administrative capacities or both to improve their implementation capabilities. This approach has clearly played an important role in the global cooperation to protect the ozone layer. The establishment of a specific fund in 1990 to pay for developing countries' incremental costs and some technology transfer can be seen as a watershed development within the regime in terms of securing support among developing countries for the Vienna Convention and Montreal Protocol on Substances That Deplete the Ozone Layer (Benedick 1991; DeSombre and Kaufman 1996; Wettestad 1999; Parson 2003). Robert Keohane and Marc Levy's 1996 volume more generally discusses the funding and technology transfer issue as the background to a number of case studies. Another institutional technique to deal with the problem of differing capabilities is the possibility of differentiated commitments. CLRTAP and the 1999 Protocol to Abate Acidification, Eutrophication, and Ground-Level Ozone is a very good example, as is elaborated later in the chapter.

A second central obstacle to effectiveness is marked *differences among the parties with regard to perceptions of the (seriousness of the) environmental problems*. An important institutional cure in this connection is the establishment of a well-functioning international knowledge-improvement effort. CLRTAP is again a very good example (as also discussed later). But improving knowledge is not enough in itself; that knowledge also has to be effectively communicated to decision makers. The promises and pitfalls of organizing a good science–politics dialog in the international regime context have been discussed in a number of studies (e.g., Andresen, Skodvin, Underdal, and Wettestad 2000; Skodvin 2000; Parson 2003). As we pointed out, a central dilemma is to find the right balance between scientific integrity and political involvement. We considered the establishment of a specific buffer body (or several bodies) to be a promising technique. I look into how the CLRTAP has handled this challenge in the fourth section.

A third central obstacle to regime effectiveness is undoubtedly *differences among the parties with regard to positions on how to deal with the problems*. An important institutional technique to beat the "law of the least ambitious program" (Underdal 1980) and to come up with strong and ambitious policies and protocols is the introduction of majority voting. CLRTAP is not a good example in this regard, nor are many other international regimes, in fact. Within international environmental politics, the most relevant example is probably the European Union (EU) (e.g., Wallace and Wallace 2000). But there are also other and less-demanding "fast-track" options (cf. Sand 1990), among them the possibility of opt-out provisions and the establishment of smaller clubs where not all regime parties participate. The North Sea cooperation is a good case in this regard (see Skjærseth 2000).

A fourth central obstacle to effectiveness is *uncertainty among the parties with regard to domestic regulatory implications and domestic abatement possibilities more generally*. A standard institutional technique is here the establishment of specific clauses on regular renegotiations of commitments. Both the North Sea regime and the ozone regime are good examples because they have developed through several stages and amendments (Skjærseth 2000; Parson 2003). Within the ozone regime, there was for instance first the 1987 Montreal Protocol. The requirements were then strengthened, and further substances were added in the 1990 London and 1992 Copenhagen amendments and then subsequently in the 1997 Montreal and 1999 Beijing amendments.

A fifth central obstacle to effectiveness is clearly *absent or diffuse knowledge regarding parties' "real" implementation and follow-up*. The general cure is the establishment of a reporting and verification system. However, a more specific institutional technique is the establishment of a specific implementation/compliance committee. CLRTAP is a good example here in that the operation of a specific Implementation

Committee from 1997 on has considerably improved reporting procedures and regime debates on implementation, as elaborated later in the chapter. The ozone regime established a specific Implementation Committee in 1990 (Parson and Greene 1995; Greene 1996; Victor 1998; Parson 2003), and this step has functioned as an important model for the climate regime (Wettestad 2005b).

CLRTAP: Steadily Improving Effectiveness Due to a Clever Design?

CLRTAP Background

In 1968, the Swedish scientist Svante Odén published a paper in which he argued that precipitation over Scandinavia was becoming increasingly acidic, thus inflicting damage on fish and lakes. Moreover, he maintained that the acidic precipitation was to a large extent caused by sulfur compounds from British and central European industrial emissions. This development aroused broader Scandinavian concern and diplomatic activity related to acid pollution. The specific background for formal negotiations on an air pollution *convention* was the East–West détente process in the mid-1970s, in which the environment was identified as one potential area for cooperation. Due to the East–West dimension, the United Nations Economic Commission for Europe (UNECE) was chosen as the institutional setting for the negotiations.

CLRTAP was signed by 33 contracting parties (32 countries and the European Commission) in Geneva in November 1979. Four main aspects of the 1979 convention may be discerned: first, the recognition that airborne pollutants were a major problem; second, the declaration that the parties would "endeavor to limit and, as far as possible, gradually reduce and prevent air pollution, including long-range transboundary air pollution" (Article 2); third, the contracting parties' commitment, "by means of exchange of information, consultation, research and monitoring, [to] develop without undue delay policies and strategies which should serve as a means of combating the discharge of air pollutants, taking into account efforts already made at the national and international levels" (Article 3); and fourth, the intention to use "the best available technology which is economically feasible" to meet the convention's objectives.

The convention did not specify any pollutants but stated that monitoring activity and information exchange should start with sulfur dioxide (SO_2). The CLRTAP has been in force since 1983 and had been ratified by 51 parties by April 2007. Moreover, it was to be overseen by an executive body that would include representatives of all parties to the convention as well as of the European Commission. Furthermore, the ECE secretariat was given a coordinating function. The institutional structure has also included several working groups, task forces, and international cooperative programs. Rooted in the convention's strong initial focus on knowledge improvement and monitoring, a specific financing protocol for the European Monitoring and Evaluation Program (EMEP) was established in 1984.

CLRTAP's Effectiveness: Medium or High?

How effective has CLRTAP been? In previous studies (Wettestad 1999, 2002b), I landed on an overall "medium" effectiveness score for the regime. This determination was based on comparative reasoning, backed up by case-specific evidence revealing both strengths and weaknesses.

In a comparative perspective, it is, on the one hand, quite likely that the regime to protect the ozone layer has achieved more significant behavioral change, which also makes problem solving possible and hence works toward a high score. On the other hand, the climate-change regime has achieved little behavioral change, and problem solving seems very far away, hence working toward a low score. CLRTAP seems more in between.

This mixed/medium position can be further substantiated in the following way: in terms of policy development through the establishment of specific protocols, it is clear that the convention's record is impressive indeed. In this connection, it is interesting to note that the initial convention did not mention subsequent protocols at all! The protocol development can be summed up as follows:

- *The 1985 Protocol on the Reduction of Sulfur Emissions.* In Helsinki in July 1985, 21 countries and the European Commission signed this legally binding protocol. It stipulated a reduction of emissions/transboundary fluxes of SO_2 by at least 30 percent as soon as possible or by 1993 at the latest, with 1980 levels as a baseline. However, some major emitter states failed to join the agreement, among them the United Kingdom, the United States, and Poland. The protocol entered into force in September 1987 and had been ratified by 25 parties by September 2010.
- *The 1988 Sofia Protocol Concerning the Control of Nitrogen Oxides.* Here, the signatories pledged to freeze nitrogen oxide (NO_x)

emissions at the 1987 level from 1994 onward and to negotiate subsequent reductions. Twenty-five countries signed the protocol, including the United Kingdom and the United States. Moreover, 12 European signatories went a step further and signed an additional (and separate) joint declaration committing them to a 30 percent reduction of emissions by 1998. The protocol entered into force in February 1991 and has been ratified by 34 parties.

• *The 1991 Geneva Protocol on Volatile Organic Compounds.* Volatile organic compounds (VOCs) are a group of chemicals that are precursors of ground-level ozone. The protocol called for a 30 percent reduction in VOC emissions between 1988 and 1999 based on 1988 levels—either at national levels or within specific "tropospheric ozone–management areas." Some countries were allowed to opt for a freeze of 1988 emissions by 1999.[3] Twenty-one parties signed the protocol in 1991, and it entered into force in September 1997. It has been ratified by 24 parties.

• *The 1994 Protocol on Further Reduction of Sulfur Emissions (Second Sulfur Protocol).* This protocol was signed in Oslo in June 1994 by 28 parties. It was based on the critical-loads approach. The aim of this approach is that emissions reductions should be negotiated on the basis of the (varying) effects of air pollutants rather than by choosing an equal-percentage reduction target for all countries involved.[4] Hence, the protocol set out individual and varying national reduction targets for the year 2000 for half of the countries and additional targets for the other half to be reached by 2005 and 2010, with 1980 as the base year. The protocol entered into force in August 1998 and has been ratified by 29 parties.

• Two protocols on transboundary air pollution by *heavy metals* and *persistent organic pollutants* (POPs) were then signed by 34 parties in Århus, Denmark, in June 1998. They have been ratified by 29 parties and entered into force in late 2003.

• *The 1999 Protocol to Abate Acidification, Eutrophication, and Ground-Level Ozone.* This protocol was signed in Gothenburg in December 1999 by 31 parties and is by far the most advanced within the regime. It covers four substances (NOx, VOCs, ammonia, and SO_2) and three environmental effects (acidification, ground-level ozone, and eutrophication). It establishes varying reduction targets for all parties involved (Wettestad 2002a). It entered into force in May 2005 and has been ratified by 25 parties.

Hence, there has been a steady development of protocols, covering more substances with regulations that are gradually becoming both binding and specific and more fine-tuned to ecological and economic variations between the countries. Moreover, national compliance with these protocols must be characterized overall as high. The work on reducing sulfur emissions has in particular been marked by substantial overcompliance. Sulfur emissions in Europe were reduced by 71 percent in the period from 1980 to 2000. And in the period covered by the Gothenburg Protocol (i.e., with a 1990 baseline), by 2006 such emissions had been reduced by 65 percent. This means that compliance to the Gothenburg Protocol had already been achieved for such emissions (UNECE 2006).

Then there is the (social) learning dimension (see Young and Levy 1999; Underdal and Hanf 2000; Social Learning Group 2001a, 2001b). It is clear that CLRTAP has been very important as a forum for organizing the production and dissemination of economic and natural scientific knowledge. Moreover, it has also been important as a forum where bureaucrats, researchers, and nongovernmental organizations (NGOs) from various countries have met and learned from each other. It is very tricky to assess the exact value of this organization and dissemination of knowledge. It has clearly been an important factor for driving the regime forward and producing scientific reports and protocols. But, as I elaborate further later, in the implementation processes other and more substantial economic and social forces have generally come to the forefront. Nevertheless, as an experienced regime analyst pointed out back in the late 1990s, "It is hard to imagine being where we are today regarding transboundary air pollution in the absence of the LRTAP process."[5] So the regime has clearly contributed to the reductions in emissions that we have witnessed and to the important steps taken toward problem solving.

Yet an assessment of "relative improvement" requires a thorough consideration of other potential causal factors than the ones related to the regime. A closer scrutiny of available national and subnational knowledge indicates that many forces other than the CLRTAP regime have been involved in bringing about policy changes and emissions reductions. Much behavioral change would probably have happened "anyway" due to energy policy changes, more fundamental economic and industrial changes, European Community/Union processes, and so on, even if the picture clearly varies between countries.[6] Take, for instance, the SO_2 and NO_x implementation processes in the United Kingdom, Germany, the Netherlands, and Norway in the 1980s and 1990s. Aided not least by knowledge produced within the context of the

Hanf and Underdal project (Underdal and Hanf 2000), it can be concluded that the majority of the initial reductions and compliance levels achieved in three of the four countries (the United Kingdom, the Netherlands, and Norway) can apparently be explained by processes that are not related primarily to environmental protection, at least with regard to sulfur reductions (Wettestad 1997, 1998; Underdal and Hanf 2000).

In the United Kingdom, industrial recession and reduced energy demand in the 1980s were important factors. Moreover, privatization and the switch from coal to gas were important factors. In the Netherlands, a gradual conversion to domestic natural gas for domestic political and financial reasons was clearly important. In Norway, much was achieved by, among other things, reducing consumption of heavy fuel oil on land (Wettestad 2004). The exception is Germany, where environmental regulations were—at least initially—the main driving forces.

Focusing more closely on the CLRTAP contribution to these processes, I have assessed that the direct, easily detectable influence has been moderate, at least for these particular countries (including Germany). In a situation without the CLRTAP, significant initial reductions would probably have taken place anyway due to other economic and political processes and domestic political pressure motivated by environmental damage. Take, for instance, Germany. It is important to remember that Germany was almost as reluctant as the United Kingdom at the CLRTAP negotiations in the late 1970s. Without the rapidly increasing concern about forest damage/*Waldsterben* in the early 1980s, both German and European acid rain politics would have looked very different today. Moreover, with regard to German and British acid rain politics, the European Community/EU decision-making arena has possibly been more important than the CLRTAP.[7]

However, there is still a possibility that CLRTAP was more important for other western European and especially eastern European countries in the 1980s and 1990s. Marc Levy suggests that countries such as Austria, Finland, the Netherlands, and Switzerland were influenced by CLRTAP in the 1980s through increased awareness of domestic acid rain damage. He also concludes that in the same period countries such as Denmark and the Soviet Union were influenced by the CLRTAP through various types of linkage effects (1993, 118–121, 122–126). In the case of Denmark, Levy maintains that Nordic Council commitments were influential. In the Russian case, emissions were reduced to further the high-politics goal of East–West détente.

With respect to the "distance to collective optimum" and hence the question of environmental optimality and problem solving, the scientific community widely felt that the targets in the first rounds of protocols were ecologically ineffective but were nevertheless steps in the right direction.[8] The adoption of the most recent and most ambitious protocol in CLRTAP history—the 1999 Gothenburg Protocol—pointed toward a considerable improvement. If it were implemented faithfully, substantial environmental improvements would be within reach (although there would still be remaining air quality problems, as I elaborate later). What is then the overall progress picture in this regard? In February 2007, *ENDS Report* ran the headline "Europe on Track to Meet Gothenburg Targets." Recent information from the UNECE (2010) gives the following implementation overview picture:

- With regard to sulfur targets, progress is good. Between 1990 and 2006, SO_2 levels dropped by 70 percent.
- As to NO_x, progress is also substantial. Between 1990 and 2006, NO_x levels dropped by 35 percent, although the target is a 41 percent cut.
- Regarding VOCs, progress is also very good. By 2006, emissions were down 41 percent and hence going beyond the 40 percent 2010 target.
- Finally, with regard to ammonia, emissions were down 22 percent and already greater than the 17 percent target.[9]

With regard to acidification, critical-loads modeling indicates that critical levels were being exceeded in around 32.5 million hectares of ecosystem area in 1990. In comparison, implementation of the Gothenburg Protocol and the National Emission Ceilings Directive reduced this area to around 4.4 million hectares by 2010. Likewise, with regard to ozone, between 1990 and 2010 there was a reduction of around 70 percent in instances where World Health Organization guidelines for protecting human health were exceeded. Yet some vulnerable areas *still* suffered from acidification in 2010 and will continue to suffer, and World Health Organization guidelines will still be exceeded in many places.

All in all, although such an evaluation is clearly highly complex and complicated, it is still tempting to give the regime a "medium" score—not least because relative improvement brought about *specifically by the regime* seems only moderate so far, and there is still a certain distance to go to reach optimal air quality and environmental conditions in Europe.[10]

Shedding Light on Medium Impact: Problems and Institutions

In this section, I first give a summary overview of how the central explanatory perspectives of problem characteristics and problem-solving capacity can shed light on the "medium" effectiveness of CLRTAP. Then I turn to the specific institutional "cures" and techniques outlined in section three.

My general assessment is that CLRTAP's perhaps not more than medium success so far has much to do with an initially strongly malign case to deal with: first, emissions of the pollutants involved stem from societally important activities related to energy production and consumption, industrial processes, and transport, and regulators have had to deal with powerful target groups.[11] Moreover, there were generally technologically complicated and quite expensive abatement options and a lack of technological breakthroughs. Third, and not least important, there was a strong asymmetry in the transboundary flow of pollutants, with some nations being net importers and some net exporters.

This situation was furthermore worsened by an asymmetrical vulnerability to air pollutants, with unfortunate combinations such as Norway and Sweden, both considerable net importers of pollutants with particularly vulnerable soil characteristics. In addition, one must not forget the East–West context that increased the need for delicate diplomatic balancing acts as well as consensual and "nonintrusive" processes. However, for understanding the degree of success the CLRTAP regime has achieved, not least symbolized by an impressive regulatory development in the 1990s and significant implementation after 2000, the catalytic event is definitely Germany's turnabout in 1982 related to the domestic *Waldsterben* (forest dieback) uproar. Germany's shift from laggard to leader represented a crucial and symbolic lasting shift in the power balance between reluctants and pushers within the regime.

The regime's problem-solving capacity can be characterized overall as moderate, although it has clearly been increasing over time. The moderate element is reflected in institutional aspects such as a limited and stable secretarial capacity (in a period where several protocols and tasks have been added to the regime) and a consensual decision-making style. However, it should be noted that the consensus requirement has been exercised with some flexibility. Reluctant countries have simply not signed the protocols and hence not held back the rest of the countries. Perhaps the most important institutional contribution to effectiveness so far has been the evolution of the scientific working groups and hence the "scientific–political complex." As indicated earlier, this complex has contributed to considerable social learning among the parties.

The strength of entrepreneurial leadership has increased over time, related primarily to the catalytic change in German air policies. German leadership added considerable political weight to the processes in the 1980s and 1990s and was exercised with continuity at the point in the regime development process where several Nordic countries' interests became much more complicated and the initial Nordic leadership coalition broke down (from the mid-1980s on). This breakdown occurred mostly because Norway found itself in a more difficult position in the NO_x and VOC negotiations, due mainly to a developing domestic oil and gas production (and related increasing emissions). The breakdown symbolizes the fact that even if procedural leadership has been strengthened over time and general problem-solving capacity improved, many countries' basic interests have continued to be complicated in this issue area, both with regard to domestic regulatory capacities and international competition.

The relevance of the more specific institutional techniques described earlier is significant here. As noted, one central obstacle to the improvement of regime effectiveness is often marked differences in the actors' capabilities to establish or implement abatement policies that would result in the establishment of a specific *funding mechanism*. Within CLRTAP, this difference has been most prominent between Western and Eastern actors. Although the funding mechanism has not played a central role in CLRTAP, it has in fact been used there also. In order to facilitate full involvement in the negotiations of countries with economies in transition, the CLRTAP Executive Body established the Trust Fund for Assistance to Countries in Transition in 1994.[12] Donor countries, mainly from western Europe, deposited money into the trust fund, the Executive Body decided on its use, and the secretariat was authorized to offer funding to one government-designated expert from each qualified country. According to well-informed sources, this arrangement contributed to wide participation in the negotiations leading up to the 1999 Gothenburg Protocol.[13]

Another institutional technique to deal with the problem of differing capabilities is *differentiated commitments*. This technique is clearly very relevant in the CLRTAP context in that CLRTAP commitments over time have become steadily more differentiated. As indicated in the earlier overview of CLRTAP policy development, the first step away from the "common cuts" approach was taken in the 1991 VOC Protocol. The differentiation was taken a significant step further in the 1994 Second Sulfur Protocol, where the concept of critical loads was used for the first time in CLRTAP's history as a foundation for policymaking. Then this whole exercise was developed considerably in the work on the 1999 Gothenburg Protocol. Here, the outcome was differing targets for the countries involved with regard to the four substances (Wettestad 2002b).

An important institutional technique for dealing with marked differences among the parties with regard to perceptions of the (seriousness of the) environmental problems is the establishment of a well-functioning *international knowledge-improvement effort.* In this connection, CLRTAP is a very good example (Wettestad 2000). The CLRTAP model has several main aspects. First, the system's basic flexibility must be noted. As Levy pointed out back in 1993, the fact that the CLRTAP has been "consistently science- and ecosystem-driven" means that working groups have progressively been organized around potential environmental damage and permitted transfrontier pollutants to enter onto the diplomatic agenda: "This accounts for the ease with which VOCs entered the agenda, as well as for the current investigations into mercury and persistent organic compounds" (111).

Second, the formally advanced (i.e., financing based on a separate, specific protocol) and well-functioning EMEP system has represented a strong scientific foundation and "core" in the development of the regime (e.g., Sliggers and Kakebeeke 2004). A third interesting element in the CLRTAP model is the establishment of a permanent negotiating forum in the Working Group on Strategies. This body may be seen as a mediating buffer between science and politics—a "not too formal" meeting place for scientists and administrators that allows the building of consensual knowledge on both scientific and political strategic matters (cf. the buffer idea put forward in Andresen, Skodvin, Underdal, and Wettestad 2000). Regime participants emphasize the flexibility in frequency of meetings and the generally much less time-consuming formalities as an advantage of the working group style of functioning compared to the Bureau of the Executive Body meetings.

To deal with differences among the parties regarding positions on how to deal with the problems, a technique utilized within CLRTAP is the "fast-track" option, including opt-out provisions and the establishment of smaller clubs where not all regime parties participate. A good example of the latter is the establishment of the 1985 Helsinki Sulfur Protocol, where the opposition of big emitters (and parties to the CLRTAP) such as Poland and the United Kingdom was not allowed to block the adoption of the protocol, and a sort of "fast-track protocol" was simply adopted without them. By the time the 1994 Second Sulfur Protocol was adopted, Poland and the United Kingdom were onboard as signatories.

An interesting institutional technique for dealing with uncertainty among the parties with regard to domestic regulatory bite and domestic abatement possibilities is the establishment of specific *clauses on regular renegotiations of commitments*. CLRTAP is a good example, where SO_2, NO_x, and VOC levels have been negotiated and renegotiated in light of improved knowledge on environmental conditions and regulatory instruments. In this process, the initial basically flat-rate reduction commitments have been replaced with the much more ambitious and differentiated commitments in the 1999 Gothenburg Protocol.

Finally, the obstacle to effectiveness created by missing and diffuse knowledge about parties' "real" implementation and follow-up can be addressed by the establishment of a reporting and verification system and by *the establishment of a specific implementation/compliance committee*. As indicated, CLRTAP is an interesting case in this regard (Wettestad 2005a, 2005b). The first step was taken in connection with the 1991 VOC Protocol, which required the parties to establish a mechanism for monitoring compliance with the protocol. This call was followed up and made more specific in the 1994 Second Sulfur Protocol. Article 7 formally established the Implementation Committee. The committee's mandate was to review implementation and compliance, including decisions on "action to bring about full compliance with the protocol."

The more specific and practical establishment of the Implementation Committee took place in 1997. At this point, the mandate was broadened to cover the review of compliance to all the CLRTAP protocols. As to environmental politics in general, the Implementation Committee established within the ozone-layer regime has clearly served as an institutional model and is an example of interinstitutional learning. Apart from "institutional diffusion" from the ozone regime, important background factors for tougher compliance procedures in CLRTAP were the fundamental changes in the East–West relationship and the greater openness in the East that took place from the early 1990s on. These changes provided a much more beneficial setting for critical follow-up discussions, at both the international and national levels, than was the case in the 1970s and 1980s. In addition, the increased regulatory sophistication over time also created an increased need for improved institutional procedures.

The CLRTAP Implementation Committee is composed of eight legal experts from the parties. At its first meeting, it decided that it would take all decisions by consensus, and any report to the Executive Body on a specific party would first be shown to "and if necessary discussed with" that party. The secretariat's important role was also indicated, and, given the committee's limited own resources, the committee was to reach its conclusions on the basis of analyses carried out by the secretariat or experts. The Implementation Committee initially concentrated on reviewing reporting procedures and practices. Part of this work consisted of publishing overview tables of reporting "scores." So a main part of its work and hence what may be characterized as its "compliance strategy" have been to increase transparency.

The committee has adopted a strategy of gradually increased highlighting of cases of noncompliance in reporting: the first instance of noncompliance is noted without highlighting the party's name; if noncompliance happens again, the party's name is revealed; and the third time around, the committee includes the party in a recommendation to the Executive Body that action be taken to achieve reporting compliance. The committee has also systematically reviewed compliance with regard to the various protocols. Experienced negotiators emphasize the value of having both technical and legal expertise represented on the Implementation Committee. All in all, the committee has considerably improved reporting procedures and regime debates on implementation.

Concluding Comments

Overall, I think the CLRTAP regime has been cleverly designed, not least with regard to its strong focus on knowledge development and its dynamic and gradual style of development, differentiating and strengthening commitments as knowledge has improved and better technologies have become available. This improvement has in turn led to a reduction in the malignancy of the problems addressed by the regime. So there has been important *internal* interaction effects.

And, over time, as increasingly analyzed within the research community (e.g., Oberthür and Gehring 2006), important *external interaction* has taken place. Probably of most significant causal importance, several relevant and related policy developments and implementations have occurred within the EU (e.g., Wettestad 2002a). This interaction effect has probably been much more important than the fact that the increasing attention to climate change in recent years has taken some political energy out of the fight against air pollution. But then again, on balance, much climate-change policy also has beneficial effects in reducing air pollution, so his particular interaction may have been on the net positive side for the fight against air pollution spearheaded by CLRTAP.

Finally, this analysis raises some main interesting topics for further research. With regard to *actors*, increasing attention has been given over time to the role of environmental NGOs in the policymaking, implementation, and effectiveness phases. Although industry's role has been given some attention, too, it can be argued that these actors should be given (even) more attention, not least in light of the increasing attention given to market-based flexible policy instruments and corporate social responsibility. After all, industry is a very important target group, controlling the effectiveness of policies to a far higher degree than NGOs.

With regard to *policy instruments*, a very interesting development is the increasing attention and weight given to so-called flexible policy instruments, including voluntary agreements and emissions trading (see, e.g., Wettestad 2005c; Skjærseth and Wettestad 2008). This instrument is also relevant to further development of air pollution policies and CLRTAP. For instance, countries such as the Netherlands are developing an emission-trading system for air pollutants covered by CLRTAP. Among other things, this development raises the question to what extent and how it may require the need for adjusted and changed regime design. For instance, existing monitoring and verification systems are geared to checking governmental actions. It is clear that emissions trading will mean a messier picture, with a flurry of transboundary transactions and multilevel games (e.g., Wettestad 2007). Computer-based registries at the national and international levels are being established to meet this challenge. But how will this system function in practice? It is also clear that the meaning of *national* environmental policies and commitments will change in such a flexible and fundamentally transnational context. These themes are interesting for further research within the CLRTAP context.

Finally, with regard to *implementation research* (as discussed in more detail in Wettestad 2005a), a wave of research was started in the mid-1990s and ended just after the turn of the millennium. Several years have passed since these studies were done, so is it perhaps time for a new wave of in-depth country case studies? In the case of CLRTAP, to my knowledge very little implementation research has been carried out with regard to VOC, heavy metal, and POP commitments. Several countries seem to have problems with their implementation of the NO_x commitments agreed to under the Gothenburg Protocol, so this topic in particular needs closer scrutiny. And detailed studies of the implementation of the Gothenburg Protocol become increasingly meaningful and important because the target date (2010) now has passed. However, synergistic interaction must be kept in mind, particularly with EU politics.

Notes

1. I thank Rolf Lidskog and Göran Sundqvist for very helpful comments on earlier drafts. I also thank Maryanne Rygg, Fridtjof Nansen Institute (FNI), for editing assistance.

2. See Miles, Underdal, Andresen, et al. 2002, chapter 1, for a further elaboration of these two seminal perspectives.

3. Among the signatories, 15 countries and the European Commission committed themselves to the regular 30 percent reduction; 4 chose the freeze option; and 3 chose the Tropospheric Ozone Management Area (TOMA) option. See Gehring 1994, 180.

4. For analyses of the negotiations and content of the 1994 sulfur protocol, see Gehring 1994 and Churchill, Kutting, and Warren 1995.

5. Communication with Oran Young, October 1997.

6. According to Marc Levy, "The sulfur protocol *probably* had significant effects on the emission reductions in seven countries, including the largest and fourth-largest emitters in Europe (USSR and United Kingdom). A protocol that affects only these seven *probably* counts as a success" (1993, 126, emphasis added).

7. For instance, the effect of the 1988 Large Combustion Plant Directive has been significant in the United Kingdom.

8. More about this topic in Wettestad 2002b.

9. See http://www.unece.org/env/lrtap/30anniversary.html.

10. For a different, more quantitative approach to assessing CLRTAP effectiveness, see Mitchell 2002; Hovi 2003; Hovi, Sprinz, and Underdal 2003.

11. All the elements summed up in this section are further elaborated in Wettestad 1999 and 2002b.

12. See Selin 2000, 137.

13. Communication with Lars Nordberg, former head of the CLRTAP Secretariat, 16 October and 30 November 2000.

References

Andresen, Steinar, Tora Skodvin, Arild Underdal, and Jørgen Wettestad. 2000. *Science and International Environmental Regimes: Combining Integrity with Involvement*. Manchester, UK: Manchester University Press.

Benedick, Richard E. 1991. *Ozone Diplomacy*. Cambridge, MA: Harvard University Press.

Churchill, R. R., G. Kutting, and L. M. Warren. 1995. The 1994 UN ECE Sulphur Protocol. *Journal of Environmental Law* 7(2): 169–197.

DeSombre, Elizabeth, and Joanne Kaufman. 1996. The Montreal Protocol Multilateral Fund: Partial Success. In *Institutions for Environmental Aid: Pitfalls and Promise*, ed. Robert Keohane and Marc Levy, 89–126. Cambridge, MA: MIT Press.

Europe on Track to Meet Gothenburg Targets. 2007. *ENDS Europe Report* 10 (February): 7.

Gehring, Thomas. 1994. *Dynamic International Regimes: Institutions for International Environmental Governance*. Peter Lang: Berlin.

Greene, Owen. 1996. The Montreal Protocol: Implementation and Development in 1995. In *Verification 1996: Arms Control, Environment, and Peacekeeping*, ed. John B. Poole and Richard Guthrie, 407–426. Boulder, CO: Westview Press.

Hovi, Jon. 2003. Regime Effectiveness and the Oslo–Potsdam Solution. *Global Environmental Politics* 3 (3): 105–108.

Hovi, Jon, Detlef Sprinz, and Arild Underdal. 2003. The Oslo–Potsdam Solution to Measuring Regime Effectiveness: Critique, Response, and the Road Ahead. *Global Environmental Politics* 3 (3): 74–97.

Keohane, Robert O., and Marc A. Levy, eds. 1996. *Institutions for Environmental Aid: Pitfalls and Promise*. Cambridge, MA: MIT Press.

Levy, Marc A. 1993. European Acid Rain: The Power of Tote-Board Diplomacy. In *Institutions for the Earth: Sources of Effective International Environmental Protection*, ed. Peter Haas, Robert Keohane, and Marc Levy, 75–133. Cambridge, MA: MIT Press.

Miles, Edward L., Arild Underdal, Steinar Andresen, Jørgen Wettestad, Jon B. Skjærseth, and Elaine Carlin. 2002. *Environmental Regime Effectiveness: Confronting Theory with Evidence*. Cambridge, MA: MIT Press.

Mitchell, Ron. 2002. A Quantitative Approach to Evaluating International Environmental Regimes. *Global Environmental Politics* 2 (4): 58–83.

Oberthür, Sebastian, and Thomas Gehring, eds. 2006. *Institutional Interaction: How to Enhance Synergies and Prevent Conflicts between International and EU Institutions*. Cambridge, MA: MIT Press.

Odén, Svante. 1968. *The Acidification of Air and Precipitation and Its Consequences in the Natural Environment*. Ecology Committee Bulletin no. 1. Stockholm: Swedish National Research Council.

Parson, Edward A. 2003. *Protecting the Ozone Layer: Science and Strategy*. Oxford, UK: Oxford University Press.

Parson, Edward A., and Owen Greene. 1995. The Complex Chemistry of the International Ozone Agreements. *Environment* 37 (2): 16–20, 35–43.

Sand, Peter. 1990. *Lessons Learned in Global Environmental Governance*. Washington, DC: World Resources Institute.

Selin, Henrik. 2000. *Towards International Chemical Safety: Taking Action on Persistent Organic Pollutants*. Linkøping, Sweden: Department of Water and Environmental Studies.

Skjærseth, Jon B. 2000. *North Sea Cooperation: Linking International and Domestic Pollution Control*. Manchester, UK: Manchester University Press.

Skjærseth, Jon B., and Jørgen Wettestad. 2008. *EU Emissions Trading: Initiation, Decision-Making, and Implementation*. Aldershot, UK: Ashgate.

Skodvin, Tora. 2000. *Structure and Agent in the Scientific Diplomacy of Climate Change: An Empirical Case Study of Science–Policy Interaction in the Intergovernmental Panel on Climate Change*. Dordrecht, Netherlands: Kluwer.

Sliggers, Johan, and Willem Kakebeeke, eds. 2004. *Clearing the Air: 25 Years of the Convention on Long-Range Transboundary Air Pollution*. Geneva: United Nations Publication.

Social Learning Group. 2001a. *A Comparative History of Social Responses to Climate Change, Ozone Depletion, and Acid Rain*. Vol. 1 of *Learning to Manage Global Environmental Risks*. Politics, Science, and the Environment. Cambridge, MA: MIT Press.

Social Learning Group. 2001b. *A Functional Analysis of Social Responses to Climate Change, Ozone Depletion, and Acid Rain*. Vol. 2 of *Learning to Manage Global Environmental Risks*. Politics, Science, and the Environment. Cambridge, MA: MIT Press.

Underdal, Arild. 1980. *The Politics of International Fisheries Management: The Case of the Northeast Atlantic*. Oslo: Universitetsforlaget.

Underdal, Arild. 1990. Negotiating Effective Solutions: The Art and Science of Political Engineering. Unpublished paper, University of Oslo.

Underdal, Arild. 1992. The Concept of Regime "Effectiveness." *Cooperation and Conflict* 27 (3): 227–240.

Underdal, Arild, and Ken Hanf, eds. 2000. *International Environmental Agreements and Domestic Politics: The Case of Acid Rain*. Aldershot, UK: Ashgate.

UNECE (United Nations Economic Commission for Europe). 2006. *Strategies and Policies for Air Pollution Abatement*. Geneva: UNECE Information Service.

UNECE (United Nations Economic Commission for Europe). 2010. *UNECE's Convention on Long-Range Transboundary Air Pollution Celebrates 30th Anniversary*. Available at http://www.unece.org/env/lrtap/30anniversary.html.

Victor, David G. 1998. The Operation and Effectiveness of the Montreal Protocol's Non-Compliance Procedure. In *The Implementation and Effectiveness of International Environmental Commitments*, ed. David G. Victor, Kal Raustiala, and Eugene B. Skolnikoff, 137–177. Cambridge, MA: MIT Press.

Wallace, Helen, and William Wallace. 2000. *Policy-Making in the European Union*. Oxford, UK: Oxford University Press.

Wettestad, Jørgen. 1997. Acid Lessons? Assessing and Explaining LRTAP Implementation and Effectiveness. *Global Environmental Change* 7 (3): 235–249.

Wettestad, Jørgen. 1998. Participation in NOx Policy-Making and Implementation in the Netherlands and Norway: Different Approaches, but Similar Results? In *The Implementation and Effectiveness of International Environmental Commitments*, ed. David G. Victor, Kal Raustiala, and Eugene B. Skolnikoff, 381–431. Cambridge, MA: MIT Press.

Wettestad, Jørgen. 1999. More "Discursive Diplomacy" Than "Dashing Design"? The Convention on Long-Range Transboundary Air Pollution (LRTAP). In *Designing Effective Environmental Regimes: The Key Conditions*, 85–125. Cheltenham, UK: Edward Elgar.

Wettestad, Jørgen. 2000. From Common Cuts to Critical Loads: The ECE Convention on Long-Range Transboundary Air Pollution (CLRTAP). In *Science and Politics in International Environmental Regimes: Between Integrity and Involvement*, ed. Steinar Andresen, Tora Skodvin, Arild Underdal, and Jørgen Wettestad, 95–122. Manchester, UK: Manchester University Press.

Wettestad, Jørgen. 2001. The 1999 Multi-Pollutant Protocol: A Neglected Break-Through in Solving Europe's Air Pollution Problems? *Yearbook of International Co-operation on Environment and Development* 9: 35–41.

Wettestad, Jørgen. 2002a. *Clearing the Air: European Advances in Tackling Acid Rain and Atmospheric Pollution.* Aldershot, UK: Ashgate.

Wettestad, Jørgen. 2002b. The Convention on Long-Range Transboundary Air Pollution (CLRTAP). In *Environmental Regime Effectiveness: Confronting Theory with Evidence*, ed. Edward Miles, Arild Underdal, Steinar Andresen, Jørgen Wettestad, Jon B. Skjærseth, and Elaine Carlin, 197–223. Cambridge, MA: MIT Press.

Wettestad, Jørgen. 2004. Air Pollution: International Success, Domestic Problems. In *International Regimes and Norway's Environmental Policy: Crossfire and Coherence*, ed. Jon B. Skjærseth, 85–111. Aldershot, UK: Ashgate.

Wettestad, Jørgen. 2005a. The Effectiveness of Environmental Policies. In *Palgrave Advances in International Environmental Politics*, ed. Michele Betsill, Kathryn Hochstetler, and Dimitris Stevis, 373–410. New York: Palgrave Macmillan.

Wettestad, Jørgen. 2005b. Enhancing Climate Compliance: What Are the Lessons to Learn from Environmental Regimes and the EU? In *Implementing the Climate Regime: International Compliance*, ed. Olav S. Stokke, Hovi Jon, and Geir Ulfstein, 209–233. London: Earthscan.

Wettestad, Jørgen. 2005c. The Making of the 2003 EU Emissions Trading Directive: An Ultra-Quick Process Due to Entrepreneurial Proficiency? *Global Environmental Politics* 5 (1): 1–24.

Wettestad, Jørgen. 2007. Monitoring and Verification. In *The Oxford Handbook of International Environmental Law*, ed. Dan Bodansky, Jutta Brunnee, and Ellen Hey, 974–995. Oxford, UK: Oxford University Press.

Young, Oran R., and Marc A. Levy. 1999. *The Effectiveness of International Environmental Regimes: Causal Connections and Behavioral Mechanisms.* Cambridge, MA: MIT Press.

3

Institutional Linkages and European Air Pollution Politics

Henrik Selin and Stacy D. VanDeveer

Much contemporary air pollution politics and policymaking occur in institutionally dense settings involving a large set of actors interacting in complex webs of connections. Scholars and policymakers are paying a great deal of attention to characteristics and consequences of linkages among the growing number of international institutions associated with environmental politics and management (Young 2002; Raustiala and Victor 2004; Oberthür and Gehring 2006; Chambers 2008; Young, King, and Schroeder 2008; Selin 2010). As the quantity and scale of human activities with significant transnational environmental effects and the number and scope of international institutions that address environmental issues continue to grow, the structures and agents associated with these institutions increasingly intersect. As such, linkage politics—the strategic use of linkages—is of growing importance in environmental regime formation and implementation. Perhaps nowhere is this clearer than in Europe, where a host of regional and domestic environmental institutions have proliferated among states and other actors actively engaged in institution-building efforts at the global level.

Institutional linkages are not only theoretically interesting, but also empirically researchable with respect to causal pathways of influence on policymaking and implementation across separate policy venues. We offer an analytical framework for studying characteristics and effects of institutional linkages on environmental governance (Selin and VanDeveer 2003). This institutionally focused framework is agnostic with respect to the relative utility and explanatory power of different larger theories, including rational choice and constructivist informed analysis. For example, using our analytical framework to study linkage politics, participants in regime development may be seen to act strategically to exploit institutional linkages in pursuit of their defined interests and preferences, or linkages may be seen to serve as vehicles for creating and diffusing norms, ideas, and knowledge across institutions. As such, the linkage research outlined here exemplifies the contemporary convergence of theoretical approaches to the study of international environmental governance outlined in chapter 1, engaging aspects of institutional approaches, network analysis, and co-productionist dynamics.

European air pollution politics around sulfur, nitrogen, volatile organic compounds (VOCs), and ground-level ozone demonstrates the complexity of institutional linkages (Zito 2000; Wettestad 2002; Selin and VanDeveer 2003; VanDeveer 2005; Tuinstra 2008). The two primary multilateral institutions for European air pollution cooperation and regulation are the Convention on Long-Range Transboundary Air Pollution (CLRTAP) and the European Union (EU). This chapter applies the analytical framework first presented in Selin and VanDeveer 2003 to map governance and actor linkages in European air policy and to assess their implications for policymaking and implementation. It argues that many aspects of European air policy cannot be properly understood without taking CLRTAP–EU linkages into consideration. The chapter pays particular attention to linkages related to the geographical scope, membership, and regulatory authority of CLRTAP and the EU.

Analyzing Institutional Linkages

Institutional linkages affect both decision-making processes across policy forums and the effectiveness of related policy instruments. The analytical framework for empirically studying institutional linkages that we offer in Selin and VanDeveer 2003 distinguishes between governance and actor linkages. Figure 3.1 illustrates this framework, using CLRTAP and EU air policy as an example. This section uses the analytical framework to highlight major arguments and insights from related scholarly work on institutional linkages. We subsequently apply this framework to map and discuss important governance and actor linkages in European air pollution politics. With attention to the co-production of knowledge and social order in scientific, technical, and political spheres, our framework can be situated among the converging approaches discussed in chapter 1.

Governance Linkages

Consistent with other scholarly work under the broad umbrella of the “new intuitionalism” literature spanning a range of social science disciplines, the term *governance linkages* refers to structural connections between immaterial components of particular international institutions

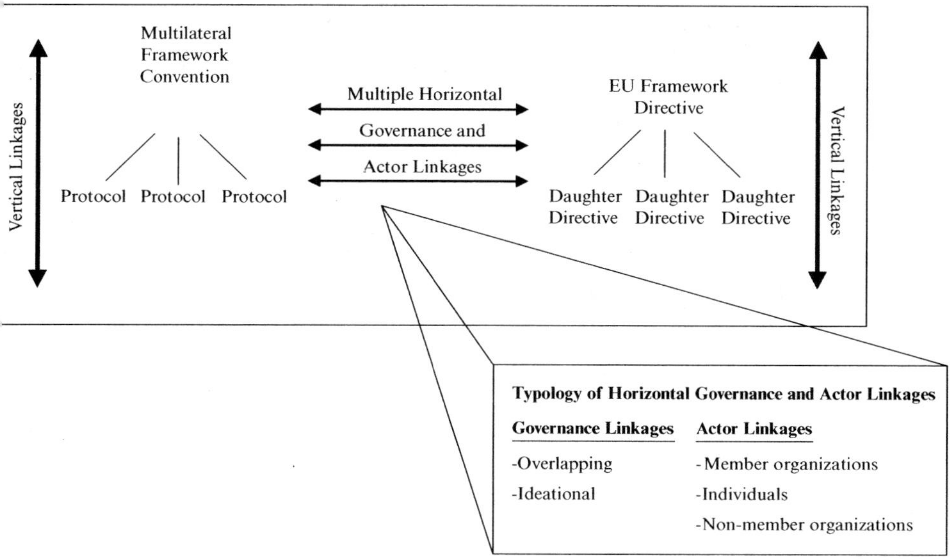

Figure 3.1
Vertical and horizontal linkages between CLRTAP and EU air pollution governance.

where institutions are interrelated sets of rules and practices that prescribe behavior, constrain activity, and shape expectations (Keohane 1989; Selin 2010). In this respect, institutions are not actors (e.g., they do not possess "agency" as generally understood) (Young, King, and Schroeder 2008). Institutions do not possess an independent capacity to act. However, they do shape actors' actions, understandings, and choices. Where institutional linkages exist, rules and activities associated with one institution exert influence on the development and performance of another institution or its participating agents or both.

Institutional governance linkages may have both synergetic and conflicting effects on management and problem solving (Rosendal 2001). Synergy arises in situations where two institutions are mutually reinforcing, which enhances overall management and problem-solving efforts in an area such as European air pollution policy. In contrast, conflict arises when the two institutions' objectives contradict each other, which may hamper the ability to govern effectively. In addition, there can be important competition across interrelated institutional venues. Then, policy developments under one institution act as a stimuli and incentive for the strengthening of regulations in the other institution. Competition across venues, however, may not benefit two institutions equally; it can enhance one institution's political and regulatory status at the expense of the other's status.

Governance linkages can operate vertically and horizontally across geographical and jurisdictional scales (Young 2002; Young, King, and Schroeder 2008; Selin 2010). For example, in a vertical linkage the operation of a regional regime may be affected by policy debates and decisions taken under a global regime. Regimes may also be embedded within broader institutions that operate within the international system or at the level of general policy arenas, including major components of public international law such as that of state sovereignty. Another common nested governance structure is the convention-cum-protocol model (Susskind 1994; Young 1996; Joyner 1998). As in the case of CLRTAP, the development of a framework convention is then followed by the creation of more detailed regulatory protocols. EU air pollution policy similarly includes the development of a broad framework directive to which regulatory requirements have been added in the form of more detailed daughter directives.

Horizontal linkages are interactions among institutions operating at a similar level of social organization (e.g., two regional institutions or two global institutions). Two types of horizontal linkages of direct relevance for CLRTAP–EU linkages are overlapping and ideational linkages (in other cases such linkages may be vertical). Overlapping linkages exist when institutions that were formed largely separately intersect on a de facto basis, impacting on each other (Young 1996). They may be functional and political (Young 1999). Functional linkages exist in biophysical and geographical terms. On European air pollution, CLRTAP and the EU cover much of the same geographical area, and emissions of a set of pollutants that are regulated under both forums are transported across national boundaries and are subject to complex interactions in the environment after deposition.

In cases of political linkages, the design and content of one institution affect the formation or operation of another (Stokke 2000; Oberthür 2001). The discussion of "interaction through commitment," "behavioral interaction," and "impact-level interaction" in Sebastian Oberthür and Thomas Gehring's 2006 edited volume on institutional interaction also addresses political linkages as understood here. In their analysis, "interaction through commitment" refers to instances where parties' acceptance of particular principles, norms, and rules in one policy forum affect stakeholders' interests and decision making in another policy forum. In cases of "behavioral interaction," behavioral changes in one policy forum affect other forums' effectiveness. In instances of "impact-level interaction," one policy instrument's ultimate governance target is affected by direct spillover effects from particular decisions or actions taken under another, functionally interdependent instrument.

Our 2003 framework divides political linkages into *norm overlaps* and *rule overlaps* (both are included in the category of interaction through commitment in Oberthür and Gehring 2006; see also Rosendal 2001). Norm overlap exists when specific policy objectives and explicit or tacit standards of appropriate behavior in particular situations carry legitimacy among participants of two or more institutions. That is, multiple institutions support consistent behavior and activities. Rule overlaps exist when similar mandatory prescriptions or proscriptions for actions have been formulated under multiple institutions. In such cases, multiple institutions contain comparable regulations. Both norm overlaps and rule overlaps are visible in European air pollution abatement as CLRTAP and the EU have increased the number of regulated substances and converged in both regulatory design and content.

Ideational linkages denote knowledge-based linkages across formally independent institutions. Such linkages are similar to the idea of cognitive interaction where stakeholders' transference of ideas shapes decisions in multiple policy forums (Oberthür and Gehring 2006). Important examples of ideational linkages include common technical and scientific concepts and practices as well as their physical expressions in the form of assessment models and reports. Organized environmental scientific assessments whereby expert knowledge is evaluated, integrated, and presented to inform decision making are increasingly utilized in international environmental cooperation (Farrell and Jäger 2006; Mitchell, Clark, Cash, et al. 2006). In other words, many aspects of the organized and usable scientific information and its representation may also function as ideational linkages between environmental policymaking venues or specific international agreements. Likewise, the use of social-economic analysis has increased in recent years—for example, in evaluating the international and domestic impact of different abatement strategies. Techniques and results of such analyses often shape policymaking in multiple forums as they serve to connect debates and outcomes across policy forums (Selin and Eckley 2003; Selin 2010).

Actor Linkages

Supplementing governance linkages, actor linkages are agent-based linkages across institutions and forums. A fundamental difference should be made between structural governance linkages that guide the behavior of actors who possess no independent ability to act and of actors or agents who have the ability to act independently and are the ones that make decisions and drive policy developments, including connections across institutions (Young, King, and Schroeder 2008; Selin 2010). It is thus critical that analysts pay close attention to how different kinds of actor linkages shape policymaking and implementation. Ways in which actor linkages can have important consequences for governance and problem solving include forum shopping, scale shopping, and other forms of purposeful use of linkages in pursuit of particular interests and policy outcomes. To this end, actors can be grouped into three major categories: *member organizations*, *nonmember organizations*, and *individuals* (see figure 3.1).

Formalized international institutions typically have state members, although membership can also include intergovernmental organizations such as the EU and United Nations specialized agencies. Also, in very few cases, such as in the Arctic Council, nongovernmental organizations (NGOs) have similar, even if not identical, participation rights as states.[1] In general, members have incentives to push for compatible norms and rules across institutions to avoid the development of conflicting goals in separate forums. Environmental policy leaders may also attempt to use institutional linkages to venue-shop for the forum most amenable to their calls for stronger policy or the venue most able to implement commonly agreed goals (Young 2002). Of course, other actors may seek to use linkages to block formulation of new policy or the implementation of existing policy.

Organizations other than formal members also influence institutional activities, often in their capacity as observers through provisions that allow for nonmembers to present viewpoints and information to member delegations. Such observers are usually environmental or scientific NGOs or private-sector associations, but other international organizations or nonmember states can also be observers. NGOs can act as catalysts of change and participate by providing scientific information, developing policy proposals, partaking in transnational coalition building, reporting from negotiations, helping monitor commitments, and mobilizing public opinion (Princen and Finger 1995; Betsill and Corell 2008). The tendency to allow NGOs more prominent roles in international cooperation arrangements and a growing importance of linkages suggest also that NGOs must spend more resources coordinating their activities and positions across institutions.

The importance of leadership is visible in much institutional development and linkage politics (Selin 2010). Facilitating leadership efforts by individual or groups of states and organizations, individual state officials and representatives of international organizations often participate in multiple international regimes and exercise leadership in a linkages context. Furthermore, individuals, often part of larger professional and personal networks, may influence multiple organizations or change institutional affiliation or both, thereby changing their roles over time. The literature on leadership and informal networks notes the importance of individuals in international cooperation when, for example, they play crucial roles in international cooperation by participating in coalition formation, agenda setting, decision making, and deal brokering, in these capacities regularly affecting processes and outcomes (Osherenko and Young 1993; Keck and Sikkink 1998; Underdal 1998).

CLRTAP and EU Air Politics: The Shift toward Brussels

As discussed in greater detail in the preceding chapters, CLRTAP was negotiated between 1977 and 1979, signed in November 1979, and entered into force in March 1983 under the auspices of the United Nations Economic Commission for Europe (UNECE) (Levy 1993). CLRTAP meetings organized by the secretariat are held at the UNECE in Geneva. As a framework convention, CLRTAP established a basis for continuing research, information sharing, and policymaking. Eight subsequent CLRTAP protocols have been negotiated to date, six of which are of relevance for addressing environmental problems associated with sulfur, nitrogen, and VOCs.[2] Table 3.1 lists these six protocols, briefly describes their major provisions, includes information on signatories and parties, and notes progress in implementation. Table 3.2 lists all 51 CLRTAP parties and presents information on the protocols' national ratification status.

The first CLRTAP pollution-specific protocols set common, time-specific reduction targets for all parties. In contrast, the multieffect/multipollution Protocol to Abate Acidification, Eutrophication, and Ground-Level Ozone, adopted in Gothenburg in 1999, sets a series of variable national emission-reduction targets for each pollutant for each state (Wettestad 2002). The Gothenburg Protocol is also based on the critical-loads concept, which was first used in the 1994 Protocol on Further Reduction of Sulfur Emissions, or Second Sulfur Protocol. The

Table 3.1
CLRTAP and Its Protocols

1979	*CLRTAP*: Framework convention, states agree to endeavor to limit and reduce air pollution or both using best available technologies and to share scientific, technical, and environmental policy information. (Adopted in Geneva, 13 November 1979; entry into force 16 March 1983; 51 parties as of 16 April 2010.)
1984	*EMEP Protocol*: Creates a multilateral trust fund for the long-term financial support of EMEP activities. (Adopted in Geneva 28 September 1984; entry into force 28 January 1988; 43 parties as of 16 April 2010.)
1985	*Protocol on the Reduction of Sulfur Emissions or Their Transboundary Fluxes (First Sulfur Protocol)*: States agree to reduce sulfur emissions or their transboundary fluxes by 30 percent from 1980 levels by 1993. All parties at the time in compliance by 1998. (Adopted in Helsinki 8 July 1985; entry into force 2 September 1987; 25 parties as of 16 April 16 2010.)
1988	*Protocol Concerning the Control of Nitrogen Oxides or Their Transboundary Fluxes (Nitrogen Oxides [NO_x] Protocol)*: States commit to freezing NO_x emissions (at 1987 or earlier levels) by the end of 1994 and to future cooperation to further reduce NO_x emissions and establish critical loads. Eighteen of the protocol's 25 parties comply with the terms of the freeze. Twelve western European states go further, aiming to reduce NO_x emissions by 30 percent by 1998. (Adopted in Sofia 31 October 1988; entry into force 14 February 1991; 34 parties as of 16 April 2010.)
1991	*Protocol Concerning the Control of Emissions of Volatile Organic Compounds or Their Transboundary Fluxes (VOCs Protocol)*: States agree to reduce VOCs emissions by 30 percent from a chosen baseline year between 1984 and 1990. Most countries choose 1988. By 2000, only seven states achieve the targeted reductions, and four more states document reductions between 16 and 21 percent. (Adopted in Geneva 18 November 1991; entry into force 29 September 1997; 24 parties as of 16 April 2010.)
1994	*Protocol on Further Reduction of Sulfur Emissions (Second Sulfur Protocol)*: Replaces the expired 1985 Sulfur Protocol. Retaining 1980 levels as a baseline, using an "effects-based" approach, and setting "target loads" based on calculated critical loads, states agree to different emissions reductions by 2000, 2005, and 2010—representing a 60 percent reduction in the difference between existing deposition levels and critical loads. By the late 1990s, 21 states either achieve their 2000 target levels or appear to be on track to do so. (Adopted in Oslo 14 June 1994; entry into force 5 August 1998; 27 parties as of 16 April 2010.)

Table 3.1
(continued)

1999	*Protocol to Abate Acidification, Eutrophication, and Ground-Level Ozone (multipollutant/multieffect protocol)*: States agree on national emission ceilings for 2010 for sulfur, NO_x, VOCs, and ammonia to ensure that critical loads and levels for acidification, eutrophication, and ground-level ozone are not exceeded over time. When the protocol is fully implemented, Europe's sulfur emissions will be cut by at least 63 percent, NO_x emissions by 41 percent, VOC emissions by 40 percent, and ammonia emissions by 17 percent compared to 1990. (Adopted in Gothenburg 30 November 1999; entry into force 17 May 2005; 25 parties as of 16 April 2010.)

Source: Adapted from Selin and VanDeveer 2003.

concept of critical loads denotes an attempt to establish a critical environmental threshold (or exposure level) below which no harmful effects occur. Emission reductions are divided among countries on a regional basis in an attempt to minimize the costs for a region as a whole while meeting set environmental and human health targets for all parts of the region (Bäckstrand 2001; Grennfelt and Hov 2005; Tuinstra, Hordijk, and Kroeze 2006).

Detailed reviews of the CLRTAP parties' compliance with protocol obligations, including the 1999 Gothenburg Protocol, conducted by CLRTAP's Implementation Committee in 2007 demonstrate that most countries comply with reporting requirements but face a number of challenges in meeting emission-reduction commitments (UNECE 2007b, 2007c). On reporting, 80 to 100 percent of parties were in broad compliance with reporting requirements between 2003 and 2006 (UNECE 2007b). Yet the quality and consistency of this information were considerably more variable when it came to what was necessary to assess parties' compliance with the various obligations under the Gothenburg Protocol. For 19 countries reporting, the Implementation Committee deemed the information too erroneous, incomplete, or irrelevant to allow for a definitive assessment of these parties' compliance. The information did demonstrate that 13 states were in noncompliance with at least one emissions requirement.

The comprehensive review of the Gothenburg Protocol also assessed progress toward critical-loads goals for acidification and eutrophication as well as critical levels for ground-level ozone (UNECE 2007b). If fully implemented, the protocol would measurably reduce emissions and make

Table 3.2
CLRTAP Parties and Protocol Signatories (S) and Ratifications (R)

CLRTAP Party	EMEP	First Sulfur Protocol	Nitrogen Oxides Protocol	VOC Protocol	Second Sulfur Protocol	Multipollutant/ Multieffect Protocol
Albania		R	R			
Armenia						S
Austria	**R**	**R**	**R**	**R**	**R**	S
Azerbaijan						
Belarus	R	R	R			
Belgium	**R**	**R**	S	S	**R**	**R**
Bosnia and Herzegovina	R					
Bulgaria	**R**	**R**	**R**	**R**	S	**R**
Canada	R	R	R	S	R	S
Croatia	R		R	R	R	R
Cyprus	**R**					**R**
Czech Republic	**R**	**R**	**R**	**R**	**R**	**R**
Denmark	**R**	**R**	**R**	**R**	**R**	**R**
Estonia	**R**	**R**	**R**	**R**		
Finland	**R**	**R**	**R**	**R**	**R**	**R**
France	**R**	**R**	**R**	**R**	**R**	**R**
Georgia						

Table 3.2
(continued)

CLRTAP Party	EMEP	First Sulfur Protocol	Nitrogen Oxides Protocol	VOC Protocol	Second Sulfur Protocol	Multipollutant/ Multieffect Protocol
Germany	**R**	**R**	**R**	**R**	**R**	**R**
Greece	**R**		**R**	S	**R**	S
Hungary	**R**	**R**	**R**	**R**	**R**	**R**
Iceland						
Ireland	**R**		**R**		**R**	S
Italy	**R**	**R**	**R**	**R**	**R**	S
Kazakhstan						
Kyrgyzstan						
Latvia	**R**					**R**
Liechtenstein		R	R	R	R	S
Lithuania		**R**	**R**	**R**	**R**	**R**
Luxembourg	**R**	**R**	**R**	**R**	**R**	**R**
Malta	**R**					
Monaco	R			R	R	
Montenegro	R					
Netherlands	**R**	**R**	**R**	**R**	**R**	**R**
Norway	R	R	R	R	R	R
Poland	**R**				S	S

Table 3.2
(continued)

CLRTAP Party	EMEP	First Sulfur Protocol	Nitrogen Oxides Protocol	VOC Protocol	Second Sulfur Protocol	Multipollutant/ Multieffect Protocol
Portugal	**R**			S		**R**
Republic of Moldova						S
Romania	**R**					R
Russian Federation	R	R	R		S	
Serbia and Montenegro	R					
Slovakia	**R**	**R**	**R**	**R**	**R**	**R**
Slovenia	**R**		**R**		**R**	**R**
Spain	**R**		**R**	**R**	**R**	**R**
Sweden	**R**	**R**	**R**	**R**	**R**	**R**
Switzerland	R	R	R	R	R	R
FYR of Macedonia	R	R	R	R	R	
Turkey	R					
Ukraine	R	R	R	S	S	
United Kingdom	**R**		**R**	**R**	**R**	**R**
United States	R		R	S		R
EU	**R**		**R**	S	**R**	**R**

Note: The 27 EU member states, the EU, and their ratifications are given in bold.
Source: Compiled from http://www.unece.org/env/lrtap.

progress toward the set critical loads and levels. Yet for all three environmental issues scientific research and modeling conducted after the signing of the protocol suggests that progress toward these goals—and toward the environmental and human health effects of meeting the goals—is likely to be considerably smaller than was believed at the time of the protocol's adoption in 1999. In particular, little progress toward reducing eutrophication has been achieved, and although incidents of extremely high ground-level ozone have been reduced, background levels continue to increase. With the approach of the protocol's 2010 deadline for emissions reductions and the knowledge that serious environmental challenges remained, CLRTAP parties agreed to launch negotiations over protocol revisions, including the setting of more stringent emissions ceilings for 2020.

Parallel to CLRTAP policymaking, the EU has become increasingly active on air pollution–abatement issues. Compared to CLRTAP's transboundary environmental focus, the EU has often focused more on local human health aspects, sometimes in collaboration with the World Health Organization. Whereas some EU air policy dates back to the 1970s, the main policy developments began in the early 1980s and greatly accelerated in the 1990s and 2000s. Current EU air policy sets standards for ambient air quality, establishes national emissions ceilings (NECs) for acidification and eutrophication, regulates emissions from stationary sources, targets emissions from road vehicles and ships, and seeks to reduce emissions of VOCs. Table 3.2 shows which of the eight CLRTAP protocols the European Commission and the separate EU member states have signed and ratified. Table 3.3 lists key EU air policy legislation and CLRTAP protocols together, illustrating the dual development of CLRTAP and EU air policies over time.

The development of the Gothenburg Protocol and subsequent efforts to implement it, in conjunction with the EU's creation of NECs for all member states, greatly accelerated EU–CLRTAP linkages (of course, these linkages were also shaped by a concurrent enlargement of EU membership to include countries that were already CLRTAP parties, discussed further later). The effort to revise the Gothenburg Protocol with more stringent emissions ceilings and potentially to add commitments to reduce short-lived climate-change forcers such as black carbon and tropospheric ozone (as well as their precursor pollutants) will likely further expand the number and scope of the already significant governance and actor linkages between CLRTAP and the EU. That is, CLRTAP remains both an important venue for scientific, technical, and political

Table 3.3
Main EU Air Legislation and CLRTAP Protocols

Year	Directives and Protocols
1970	EC Directive 70/220: Set initial community standards for spark-ignition engines for carbon monoxide and hydrocarbons. NO_x added in 1978. All standards tightened through 1987.
1972	EC Directive 72/306: Set early limits on the opacity of emissions from diesel engine vehicles.
1975	EC Directive 75/716: Set initial community standards for the sulfur content in fuel. Standards for lead and VOC emissions added subsequently. All standards tightened progressively through 1996.
1979	UNECE CLRTAP.
1980	EC Directive 80/779: Set initial community air quality limit values on sulfur dioxide and particles.
1984	EC Directive 84/360: Framework directive on emissions from industrial plants.
1984	CLRTAP EMEP Protocol.
1985	CLRTAP Sulfur Protocol.
1988	EC Directive 88/76: Revoked principle of optional harmonization. EC Directive 88/77 and 88/436: Set initial community standards for diesel engines. EC Directive 88/609, Large-Combustion Plant Directive: Set both technology-based emission limits and overall percentage reductions.
1988	CLRTAP NO_x Protocol.
1990	EC Directive 90/313: Established requirement of public access to environmental information. EC Directive 90/1290: Established European Environmental Agency to collect and provide environmental information to the European Community and member states. Became operational in 1992.
1991	EC Directive 91/441: Set limits for road vehicle emissions, including VOCs.
1991	CLRTAP VOC Protocol.
1992	EU Directive 92/72: Required ozone monitoring and established air quality standards for ozone.
1993	EU Directive 93/12: Set limit values for sulfur content of gas oils.
1994	CLRTAP Sulfur Protocol on Further Reductions
1996	EU Directive 96/61, Integrated Pollution Prevention and Control Directive: Established multimedia permitting system and strengthened implementation mechanisms. EU Directive 96/62, Air Quality Framework Directive: Set air quality standards and requires monitoring and notification of the public when standards are not met. Intended to beget daughter directives.

Table 3.3
(continued)

Year	Directives and Protocols
1999	CLRTAP Multipollutant/Multieffect Protocol
1999	EU Directive 99/30, daughter directive to 96/62: Set limit values for sulfur, nitrogen, particulate matter, and lead in ambient air (replaced by Directive 08/50).
1999	EU Directive 99/32: Set maximum permitted concentration for sulfur in heavy fuel oil.
2000	EU Directive 2000/69, daughter directive to 96/62: Set limit values for carbon monoxide and benzene (replaced by Directive 08/50).
2001	EU Directive 2001/80 on large-combustion plants: Set emission limits for sulfur dioxide, NO_x, and dust.
2001	EU Directive 2001/81 on National Emission Ceilings: Set ceilings for acidifying substances and ozone-forming pollutants for each member state.
2002	EU Directive 2002/3, daughter directive to 96/62: Set limit values for ground-level ozone (replaced by Directive 08/50).
2004	EU Directive 2004/107, daughter directive to 96/62: Set target levels for arsenic, cadmium, mercury, nickel, and polycyclic aromatic hydrocarbons in ambient air.
2008	EU Directive 2008/50 on Ambient Air Quality and Cleaner Air for Europe: Merged the first three daughter directives to 96/62 in a single piece of legislation and set new standards for fine particulates.

Source: Adapted from Selin and VanDeveer 2003.

air pollution–abatement work and an influential part of European air pollution linkage politics even as the EU has greatly expanded its own air pollution work.

In 1996, the EU adopted a framework directive on ambient air quality standards (96/62). In 1997, the European Commission launched a comprehensive Acidification Strategy. Pursuant to this strategy, the EU ratified the 1994 CLRTAP Second Sulfur Protocol, strengthened controls on sulfur content of certain fuels through the 1998 auto-oil regulation, and adopted a new directive on sulfur in liquid fuels in 1999. The EU also reached an agreement on a 2001 framework directive (01/81), setting NECs for each member state for sulfur, nitrogen, and ammonia. These targets were to be met by 2010, and activities were to be coordinated with the EU's strategy for reducing levels of ground-level ozone. In addition, the EU in 2001 adopted an updated directive (01/80) that strengthened earlier regulations on emissions from large-combustion plants under a revised directive that was adopted in 1988 (88/609).

Beginning in the late 1990s, the EU developed several pollution-specific daughter directives to the 1996 framework directive on air quality standards (see table 3.3). These daughter directives set mandatory limit values for the concentrations of a multitude of pollutants in air: the first daughter directive (99/30) set standards for sulfur dioxide, nitrogen dioxide, particulates, and lead; the second (00/69) set standards for carbon monoxide and benzene; the third (02/3) addressed ground-level ozone; and the fourth (04/107) targeted levels of arsenic, cadmium, mercury, nickel, and polycyclic aromatic hydrocarbons in ambient air. This relatively recent expansion of EU air policy has in part shifted the locus of European air pollution abatement from Geneva to Brussels. The national NEC standards, however, differ somewhat from those set under the Gothenburg Protocol.

The European Commission launched its Clean Air for Europe (CAFE) program in 2001 with the support of other EU organizations and member states. The aim of the CAFE program was to develop a long-term, strategic, and integrated EU air pollution policy by gathering the increasingly complex web of major air policies developed over the past decades within a single, integrated policy program. Early activities carried out under the CAFE program included a review of the implementation and effectiveness of the major air quality directives and programs in all member states; the review and update of European air quality thresholds and national emission ceilings; efforts to improve regional and domestic air quality monitoring and modeling; an increase in dissemination of public information on air quality and pollution standards; and the setting of priorities for further action on assessment and pollution abatement at the EU level.

Work under the CAFE program led to the formulation of the EU Thematic Strategy on Air Pollution in 2005. Consistent with this strategy, the European Parliament and the EU member states in 2008 adopted a directive on ambient air quality and cleaner air for Europe (08/50). This directive merged the first three daughter directives of the 1996 framework directive together with a 1997 Council Decision on information exchange into a single piece of legislation without making any changes to the air quality objectives set out in those daughter directives. The 2008 directive also set new air quality objectives for fine particulates. In addition, the European Commission, in close collaboration with other EU organizations and member states, is working to integrate air policy measures with relevant climate-change and energy policy initiatives, which have been the focus of extensive policymaking in the 2000s.

Although the European Commission announced in its 2005 Thematic Strategy on Air Pollution that the NECs needed to be cut further to meet regional and local human health and environmental goals, many member states continued to struggle to meet their requirements for 2010. An analysis conducted by the European Environment Agency (EEA) based on final data for 2005 and preliminary data for 2006, showed that 16 out of 27 member states were set to exceed national limits for at least one of the four key pollutants covered by the NEC Directive (Ågren 2009). This conclusion is consistent with findings by the CLRTAP Implementation Committee in 2007, and the EU emissions ceilings and reduction commitments are generally more stringent than those under CLRTAP. The renegotiation of the Gothenburg Protocol may also change emissions ceilings for some participating countries. In addition, EU member states are involved in discussions under CLRTAP on the possible creation of a hemispheric pollution treaty focusing on ozone and fine particulates (UNECE 2007a).

Institutional Linkages in European Air Pollution Abatement

A host of major governance and actor linkages shape the development and implementation of European air pollution policy.

Governance Linkages

There are several overlapping linkages between CLRTAP and the EU. Functional overlaps are seen in the overlapping geographical areas and sets of pollutants covered by CLRTAP and the EU. Throughout the 1980s, CLRTAP was the policy forerunner, serving as a model for EU members pushing for expanded EU air regulations. In the 1990s, both CLRTAP and the EU expanded their regulatory scope by increasing the number of pollutants and emission sources they regulate, also setting more ambitious reduction targets. In this process, the EU closed some of the gap between itself and CLRTAP (Wettestad 2002). The European Commission also planned to launch the NEC Directive ahead of parallel work in CLRTAP on the multipollutant/multieffect protocol. However, this ambition failed, in part as a result of the resignation of the European Commission following a corruption scandal. Despite growing CLRTAP–EU linkages, there are differences in policy outcomes. For example, CLRTAP and EU national emission ceilings differ slightly for countries that are members to both. The revisions of the Gothenburg Protocol and the NEC Directive offer opportunities to close these gaps, but adding more attention to climate-change drivers may also create new EU–CLRTAP commitment differences.

Political overlaps in the form of shared norms and rules between CLRTAP and EU air policy are in part the result of conscious strategic linking by members to both, fueled by their interests in harmonizing activities and regulations as well as the European Commission and environmental leader states' interests in strengthening pan-European air pollution policy (Wettestad 2002). Shared CLRTAP–EU norms include emission-reduction goals for transboundary pollutant transport based on similar reduction methods. Such reduction measures include the use of critical-loads approaches to calculate desirable emission reductions, the setting of country-based emissions ceilings, and the use of best-available techniques and emission-limit values for specific emissions sources. Common CLRTAP–EU rules, including specific emissions-reduction requirements, have also increased over time, as seen in similarities between the CLRTAP 1999 Gothenburg Protocol and the EU NEC Directive.

The use of similar regulatory approaches in CLRTAP and the EU has created multiple ideational linkages, including the concurrent use of critical loads for abating acidification, eutrophication, and ground-level ozone based on the Regional Acidification Information System (RAINS) atmospheric transport and deposition model. The critical-loads concept and the RAINS model were developed for the CLRTAP negotiations by researchers and modelers at the International Institute of Applied Systems Analysis (IIASA) and first applied to policy in the CLRTAP Second Sulfur Protocol (Bäckstrand 2001; Castells and Ravetz 2001; Tuinstra, Hordijk, and Kroeze 2006). Technical and scientific work on critical loads has been used in both CLRTAP and the EU since the middle of the 1990s, and CLRTAP and the EU have formulated policy based on similar technical and scientific assumptions and environmental assessments. Furthermore, a substantial portion of funding for IIASA's air pollution work comes from the EU. The CAFE process built upon earlier scientific work under CLRTAP, and the European Commission draws heavily from results by the IIASA's integrated assessment teams in its ongoing work on the NEC Directive (Tuinstra 2007, 2008).

Actor Linkages

Actor linkages shape air policy processes and outcomes under both CLRTAP and the EU. The EU (or, formally, the European Community) is a party to CLRTAP, represented by the European Commission at technical and political meetings. The European Commission has the competence to negotiate in CLRTAP on behalf of all member states on issues where there is common EU legislation in place. The European Commission was not very active in CLRTAP during the 1980s (Zito 2001). However, as EU air policy developed and the impact of gradually more stringent CLRTAP policies on EU legislation grew more significant, the European Commission came to play a much more active role in CLRTAP in the 1990s, including working to coordinate member states' positions in CLRTAP negotiations.[3] These efforts were intended to increase EU influence by putting forward common positions and to ensure that member state positions in CLRTAP were compatible with evolving EU policy (Wettestad 2001; Zito 2001).

The need to enhance cooperation and coordination with CLRTAP has been highlighted in numerous internal EU discussions, although the European Commission has stressed that "such co-operation must not lead to any dilution of Community competence or control over EU policy in this area" (Commission of the European Communities 2001, section 5.9.1). The increasing involvement of the European Commission in CLRTAP, however, has also given rise to political conflicts between the European Commission and individual member states. For example, the European Commission refused to sign the 1999 CLRTAP multipollutant/multieffect protocol on the grounds that it did not cut emissions enough. European Commission representatives did not want to signal that they accepted what they saw as weak emission-reduction targets. However, all EU member states signed the protocol, many of them arguing that it was a first step that could be strengthened later and that it was better to reach agreement on the proposed reduction rather than have no agreement at all. By 2009 and 2010, many EU air pollution policy leader states were positioned as the driving forces behind the revision of the Gothenburg Protocol with more stringent goals for 2020.

Informal cooperation between the European Commission (and in particular DG Environment) and the CLRTAP Secretariat was established in the mid-1990s in connection with the European Commission's early work on the EU Acidification Strategy. In 2001, the CLRTAP Executive Body accepted a formal invitation from the commission to participate in the Technical Analysis Group to coordinate the technical analysis work carried out within the CAFE program. It also agreed to set up a CLRTAP–EU group to enhance coordination of developing strategies and work programs. Commission representatives, moreover, participate in meetings of the smaller CLRTAP Bureau of the Executive Body. Such explicit CLRTAP–EU coordination is likely to further increase linkages across the two forums and increase incentives for actors to use linkages strategically to pursue policy goals.

A second important actor linkage between CLRTAP and the EU is, of course, their overlapping national membership. Since the early 1990s, the EU membership has increased from 12 to 27 countries. All 27 EU member states are also among the 51 parties to CLRTAP. In general, those states typically regarded as policy leaders and laggards in CLRTAP play similar roles in community air policy. The Scandinavian countries and the Netherlands, joined by Germany after 1982, have been the traditional environmental policy leader states in CLRTAP (Levy 1993). Early political resistance to stronger air pollution policy and regulations came primarily from the United Kingdom, supported by Italy, Greece, and Ireland in both CLRTAP and the EU. When Spain and Portugal, who were laggards in CLRTAP, joined the EU in 1986, the group of relatively less affluent and less green EU members grew.

The accession of Austria, Sweden, and Finland in 1995 significantly affected EU air policy and increased linkages between CLRTAP and the EU (Wettestad 2002). These countries, together with Denmark, Germany, and the Netherlands, pushed hard for more stringent EU air pollution policy in the late 1990s to bring it in line with the more ambitious but less enforceable CLRTAP agreements. For example, EU environmental leaders engaging in forum shopping pushed to have an NEC Directive finished before the signing of the 1999 CLRTAP multipollutant/multieffect protocol. They hoped that higher EU standards would set a precedent that could be used to strengthen the CLRTAP agreement. Because the NEC Directive is more ambitious than the national targets formulated under the CLRTAP Gothenburg Protocol, EU members face larger reductions than if there had not been an NEC Directive.

Review of the NEC Directive was directly linked with the concurrent review of the Gothenburg Protocol in the 2000s. The CLRTAP technical support for this review rests on the European Monitoring and Evaluation Program (EMEP), carried out by IIASA and the Norwegian Meteorological Institute. This CLRTAP work is partially funded by the European Commission. However, the overlapping assessments do not necessarily produce the same policy outcomes in CLRTAP and the EU, suggesting that institutional and membership differences between the two venues influence activities. Although the use of common CLRTAP and EU assessments helps to reduce overall European costs of assessments by avoiding duplication, issues of cross-institutional legitimacy have to be taken into consideration during the assessment work because the CLRTAP and EU memberships differ. States that participate in both institutions are also likely to have more influence on policy and assessment than those participating in only one. In both venues, leaders in air pollution abatement are generally found among the most active and engaged member states.

Leader states on European air policy often exercise influence by sponsoring technical and scientific activities around particular policy issues, as they seek to shape the direction of policy and technical debates. CLRTAP and EU bodies have institutionalized connections to the same organizations for data gathering and calibration as well as for scientific and technical assessment and modeling, with ramifications for policy outcomes. EU incorporation of the critical-loads concept and the RAINS model was facilitated by previous acceptance of these ideas by EU member states and the European Commission within CLRTAP. Environmental policy leaders perceived the use of RAINS and critical loads to be in their interests, and other EU member states had already accepted their use. Furthermore, the RAINS model and the critical-loads approach had established scientific and technical credibility, and any attempt to develop alternatives would be time consuming and costly.

Part of the surge in ratification and implementation efforts for CLRTAP and its protocols after the mid-1990s is attributable to the drive by many eastern European states to join the EU (Carmin and VanDeveer 2005; VanDeveer 2005, 2006). In turn, the European Commission and the EU member states pushed EU candidates both to harmonize their domestic air pollution policies with EU legislation and to ratify CLRTAP agreements. This process continued in the 2000s, as countries such as Albania, Croatia, and Macedonia stepped up CLRTAP-ratification efforts as they pursued EU accession. As the European Commission and EU member states have repeatedly stated, such harmonization is a necessity for EU membership. At the same time, the commission recognized that "CLRTAP is the main forum through which the European Community and the Member States can influence and promote emission reductions in non-member countries. Further action in some of those countries has proven generally to be highly cost-effective" (Commission of the European Communities 1999, 39).

Targeted EU support of poorer southern and eastern countries has repeatedly assisted new members in improving domestic levels of environmental protection. Accessions have also not halted progressive environmental policymaking in the EU, in part because the EU has worked to raise domestic regulations in the new member states (Carmin and VanDeveer 2005; Skjærseth and Wettestad 2007). Supported by EU programs and member states, efforts to expand the ratification and implementation of CLRTAP protocols continue apace. For example, in 2007 the Dutch government launched a program to support the ratification and implementation of the three most recent CLRTAP protocols in five non-EU Balkan countries (Albania, Bosnia, Macedonia, Montenegro, and Serbia). The program, developed and implemented with participation of the CLRTAP Secretariat, also serves to support the Balkan countries' efforts to harmonize their domestic policies with EU environmental policy, serving their aspirations for EU membership.

Since the 1990s, CLRTAP, with the support of the European Commission and many western European EU member states, has invested heavily in programs to engage and develop technical and administrative capacity in the eastern European, Caucasus, and Central Asian (EECCA) countries. In 2005, the CLRTAP Executive Body launched an action plan for EECCA countries. The UNECE also developed the project Capacity Building for Air Quality Management and the Application of Clean Coal Combustion Technology in Central Asia. These efforts helped to drive resources (human, economic, and informational) toward the ratification, implementation, and compliance of CLRTAP protocols in the EECCA countries. They also shifted a greater portion of CLRTAP's attention to the countries east of the EU border (Sliggers 2007).

Nonmember organization linkages between CLRTAP and the EU consist of a multitude of NGOs, business organizations, and research organizations active in both forums. The role of environmental NGOs in European air pollution politics is similar to their role in other issue areas and forums. In addition to policy advocacy, they act as awareness raisers and "watchdogs," communicating "bad behavior" to national NGOs, the media, and the public. The most active environmental NGO on air pollution issues has been the Air Pollution & Climate Secretariat (originally named the Swedish NGO Secretariat on Acid Rain).[4] It has been an effective ally to leader states in CLRTAP and the EU. For example, some of the early material and information that the Air Pollution & Climate Secretariat distributed reached domestic groups in countries that would not have been accessible for distribution of materials through more official government channels (Hägerhäll 1993).

Air Pollution & Climate Secretariat staff carefully calculate where and when to devote resources in the two institutions. The secretariat's credibility and legitimacy were developed over its decades-long engagement with CLRTAP activities. Starting in the 1990s and accelerating in the early years of the twenty-first century, however, the secretariat has shifted increasing amounts of its resources to EU bodies and debates on air pollution and climate-change issues—a shift in focus similar to that found among IIASA's experts. This change reflects the increasingly strict and legally binding EU regulatory system. The secretariat's widely read signature publication *Acid News* frequently compares progress (or lack thereof) in both venues, publicizes noncompliance by states in both, and covers the details of European Commission legal actions against noncompliant member states. In other words, like environmental leader states, the secretariat is attracted to the EU because of the importance of environmental policy in Brussels and the binding and increasingly enforceable nature of EU environmental policy.

The European Environmental Bureau (EEB), established in 1974, acts as a federal umbrella organization. The office in Brussels links the activities and lobbying efforts of more than 140 domestic NGOs from 31 countries. Air policy is one of the EEB's main issue areas, and many domestic environmental and public-health NGOs are also highly active in air pollution and air quality issues. Whereas the EEB's regional lobbying work focuses primarily on assessment and policymaking efforts within and among the main EU organizations, the EEB also attends and advocates at major CLRTAP meetings, including the annual Executive Body meetings. Likewise, leading industry organizations such as the European Petroleum Industry Association and the European oil companies' association for environment, health, and safety in refining and distribution have traditionally been more active in trying to influence policy developments in the EU than under CLRTAP.

Both CLRTAP and the EU also make extensive use of IIASA's expertise and models as well as the network of experts surrounding IIASA's activities. Both also use EMEP data to grapple with the complex challenges of combating tropospheric ozone pollution, acidification, and eutrophication. Although most European states have domestic monitoring systems, EMEP attempts to build and maintain a regional European monitoring and environmental data-analysis system—with some geographic gaps remaining in the system's coverage. EMEP is essentially the only pan-European monitoring and data-gathering analysis institution of this kind, and CLRTAP parties periodically review and seek to improve its activities. EMEP was established under CLRTAP in the first protocol to the convention, and this investment has proved invaluable to air pollution analysts and policymakers over the system's 30-plus years.

The EEA, which was established by a regulation in 1990 and became operational in 1994, coordinates the European Environment Information and Observation Network. The EEA was the first EU body to admit eastern European candidate countries as members in the 1990s. It continues to work with several non-EU members, including Norway and Turkey, whose nationals are employed at the EEA. The EEA does not formulate policy, but it provides member states and the European Commission with information on the state, pressures, and sensitivities of the European environment. The EEA's air pollution and CAFE-related activities support the review of existing EU legislation, including calculation of national emission ceilings. Its work with both member and candidate states have not only aided the development of EU air policy but also indirectly benefited activities under CLRTAP by generating data and building awareness among a host of CLRTAP parties.

The building of important personal networks is also quite common in European air pollution politics, where individuals can play significant roles in scientific bodies and political committees (Björkbom 1999). Many countries send the same delegates to CLRTAP and EU scientific and political meetings. Such individuals, in particular those from the environmental pusher states, often play important leadership roles in formulating policy (Bäckstrand 2001; Selin 2003). In such webs of interaction, informal networks made up of people from intergovernmental organizations, states, and NGOs can be influential vehicles for the diffusion of knowledge, ideas, policy proposals, and individual career advancement across forums. Individuals frequently move among various European air pollution institutions and organizations, taking their experience, knowledge, ideas, and personal contacts with them.

The strategic use of national experts is common in both CLRTAP and the European Commission. CLRTAP assessments and development of policy proposals are dependent on financing by member states and are carried out by national experts. Environmental leader states frequently place national civil servants in strategic political positions within the European Commission to work on predetermined issues and projects as a way to influence assessments, reports, and policy outcomes. In particular, national experts from Sweden, Norway, Germany, and the United Kingdom were very active on the development of the EU acidification and tropospheric ozone strategies and the NEC Directive (Zito 2001).[5]

Such work included explicit efforts to harmonize activities and policies between CLRTAP and the EU on air pollution issues.

There are also close connections between CLRTAP and the EU through the use of overlapping assessment models (Tuinstra 2008). In particular, the application of the RAINS model has led to the increased inclusion of IIASA alumni and other RAINS model users in EU policymaking circles. Such experience is an asset to individuals seeking to participate in policymaking. Individuals with a history of working or training at IIASA now populate many EU bodies, state administrative bodies, research institutes, EMEP bodies, and universities. Many technical and scientific assessments have been produced by such networks and evaluated at international workshops and seminars and have been crucial in designing politically acceptable solutions during negotiations. The formalization of closer cooperation between CLRTAP and the EU beginning in the late 1990s facilitated such interactions. This dominance of IIASA-connected experts, ideas, and assessment tools is not without critics or limitations, but it has been an important force linking and integrating CLRTAP and EU policymaking.

Institutional Linkages and the Future of European Air Pollution Politics

Institutional linkages are an integral part of European air pollution politics, influencing the behavior of actors as well as policymaking and implementation. Governance and actor linkages between CLRTAP and EU air policy have shaped the formulation of gradually more ambitious policy in both forums. Many governance linkages stem from the fact that CLRTAP and the EU target many of the same pollutants, using similar policy approaches. Furthermore, many states, organizations, and individuals link scientific and technical activities and policymaking processes and outcomes across CLRTAP and the EU. Linkages emerge both from strategic action by actors and from the iterated interaction among actors through networks and co-production processes. The dramatic political changes in Europe since the late 1980s followed by a series of EU enlargements in 1995, 2004, and 2007 have importantly resulted in a shift in the center of gravity for pan-European air pollution politics from CLRTAP meetings in Geneva to EU deliberations in Brussels.

European air pollution politics demonstrates how institutional linkages create pathways for influence across policy forums as key actors engage in linkage politics and forum shopping. That is, environmental leaders use linkages and the desire to harmonize standards and regulations between CLRTAP and the EU as they attempt to strengthen policy in one or both of those policy forums. This behavior has been shaped by the enlargement of the EU, from 12 members in 1994 to 27 in 2007, as more countries seek complimentary policies. Many efforts to link assessments and policymaking between CLRTAP and the EU have also been supportive of air pollution–abatement efforts in both forums. This finding is consistent with other comparative work on interactions across international environmental institutions, which found more evidence of synergetic results than of disruptive outcomes (Oberthür and Gehring 2006).

European air pollution linkage politics continues, as many countries struggle to meet both their CLRTAP and EU emission-reduction targets even as leading states and NGOs push for more stringent goals in the years to come. As CLRTAP participants review implementation progress and negotiate revisions to the 1999 Gothenburg Protocol, such work is closely intertwined with the European Commission's efforts to revise the NEC Directive and to set stricter standards for ambient air quality. Paralleling the situation from a decade earlier during the creation of the Gothenburg Protocol and the NEC Directive, the Gothenburg Protocol revision is likely to be completed before the NEC revision (set for 2013). This situation holds the promise of capturing synergistic effects across the two forums as well as the risk to create inconsistent or duplicative commitments. Many states that are members of both the EU and the CLRTAP would like to see the CLRTAP and EU negotiations move in parallel, thereby giving these states another forum to attempt to influence the European Commission. Leader states are likely to continue to venue-shop for opportunities to achieve the most stringent agreements possible as actors in both venues draw on overlapping bodies of knowledge.

The revisions of the Gothenburg Protocol and the NEC Directive are not complicated merely by overlapping membership and states' interest in pursing their policy preferences for emissions ceilings in each forum, but also by an expansion of the policy agenda. Growing concern among air pollution experts about ship-based emissions, fine particulate concentrations, and climate change drivers all intersect with more traditional concerns about land-based emissions of sulfur, nitrogen oxides, VOCs, and ammonia. Linkages between climate change and other air pollution issues are growing as the EU seeks to implement regional climate change–mitigation goals. The possible development of a hemispheric pollution treaty would furthermore create a new set of linkages because it would involve the establishment of a third multilateral institution with strong assessment and policy linkages to CLRTAP and the EU.

The growing quantity, scope, and stringency of climate change–related air pollution and energy policy goals and policy instruments should also be expected to further complicate and expand linkage politics across air pollution policy venues and issues. European and North American climate and energy politics, like aspects of transatlantic chemicals regulations, already show signs of the increasing importance of institutional linkage (Selin and VanDeveer 2006, 2009; Schreurs, Selin, and VanDeveer 2009). If climate change policymaking efforts accelerate on both sides of the Atlantic in the coming years, the linkages of CLRTAP and EU air pollution policymaking to entities beyond the European space seem likely to increase as well (Selin and VanDeveer 2010). As CLRTAP moves further into climate change–related pollution issues and the EU continues to develop its comprehensive climate change work, European and transatlantic institutional density and linkages may become even more complex and influential.

In the wake of the many important air policy expansions in CLRTAP and EU since the 1990s, there has been a greater focus on compliance issues in recent years. As many countries struggle to meet their emission-reductions targets and pollution-prevention mandates, many proponents of strong pollution controls—including the European Commission, the European Parliament, environmental leader states, and environmental NGOs—believe that comprehensive monitoring and enforcement are critical to ensuring good air quality. Because the EU has much stronger legal and political mechanisms for implementation review and enforcement, including ultimately the European Court of Justice, the partial shift in focus from policymaking to implementation will serve to further increase the relative importance of the EU over CLRTAP. This shift will be even clearer if the EU continues to expand its membership to countries that are also CLRTAP parties.

Notes

1. Six indigenous peoples groups are designated Permanent Participants in the Arctic Council in collaboration with the eight Arctic states. This designation gives them full participation rights in all meetings and activities under the Arctic Council, but they lack the full voting rights of the Arctic states. See Selin and Selin 2008.

2. In addition, CLRTAP has also negotiated protocols on heavy metals and persistent organic pollutants (Selin 2010).

3. The authors' observation of European Commission staff practices at CLRTAP negotiations, 1996–1998. For the commission's view, see Commission of the European Communities 2001, section 5.9.1.

4. The secretariat was established in connection with the 1982 Stockholm Conference on the Acidification of the Environment. Four (later expanded into five) Swedish NGOs set it up to provide information on the acid rain problem and to promote abatement action. The Swedish Environmental Protection Agency funded the secretariat until 1992, when the Ministry for the Environment overtook the role as prime financier. See Bäckstrand and Selin 2000, 96.

5. Norway is not an EU member but acts as an observer in Brussels and has the right to have national experts working within the European Commission.

References

Ågren, Christer. 2009. 16 States Exceed Emission Limits. *Acid News* 1 (March): 1 and 4–5.

Bäckstrand, Karin. 2001. *What Can Nature Withstand? Science, Politics, and Discourses in Transboundary Air Pollution Diplomacy*. Lund, Sweden: Lund University Press.

Bäckstrand, Karin, and Henrik Selin. 2000. Sweden—A Pioneer of Acidification Abatement. In *International Environmental Agreements and Domestic Politics: The Case of Acid Rain*, ed. Arild Underdal and Kenneth Hanf, 87–108. Aldershot, UK: Ashgate.

Betsill, Michele, and Elisabeth Corell, eds. 2008. *NGO Diplomacy: The Influence of Nongovernmental Organizations in International Environmental Negotiations*. Cambridge, MA: MIT Press.

Björkbom, Lars. 1999. Negotiations over Transboundary Air Pollution: The Case of Europe. *International Negotiation* 4 (3): 389–410.

Carmin, JoAnn, and Stacy D. VanDeveer, eds. 2005. *EU Enlargement and the Environment*. New York: Routledge.

Castells, Nuria, and Jerry Ravetz. 2001. Science and Policy in International Environmental Agreements: Lessons from the European Experience on Transboundary Air Pollution. *International Environmental Agreement: Politics, Law, and Economics* 1 (4): 405–425.

Chambers, W. Bradnee. 2008. *Interlinkages and the Effectiveness of Multilateral Environmental Agreements*. Tokyo: United Nations University Press.

Commission of the European Communities. 1999. *Proposal for a Directive of the European Parliament and the Council on National Emission Ceilings for Certain Atmospheric Pollutants/Proposal for a Directive of the European Parliament and the Council Relating to Ozone in Ambient Air*. COM (1999) 125. Brussels: Commission of the European Communities, 9 June.

Commission of the European Communities. 2001. *Communication from the Commission. The Clean Air for Europe (CAFE) Programme: Towards a Thematic Strategy for Air Quality*. COM (2001) 245. Brussels: Commission of the European Communities, 4 May.

Farrell, Alexander E., and Jill Jäger. 2006. *Assessments of Regional and Global Environmental Risks: Designing Processes for the Effective Use of Science in Decisionmaking*. Washington, DC: Resources for the Future.

Grennfelt, Peringe, and Øystein Hov. 2005. Regional Air Pollution at a Turning Point. *Ambio* 34 (1): 2–10.

Hägerhäll, Bertil. 1993. The Evolving Role of NGOs. In *International Environmental Negotiations: Process, Issues, and Context*, ed. Gunnar Sjöstedt, Uno Svedin, and Britt Hägerhäll Aniansson, 50–76. Stockholm: Swedish Council for Planning and Coordination of Research.

Joyner, Christoffer C. 1998. *Governing the Frozen Commons: The Antarctic Regime and Environmental Protection*. Columbia: University of South Carolina Press.

Keck, Margaret E., and Kathryn Sikkink. 1998. *Activists beyond Borders: Advocacy Networks in International Politics*. Ithaca, NY: Cornell University Press.

Keohane, Robert O. 1989. *International Institutions and State Power: Essays in International Relations Theory*. Boulder, CO: Westview Press.

Levy, Marc. 1993. European Acid Rain: The Power of Tote-Board Diplomacy. In *Institutions for the Earth: Sources of Effective International Environmental Protection*, ed. Peter M. Haas, Robert O. Keohane, and Marc A. Levy, 75–132. Cambridge, MA: MIT Press.

Mitchell, Ronald B., William C. Clark, David W. Cash, and Nancy M. Dickson, eds. 2006. *Global Environmental Assessments: Information and Influence*. Cambridge, MA: MIT Press.

Oberthür, Sebastian. 2001. Linkages between the Montreal and Kyoto Protocols: Enhancing Synergies between Protecting the Ozone Layer and the Global Climate. *International Environmental Agreement: Politics, Law, and Economics* 1 (3): 357–377.

Oberthür, Sebastian, and Thomas Gehring, eds. 2006. *Institutional Interaction in Global Environmental Governance: Synergy and Conflict among International and EU Policies*. Cambridge, MA: MIT Press.

Osherenko, Gail, and Oran R. Young. 1993. The Formation of International Regimes: Hypotheses and Cases. In *Polar Politics: Creating International Environmental Regimes*, ed. Oran R. Young and Gail Osherenko, 1–21. Ithaca, NY: Cornell University Press.

Princen, Thomas, and Matthias Finger. 1995. *Environmental NGOs in World Politics*. London: Routledge.

Raustiala, Kal, and David G. Victor. 2004. The Regime Complex for Plant Genetic Resources. *International Organization* 58 (2): 277–309.

Rosendal, G. Kristin. 2001. Impact of Overlapping International Regimes: The Case of Biodiversity. *Global Governance* 7 (2): 95–117.

Schreurs, Miranda, Henrik Selin, and Stacy D. VanDeveer, eds. 2009. *Transatlantic Environment and Energy Politics: Comparative and International Perspectives*. Aldershot, UK: Ashgate.

Selin, Henrik. 2003. Regional POPs Policy: The UNECE/CLRTAP POPs Agreement. In *Northern Lights against POPs: Combatting Toxic Threats in the Arctic*, ed. David L. Downie and Terry Fenge, 111–132. Montreal: McGill-Queens University Press.

Selin, Henrik. 2010. *Global Governance of Hazardous Chemicals: Challenges of Multilevel Management*. Cambridge, MA: MIT Press.

Selin, Henrik, and Noelle Eckley. 2003. Science, Politics, and Persistent Organic Pollutants: Scientific Assessments and Their Role in International Environmental Negotiations. *International Environmental Agreement: Politics, Law, and Economics* 3 (1): 17–42.

Selin, Henrik, and Noelle Eckley Selin. 2008. Indigenous Peoples in International Environmental Cooperation: Arctic Management of Hazardous Substances. *Review of European Community & International Environmental Law* 17 (1): 72–83.

Selin, Henrik, and Stacy D. VanDeveer. 2003. Mapping Institutional Linkages in European Air Pollution Politics. *Global Environmental Politics* 3 (3): 14–46.

Selin, Henrik, and Stacy D. VanDeveer. 2006. Raising Global Standards: Hazardous Substances and E-Waste Management in the European Union. *Environment* 28 (10): 6–17.

Selin, Henrik, and Stacy D. VanDeveer. 2009. *Changing Climates in North American Politics: Institutions, Policymaking, and Multilevel Governance*. Cambridge, MA: MIT Press.

Selin, Henrik, and Stacy D. VanDeveer. 2010. Multilevel Governance and Transatlantic Climate Change Politics. In *Greenhouse Governance: Addressing American Climate Change Policy*, ed. Barry G. Rabe, 336–352. Washington, DC: Brookings Institution Press.

Skjærseth, Jon B., and Jørgen Wettestad. 2007. Is EU Enlargement Bad for Environmental Policy? Confronting Gloomy Expectations with Evidence. *International Environmental Agreement: Politics, Law, and Economics* 7 (3): 263–281.

Sliggers, Johan. 2007. Fifteen Ratifications on the Way. *Acid News* 1 (March): 22.

Stokke, Olav Schram. 2000. Managing Straddling Stocks: The Interplay of Global and Regional Regimes. *Ocean and Coastal Management* 43 (2–3): 205–234.

Susskind, Lawrence E. 1994. *Environmental Diplomacy: Negotiating More Effective Global Agreements*. Oxford, UK: Oxford University Press.

Tuinstra, Willemijn. 2007. Preparing for the European Thematic Strategy on Air Pollution: At the Interface between Science and Policy. *Environmental Science & Policy* 10 (5): 434–444.

Tuinstra, Willemijn. 2008. European Air Pollution Assessments: Co-Production of Science and Policy. *International Environmental Agreement: Politics, Law, and Economics* 8 (1): 35–49.

Tuinstra, Willemijn, Leen Hordijk, and Carolien Kroeze. 2006. Moving Boundaries in Transboundary Air Pollution Co-Production of Science and Policy under the Convention on Long-Range Transboundary Air Pollution. *Global Environmental Change* 16 (4): 349–363.

Underdal, Arild. 1998. Leadership in International Environmental Negotiations: Designing Feasible Solutions. In *The Politics of International Environmental Management*, ed. Arild Underdal, 101–127. Dordrecht, Netherlands: Kluwer Academic.

UNECE (United Nations Economic Commission for Europe). 2007a. *Executive Summary of the Interim Report of the Task Force on Hemispheric Transport of Air Pollution.* ECE/EB.AIR/GE.1/2007/13. Geneva: UNECE, June 22.

UNECE (United Nations Economic Commission for Europe). 2007b. *Review of the 1999 Gothenburg Protocol.* ECE.EB.AIR 2007/10. Geneva: UNECE, September 28.

UNECE (United Nations Economic Commission for Europe). 2007c. *The Tenth Report of the Implementation Committee.* ECE/EB.AIR/2007/3. Geneva: UNECE, September 26.

VanDeveer, Stacy D. 2005. European Politics with a Scientific Face. In *Assessments of Regional and Global Environmental Risks*, ed. Alex Farrell and Jill Jaeger, 25–63. Washington, DC: Resources for the Future.

VanDeveer, Stacy D. 2006. Assessment Information in European Politics: East and West. In *Global Environmental Assessments: Information and Influence*, ed. Ronald B. Mitchell, William C. Clark, David W. Cash, and Nancy Dickson, 113–150. Cambridge, MA: MIT Press.

Wettestad, Jørgen. 2001. Designing Effective Environmental Regimes: The Conditional Keys. *Global Governance* 7 (3): 317–341.

Wettestad, Jørgen. 2002. *Clearing the Air: European Advances in Tackling Acid Rain and Atmospheric Pollution.* Aldershot, UK: Ashgate.

Young, Oran R. 1996. Institutional Linkages in International Society: Polar Perspectives. *Global Governance* 2 (1): 1–24.

Young, Oran R. 1999. Institutional Interplay: Exploring the Vertical Dimension. Paper presented at the Open Meeting of the Human Dimensions of Global Environmental Change Research Community, Shonan Village, Japan, June.

Young, Oran R. 2002. *The Institutional Dimensions of Environmental Change: Fit, Interplay, and Scale.* Cambridge, MA: MIT Press.

Young, Oran R., Leslie A. King, and Heike Schroeder, eds. 2008. *Institutions and Environmental Challenges: Principal Findings, Applications, and Research Frontiers*. Cambridge, MA: MIT Press.

Zito, Anthony. 2000. *Creating Environmental Policy in the European Union*. London: MacMillan.

Zito, Anthony. 2001. Epistemic Communities, Collective Entrepreneurship and European Integration. *Journal of European Public Policy* 8 (4): 583–603.

4

Transboundary Science for Transnational Air Pollution Policies in Europe

Bernd Siebenhüner

The problem of air pollution has been at the center of most industrialized countries' environmental policymaking from early on. Damages to human health, buildings, crops, and forests through emissions of toxic gases and particles have raised political concerns since the nineteenth century and have been addressed in numerous national regulations. The problem of air pollution, however, does not remain within national borders but is transboundary in nature. This fact poses a particular challenge to political processes and the political approaches to combat air pollution with regard to international coordination and agreements.

Within Europe, several levels of institutions and approaches to address transboundary air pollution have occurred over the past 30 years. The transboundary dimension of air pollution is a particular challenge for this continent because emission sources are numerous and widespread over countries that are relatively small and in close vicinity—at least when compared to North America. Only in Southeast Asia and East Asia does a similar problem structure of transboundary air pollution exist at the moment.

In addition to the transboundary dimension, air pollution also requires much scientific knowledge on the definition of the problems, the identification of dangerous substances, the monitoring of possible damages, the understanding of causal mechanisms, and the analysis of policy responses and their impact on ecosystems. Given the complexity of ecological systems with their interconnectedness, numerous causal mechanisms, synergies, and accumulation effects between different substances and abrupt system changes, scientific knowledge is limited and has to cope with paramount uncertainties that can be only partially eliminated by rigorous research and monitoring.

This chapter sets out to investigate how European policy approaches of the past 30 years have addressed these problems—specifically, the

Convention on Long-Range Transboundary Air Pollution (CLRTAP) and the European Union's (EU) air pollution policy. Both have somewhat different institutional contexts and address air pollution from different angles. The chapter compares both approaches with regard to the interface between scientific knowledge and political decision making and what effects these designs have had in the political realm.

The second section develops the analytical framework for the subsequent analysis, and the third section studies the evolution of the CLRTAP and its major concepts and achievements. The fourth section addresses the EU's air pollution policies, and the final section discusses and compares both approaches on the basis of the central criteria.

Designing Transnational Science–Policy Interfaces

Designing the interface between science and political decision making poses a significant challenge to both realms and must be thorough to function effectively. Different designs of this interface will influence the overall effects that scientific knowledge has in political processes (Mitchell, Clark, Cash, et al. 2006).

The chapter seeks to study effects of science actors and scientific knowledge in a broader sense that is not related to a particular ideal point of reference, as the notion of effectiveness suggests. Effects are understood rather as changes in the behavior of decision makers, mainly governments and international bodies, such as the EU. This notion of effects is generally known as "outcome" in policy studies and is to be distinguished from the actual impacts on target indicators, such as improvement of environmental parameters. It follows a line of reasoning in recent works on international institutions that have generally focused on changes of actor behavior instead of on environmental improvement (Keohane, Haas, and Levy 1993). Science actors such as epistemic communities (in the sense of Haas 1992), international scientific organizations, and individual research institutions can have effects on political decision-making processes by changing other actors' knowledge and belief systems and by influencing policy agendas and advancing specific solutions.

In this context, social studies of science and technology have stressed the need for a perspective on the underlying norms of both realms and the way each intrudes into the other realm. As opposed to the idea of pure science, which is regarded as a completely independent provider of knowledge governed by its own rules and quality-assurance mechanisms, this field of study regards science as being influenced by political needs and interests, in particular in those fields of knowledge where uncertainty is high (Jasanoff 1987, 1990). In this situation, clarification and transparency about which roles are to be taken by particular individuals and organizations are necessary for a functional organization in the science–policy interface.

Kai Lee (1993), for instance, argues that actors can have only one role in this interface: they are either politicians or scientists.[1] However, he acknowledges the existence of roles that lie in between both ends of the spectrum, such as administrators organizing scientific knowledge for the purposes of political decision making and professional analysts who build on their scientific knowledge in their practical activity in society, comparable to doctors or engineers. Recent case study–based research has led to the different conclusion that individuals might fill in several roles in this spectrum—sometimes in a very effective way (Farrell, VanDeveer, and Jäger 2001).

The literature on boundary organizations, with its particular focus on organizations at the science–policy interface, also builds on the notion of the existence of a dividing line between the scientific and the political. This line, however, constantly remains under negotiation between the individuals and organizations involved (Gieryn 1996; Guston 1999). In these boundary negotiations, it is decided who is entitled to deal with which kinds of issues. The outcome might be, for example, that the scientists have to surrender to political imperatives or that the policymakers are exclusively dependent on scientists and their recommendations. It is clear that there are no objectively given boundaries between the two spheres and that this line cannot be drawn with sufficient clarity due to various overlaps and the connectedness between scientific knowledge and political and social implications. Thus, from this analytical perspective, the chapter focuses on this intersection and the organization of the science–policy interface and its relation to the political outcomes of the particular policy processes. It distinguishes between three factors that influence the political effects of scientific knowledge: interaction, participation, and reflexive mechanisms.

First, there are different options for how to design *interactions* between scientists and policymakers within assessment processes. In one extreme case, interaction may be limited to an absolute minimum by isolating scientists from the policy process. In the other extreme, there would be an intense and thoroughly crafted collaboration between individuals from both fields (Farrell and Jäger 2006). The interaction might take place either in formalized settings with clearly defined individual roles or in loose forms of cooperation maintained mostly through personal engagement among certain individuals. Hence, the influence of national governments hinges largely on the design of these forms of interaction (Farrell, VanDeveer, and Jäger 2001). According to the theory of boundary organizations, each individual entity or group, irrespective of its position on this spectrum, is well advised to maintain its self-identity and protect its sources of legitimacy and credibility. Whatever the formal interaction structures might look like, there is nearly always the opportunity for informal communication between the two groups. The underlying hypothesis is that more frequent and more intense communication leads to better-informed policy outcomes and more observable political effects.

Second, answering the questions of which individuals and organizations are allowed to play a role in the science–policy interaction and when and how is essential in determining the interaction's political impact. *Participation* in different phases of an interaction can vary substantially; different actors might be involved in problem identification, generation of new knowledge, or final communication of the results. Moreover, there is a spectrum of options regarding how to arrange participation. On the one hand, broad participation of many different individuals and actor groups such as governments, scientists from many related fields, nongovernmental organization (NGO) representatives, and businesspeople might be valuable for ensuring legitimacy and policy impact. On the other, an exclusive participation of scientists might be of great value for maintaining credibility (Biermann 2001, 2002; Farrell, VanDeveer, and Jäger 2001). However, generally speaking, there should be a link between the environmental problem at hand and the participation of the political entities that are affected by the problem or involved in its causation or both (Young 2002; Young, King, and Schroeder 2008). In transboundary science–policy processes, those nations' governments whose national territories or people are affected by the problem in one way or another might have a strong interest in participating. They might be major contributors to damaging emissions, or they might be victims of these emissions or of their effects, such as sea-level rise. Thus, the researchable hypothesis is that the participation of more stakeholders in science–policy interactions leads to more observable effects in political decision making.

Third, *reflexive mechanisms* should make use of past experiences by reflecting on them and turning them into action of any kind. As Micha

Popper and Raanan Lipshitz (2000) describe, reflexive mechanisms are structural elements in organizations for reflecting on experiences and for initiating and implementing learning processes. To be effective in bringing to bear scientific and nonscientific knowledge in air pollution policies, learning mechanisms such as learning laboratories, evaluation units, and workshops that involve both scientists and political decision makers need to include these different forms of knowledge. . Reflexive mechanisms might be very informal, as in personal communication among participants in assessments or they might be highly formalized and sophisticated, as in institutionalized committees with a distinct set of rules of procedure (Siebenhüner 2002, 2003). Their existence and specific forms can be expected to have an effect on the actual policy outcomes of the science–policy interactions through individual or collective learning. However, impact on the behavior of the actors and organizations involved is not guaranteed by the fact that reflexive mechanisms are in place.

The CLRTAP

The CLRTAP is one of the main international efforts to combat acidification and other damages to ecosystems, buildings, and human health in Europe and North America. Acknowledging the fact that the pollutants travel thousands of kilometers through the air, the convention addresses the problem of air pollution on an international scale. Signed in 1979 and entered into force in 1983, it was the first international legally binding instrument to deal with this problem. Since then, eight protocols on different pollutants and procedural matters have been signed under the auspices of the United Nations Economic Commission for Europe (UNECE). Almost all European countries, including Russia and the European Community as a supranational body, as well as the United States and Canada on the North American continent participated.

The convention has set up a multilayer organization to arrange for the various countries' participation and to include scientific assessments on the numerous technical and scientific questions of air pollution.[2] The Working Group on Strategies, including representatives from all the parties of the convention, has negotiated all the protocols. It prepared the documents for the final decision making by the Executive Body. The Working Group on Effects has carried out scientific assessments of the effects of air pollution, but assessments on methodological questions regarding modeling and monitoring have been organized in the form of task forces that report to the Executive Body. The whole organizational structure and the institutions of scientific assessment have evolved over the three decades since the signing of the convention.[3]

With regard to conceptual changes in the convention process, one can distinguish three major overlapping phases. This distinction does not focus on the advancements in the negotiations and the different protocols established under the convention, but in particular on how the convention process dealt with scientific knowledge and the uncertainties involved.[4]

Setting Up the Organizational Structure (1979–1988)

After the convention was signed, several years passed before the organizational structure with the two main working groups and various task forces was established. The convention foresaw the establishment of the Executive Body and the secretariat for administrative support. It put the Executive Body in charge to create "working groups to consider matters related to the implementation and development of the present Convention" and also stated the "desirability of establishing a framework for a cooperative environmental monitoring programme" (UNECE 1979, Articles 9 and 10). As a first important step toward setting up an effective organizational structure for the assessments, the first protocol under the convention ensured funding for the European Monitoring and Evaluation Programme (EMEP).[5] By collecting, processing, and analyzing air pollution data, EMEP was the precondition for both political negotiations and scientific advice. EMEP grew out of an Organisation for Economic Co-operation and Development research program on long-range transboundary air pollution that existed from 1973 to 1975. This earlier assessment was effective inasmuch as it demonstrated that air pollutants have the potential to travel very long distances. Although it had a large number of uncertainties in its final documents, it convinced many European policymakers of the need for pollution control (VanDeveer 1998). Nevertheless, due to its limited scope and design it left some room for improvement.

Also in a very early stage in the convention process, the Working Group on Effects came into being to provide the information on which the negotiations for further protocols were based. As phrased in the convention, its task was to conduct research on "effects of sulphur compounds and other major air pollutants on human health and the environment, including agriculture, forestry, materials, aquatic and other natural ecosystems and visibility, with a view to establishing a scientific basis for dose/effect relationships designed to protect the environment" (UNECE

1979, Article 7d). On each of these areas where effects of air pollution were observable, special international programs were established under the supervision of the Working Group on Effects.[6]

EMEP and the Working Group on Effects were dedicated to providing information for the political process and were seen as necessary to convince policymakers to take action, but they were hardly used in the process of designing the first series of protocols. Although the convention considered scientific information pivotal to its processes, the first negotiations and the resulting protocols were driven more by the political will to reduce emissions than by the consideration of all related scientific information: "There was a link between policy and science but it wasn't very strong," stated one CLRTAP secretary.[7] Because the first three technical protocols were based on a flat-rate approach that pursued fixed rates of emission reductions without clear links to effects, scientific information about effects was peripheral. However, this information was important for the political process in three ways. First, it helped to put pollution control on the political agenda and to raise policymaker's awareness of the problems caused by air pollutants. Second, it helped to identify major sources of air pollution. And, third, it shaped the way uncertainty was handled in the negotiation processes as carried out in the Working Group on Strategies (Eckley 1999).

The last working group that came into being was the Working Group on Abatement Technologies. It focused on technologies and industrial processes to reduce emissions. Like the Working Group on Effects, it consisted of scientists working for governments or civil servants who were responsible for science. Therefore, it can be seen as a scientific assessment body because it was in charge of providing scientific and technical information for the preparation of protocols to the Working Group on Strategies and the Executive Body.

The assessment bodies in this phase utilized international workshops as means to facilitate the uptake of scientific knowledge in the convention process. The workshops were organized around various specific questions that came out of the task forces and working groups and included scientific expertise from outside of the convention process. Thus, the set of participating external scientists varied from one workshop to the other. In the workshops, external scientists from the participating countries brought in their detailed and highly specific knowledge and were asked to simplify it and to make it more accessible for policymakers and integrative modeling approaches.[8] They thereby helped to develop the integrated assessment models that became crucial in the second phase.

Moreover, the workshops were a means to ensure scientific credibility by including high-ranking scientists. However, the organization of a workshop is a lengthy process. The idea to organize such an event had to be approved by the specific working group and the Executive Body in their annual work plans. In total, it took a year or even longer from the idea to the actual event.

Introduction of the Critical-Loads Concept and the RAINS Model (1988–1999)

It was already in the 1988 Protocol Concerning the Control of Nitrogen Oxides that the critical-loads concept was mentioned as a possible conceptual framework for future protocols under the convention. Emerging from the strong political will in Scandinavia to avoid threshold effects in lakes, this concept implied that scientific knowledge was included more thoroughly into the process (Bull 1992). Whereas the so-called first-generation protocols simply adopted a flat-rate approach using a fixed percentage of emission reductions, the second-generation protocols starting with the sulfur protocol of 1994 were effects based and adhered to the critical-loads approach. According to Peter Haas and David McCabe, "the [critical-loads] concept was virtually revolutionary in diplomacy because it assigned differential national obligations based on the carrying capacity of vulnerable ecosystems rather than [on] a politically equitable (and arbitrary) emission cut" (2001, 327).

What was the critical-loads concept about that it managed to revolutionize the political as well as the scientific assessment processes of the convention? As phrased in the Nitrogen Oxides Protocol, "Critical load means a quantitative estimate of the exposure to one or more pollutants below which significant harmful effects on specified sensitive elements of the environment do not occur according to present knowledge" (UNECE 1988, Article 1). The 1991 Protocol Concerning the Control of Emissions of Volatile Organic Compounds also included damaging effects on human beings into this definition. In essence, this concept focused on the regional damages and approached pollution control from this end in order to protect ecosystems and human health effectively (Bull 1992, 1995). By contrast, previous pollution-abatement strategies concentrated on overall emission reductions at all sources largely irrespective of the actual damages of specific emissions in certain regions. Whereas those strategies were more or less a matter of political negotiation over how many emissions have to be reduced, in the critical-loads approach the different emission-reduction requirements could be identified for different regions. This approach allowed for greater cost effectiveness. At the same time, effective pollution control under the critical-loads approach required reliable scientific input on the regional effects and emission-reduction requirements. The critical-loads concept thus helped to generate significant effects on political decisions.

The International Cooperative Program for Mapping Critical Loads and Levels and Their Exceedances was put in charge of collecting the available regional data on damages to forests, crops, natural vegetation, soils, surface and groundwater, and materials based on data about the environmental sensitivities of these ecosystems. It subsequently produced maps on the state of ecosystems with a resolution of 150 square kilometers. These maps indicated the exceedance rates against the commonly agreed 5 percentile damage of bearable emissions.[9] Scientists from the national focal centers put in charge by the national governments carried out data collection and mapping exclusively.

In the CLRTAP, the crucial means to link these scientific data systems with the political processes under the critical-loads approach was the Regional Acidification Information System (RAINS) model—an integrated assessment model developed by the International Institute of Applied Systems Analysis (IIASA).[10] It allowed for the analysis of alternative strategies to reduce acidification and other damaging impacts on a regional scale, and in this respect it was a "scenario-generating device" (Alcamo, Shaw, and Hordijk 1990, 1). The model integrates information on

- projections of future economic, agricultural, and energy development in the participating countries
- present and future emissions of air pollutants resulting from these activities
- options for reducing emissions and the costs of these measures
- atmospheric dispersion characteristics of these pollutants, and
- environmental sensitivities of ecosystems especially toward acidification.

Therefore, it can be seen as an approach to integrate the scientific and technical information generated by the various task forces and working groups under the convention in order to provide highly policy-relevant information. Because the critical-loads approach required significant scientific informational input to the political negotiation processes to determine the relevant emission-reduction targets and to decide about the best abatement strategies to reach these targets, the RAINS model quickly gained great prominence; although alternative models were available, they could not catch up with RAINS model's advancements (Lidskog and Sundqvist 2004).[11]

At the beginning of the development of the RAINS model in 1991, the IIASA organized a workshop and invited one negotiator and one technical expert from each delegation to the convention. The workshop has been crucial for the nonscientific experts involved to capture the difficulties of the critical-loads concept and to deal with the complexity of the model and the issues of air pollution in general.[12]

The use of the RAINS model in the negotiations on further emission reductions, in particular in the preparation of the Protocol on Further Reduction of Sulfur Emissions (Second Sulfur Protocol) signed in 1994, gave a boost to the applicability of scientific information in political processes. In addition, negotiators accepted the model as scientifically credible and politically legitimate in its development (Patt 1998). Thus, the introduction of the critical-loads concept was a significant factor in securing the effects of scientific knowledge in the political processes. It led to changes in the general perception of the role of science in the political process.

Restructuring after the Gothenburg Protocol (1999–2008)

The Protocol to Abate Acidification, Eutrophication, and Ground-Level Ozone signed in Gothenburg in 1999 can be perceived as a "third-generation" pollution-control protocol because it takes into account the interdependencies of environmental effects such as acidification, eutrophication, and ground ozone accumulation and the related pollutants based on the critical-loads approach in order to achieve effective reductions of environmental damages at least cost (Jagusiewicz 1999). Given the current state of knowledge on the causes and damages of air pollution, the Executive Body decided at this time that so far no further protocols would be negotiated under the CLRTAP.[13] As a consequence, some major restructuring took place in order to redirect the organizational setup toward the reversed focus on implementation and review issues (UNECE Executive Body 1999).

The process of restructuring also affected the scientific bodies and their work. The Working Group on Abatement Techniques was dissolved because its main task was the preparation of technical annexes to the protocols; the Working Group on Strategies was renamed in order to extend its responsibility to the review of the current protocols and for possible revisions and initiatives. In addition, the focus on health effects was introduced in collaboration with the World Health Organization (WHO). Moreover, the Task Force on Integrated Assessment Modeling was moved from the Working Group on Strategies across to EMEP to link it more closely to the scientific activities. The integrated assessment modeling was thereby moved away from the political processes, resulting in possible decreases in the relevance of its work for the political decision makers. In general, the political influence of the integrated assessment modeling in the CLRTAP process decreased after the conclusion of the 1994 Second Sulfur Protocol, whereas the scientific orientation of the modeling was strengthened.[14]

However, the restructuring did not end the reflection over future strategies and pathways to combat transboundary air pollution. The topic was discussed at a workshop in 2000 in Sweden to facilitate further exchange of new knowledge and learning—in particular because it covered topics central to the development of relevant scientific knowledge such as the sector integration, the linkages between global models and regional or national scales, the role of economic models, and advancements in integrated assessment modeling (Nordic Council of Ministers 2000). New directions for the scientific assessment under the convention included the integration of health aspects and the narrowing of the monitoring grids from 150- to 50-square-kilometer grids. These efforts led to the assessment's greater relevance to regional users as well as to users from other communities such as the public health community. In 2007, a general review of the Gothenburg Protocol revealed significant emission reductions in most European countries that signed the protocol (see Wettestad, chap. 2 in this volume). Nevertheless, many non-EU countries still lag behind in their emission-reduction policies. The review also demonstrated that monitoring and abatement efforts remain largely separated in the different pollutant categories, and their interaction and multiple effects are still uncertain to a large extent. Scientific research and monitoring can thus help little to advance with this particular target of the Gothenburg Protocol (UNECE 2007).

The Overall Picture

With the successful adoption of several protocols and implementation procedures, the CLRTAP has been fairly effective in political terms. However, this assessment does not include the actual implementation of the convention and its protocols, which is under current debate. From the perspective of the science–policy interaction, the processes have been successful, and the question remains how much the particular design of the science–policy interactions contributed to this success.

Interaction

In the CLRTAP setup, there is a continuous exchange of individual negotiators and scientists between the convention's different political bodies, in particular the Executive Body and the Working Group on Strategies (and Review). Most Executive Body members also take part in the working group meetings and allow for a transfer of knowledge among theses bodies. Second, the exchange of individuals between the different working groups and even between the task forces and international cooperative programs (ICPs) under one working group is rare, although there is an intense exchange of individuals and knowledge between each working group and its affiliated task forces and ICPs because the latter two have to report to the working group.

Most of the scholars examining the CLRTAP assessments highlight the continuous, iterative communication among scientists and negotiators as one of the crucial preconditions for the successes of the convention process (VanDeveer 1998; Eckley 1999; Tuinstra, Hordijk, and Kroeze 2006).[15] Formal communication relationships exist mainly between the working groups and their subordinate task forces and ICPs as well as between the working groups and the Executive Body—mediated mostly through the Working Group on Strategies and Review. Because the working groups and task forces constantly exchange people who attend meetings of other groups or of the Executive Body, intense interaction is possible.[16] This structure allows the flow of scientific and technical information into the negotiation processes. It is rendered highly effective through the informal communication among the members of the core group of active scientists and negotiators. Because there are personal relationships among the people of this core group, informal communication plays an important role in the interaction between negotiators and scientists as well as between the two groups. In the negotiations about the Second Sulfur Protocol, the scientists from the Task Force on Integrated Assessment Modeling even confronted the negotiators with an ultimatum in order to move the negotiations forward. After having informed the political decision makers about all the possible scenarios for emission reductions and their likely outcomes, the scientists refused to carry out any further modeling until the negotiators agreed on a clear emission target for the protocol. They finally succeeded when the negotiators found some compromise on the issue, and the negotiations continued.[17] This example demonstrates the strong impact that scientists had on the actual decision making, which became possible through the good informal communication networks and the trust that negotiators had in the scientists, in particular the IIASA modelers. In addition, the critical-loads concept granted scientific assessments a stronger role in the negotiation processes. This influence became bracketed in part when the Gothenburg Protocol's multieffects approach was introduced, which highlighted the uncertainties about the interactions of pollutants and thereby limited the perceived strength of scientific information in the political process.

Participation

One of the main success factors for the CLRTAP and its assessments has been the continuity of a large percentage of its personnel, especially in the first decade of its existence. Newly acquired technical and procedural knowledge could thereby be kept inside the process and passed on through individuals. However, in the early 1990s, many people, in particular political officials, left their work on the convention and turned to the newly hot topic of climate change, which attracted significantly more attention at that time.[18] Nevertheless, those individuals who were key to the advancements of the political processes and most of the scientists remained within the framework of the convention, thus contributing to the successful input of scientific and technical information into the political processes. In addition, the secretariat remained at the same location and had a rather small staff, which allowed for the uninterrupted continuation of processes and for the accumulation of valuable procedural knowledge.

Decision making in the CLRTAP process has been highly participatory in terms of the involvement of parties to the convention—at least as far as the Executive Body and the working groups are concerned. In comparison to other UN processes, the diversity of participants in the CLRTAP process is lower because no developing countries are participating. Andrzej Jagusiewicz points out that the Executive Body "has always been sensitive to new scientific evidence, to innovative approaches to solving the problem and to the need to protect human health and ecosystems effectively and cost-effectively" (1999, 17).

The CLRTAP was one of the few fields of East–West cooperation in the 1980s and 1990s and thus was given priority among policymakers who were interested in this kind of bilateral communication and collaboration. However, in this period the East–West conflict also slowed down the processes and put up barriers to the exchange of experts and scientific knowledge. In this regard, participation during the times of intense East–West conflict limited the political effects of scientific data and information in decision making.

Reflexive Mechanisms
The CLRTAP has over time utilized mainly three means to facilitate learning: international workshops, critical discussions within the Executive Body, and external reviews. The workshops were initially organized around various specific questions that came out of the task forces and working groups and included scientific expertise from outside the convention process. Thus, the set of participating external scientists varied from one workshop to the other. In the workshops, when external scientists from the participating countries brought in their oftentimes detailed and highly specific knowledge, they were asked to render it more accessible for policymakers and integrative-modeling approaches. They thereby helped to develop the integrated assessment model that became crucial in the convention's second phase. Moreover, the workshops were a means to ensure scientific credibility by including high-ranking scientists. Because they were international in terms of participation, legitimacy was ensured as well. The aftermath of each workshop was well designed to allow for the integration of its outcomes into political processes—namely, the negotiations. Members of the working groups and the secretariat actively strove to use the outcomes for the political processes and to maintain momentum after the end of each workshop.

Whereas the workshops provided rather indirect opportunities for reflection on procedural issues, occasional evaluation and discussion in the Executive Body were more focused on the development of the organizational structure, its functioning, its effectiveness, and the assessment process. However, these discussions took place more on an ad hoc basis and were for the most part not based on thorough evaluations. In most cases, country representatives demanded them with the aim to reduce the costs of the CLRTAP process.

External review was another means to install reflexive elements in the assessment process that proved to be successful in identifying problem areas and triggering the required changes. It was crucial that these reviews were conducted by highly respected individuals or bodies from outside the process in order to maintain their independence, their credibility, and their actual effect in the political processes.

The EU's Air Pollution Policies

Air pollution policies have a similarly long tradition in the EU, as in the case of the CLRTAP. As a supranational body, the early European Community viewed the cross-national coordination and regulation among the member states as one of its key tasks. Therefore, transboundary air pollution policies fell to its responsibility and later to the EU's. In contrast to the UNECE, the institutional home of the CLRTAP, the EU's air pollution policies remain restricted to the member states. Its effects are therefore limited in scope when compared to the CLRTAP. However, because the EU is also an official party to the CLRTAP convention, significant overlap between the two is ensured.

Over the past 30 years, the EU has developed and enforced numerous pieces of legislation in addition to the CLRTAP protocols to combat transboundary air pollution (see Selin and VanDeveer 2003; Selin and VanDeveer, chap. 3 in this volume). As in the case of the CLRTAP, the member states have to implement regulatory activities, leaving mostly coordination, negotiation, and monitoring responsibilities for the EU. However, effective sanctioning mechanisms exist in the union that threaten noncomplying member states with lawsuits in the European Court—a mechanism that does not existent in the CLRTAP context.

How did these policies evolve, and what crucial steps with regard to the challenge of transboundary coordination and the inclusion of scientific knowledge can be identified? Three major phases can be distinguished in this process in interaction with the CLRTAP agreements.

Early Approaches (1972–1996)

The European Community's environmental policy began in 1972 when the heads of state and government announced a common environmental policy as a general objective for the community. However, the formulation remained institutionally rather vague, and concrete implementation of air pollution policies on the European level was slow until the 1980s (Wettestad and Farmer 2001). As a party to the CLRTAP, the European Community was also actively involved in its negotiations. Even though the community adopted several regulations on particular air pollutants—for example, the influential Large-Combustion Plant Directive of 1988—until the mid-1990s most of the initiative in air pollution policy was left to the CLRTAP process.

In 1987, the Single European Act was enforced, detailing the objectives, principles, procedures, and concrete legal foundations for the community's environmental policy. In this document, prevention was named for the first time as a guiding principle of European environmental policy. According to Article 130r, the objectives in environmental and health protection and in efficient resource utilization should be based on the following principles:

- Preventive action should be taken.
- Environmental damage should be seen as a priority rectified at source.
- The polluter should pay.

These principles established a novel guideline for policymaking in the environmental domain. Whereas previous environmental policies were enacted on the basis of observed damages, the preventive approach allowed for policy measures to be taken in advance. However, prevention required significant scientific knowledge to assess the causal chains and the possible resulting damages. Given the persistent uncertainties in scientific knowledge on environmental conditions and the related causal chains, clear standards for the political treatment of scientific uncertainties were needed but not developed in the Single European Act (Siebenhüner, Antes, and Zabel 1997).

With regard to the institutional setting of scientific knowledge in the community air pollution policies, the Directorate General XI was established in 1982 as the central body where scientists could be included in the policymaking processes. The Commission of the European Communities managed to integrate regulatory powers with scientific expertise to set agendas, outline policies, and monitor implementation through the inclusion of expertise from the member states and from the CLRTAP process. However, most legislation was based on in-house expertise of commission staff and not directly drawn from outside institutions (Wettestad and Farmer 2003).

The 1992 Maastricht Treaty on the European Union, formed to regulate the establishment of an economic and monetary union, contained some significant modifications of Article 130r. Here it is stated that the common environmental policy should be based on "the precautionary principle and on the principles that preventive action should be taken, that environmental damage should as a priority be rectified at source, and that the polluter should pay" (European Commission 1992, Article 130r). In addition to the formulation of 1987, the precautionary principle is explicitly mentioned in this document, which gives a better orientation for decision making in the situation of scientific uncertainties.

It allows for preventive measures to be taken even if no full scientific proof of causal mechanisms between certain pollutants and possible or apparent damages has been found (O'Riordan and Jordan 1995).

Until 1992, all decisions required unanimity among the member states, granting veto powers to every national government. The 1992 Maastricht Treaty on the European Union brought a partial change to this rule when it allowed for a qualified majority voting system in many environmental policy issues (Selin and VanDeveer 2003).[19] This mechanism gave the Commission of the European Communities (European Commission) and the European Council more room for maneuver, while at the same time the European Parliament was also given a more prominent role in the community legislation. Through this change, the supranational level was partially strengthened, while national interests have less influence in transboundary policymaking. Air pollution policies profited from this change because the influence of national industries on their government's environmental policies were leveled out.

EU Air Quality Framework Directive (1996–2001)

Since the early 1990s, the European Commission perceived the need for a consistent and integrated approach to combat transboundary and domestic air pollution. Endowed with additional regulatory powers, the commission prepared an air quality framework directive that aimed at the harmonization of national air quality measurements and policies. Its objectives were, first, to establish common monitoring systems and, second, to establish equal standards for emission thresholds for the different air pollutants (Wettestad and Farmer 2001). The resulting framework directive was enforced in 1996 and left most detailed regulations to subsequent daughter directives.[20]

The relationship between the EU and the member states is generally influenced by the choice of instruments on the EU level. Whereas the EU's most commonly used instruments in environmental policy are *directives*, which the member states have to implement in national laws within two years, *regulations* are directly enforced in all member states. The directive was the dominant tool to fix threshold values in all EU air pollution policies. It leaves some flexibility to the member states—a flexibility that has nevertheless often been used to delay and undermine EU decisions.[21]

However, in parallel to other environmental policy areas, air pollution policies since the early 1990s have been characterized by an ongoing process of Europeanization with a shift of competences from the member states to the community. National environmental policies have had to line up with the policies made in Brussels. Countries with a long tradition in environmental policy, such as Germany, have had to modify established and sometimes even older regulations and concepts with enormous administrative efforts. Therefore, high environmental standards have often been hindered both by those member states with an elaborated environmental policy and by those with very low standards who fear for their competitive advantages. Moreover, specific national features in general—such as the degree of economic development, the state of the environment, geographic and demographic features, infrastructure—led to different priorities and may explain the large extent of compliance deficits in meeting EU regulations.

With regard to the use of scientific knowledge in the preparation and implementation of the EU Air Quality Framework Directive, patterns established in the first period changed significantly. Whereas previous legislation drew mainly on internal expertise and data from national research and monitoring centers, the framework directive built largely on advice from the WHO. The publication of WHO guidelines for health thresholds and its recommendations had already been taken up in a monitoring directive of 1992. In the further preparation of the Air Quality Framework Directive, WHO experts were contacted to comment on "limit values." The experts from the WHO's European Regional Centre and the European Commission concluded an agreement to fund collaboration on air quality guidelines for Europe that were adopted in 1996 (European Commission 2001a). These guidelines were in large part directly taken up in the Air Quality Framework Directive.

In the process of the preparation of the daughter directives, the scientific basis of the policymaking process was enlarged. Groups of experts from the European Environment Agency, WHO, the European Commission, the member states, industry, and environmental NGOs were involved in the development of emission standards for the different pollutants. Among these participants, the WHO can claim a dominant role because the justification for the adopted standards in the daughter directives was taken from the WHO guidelines (Wettestad and Farmer 2003).

Clean Air for Europe Program (2001–2008)

As part of the sixth environment action program of 2001, the European Commission and the European Council launched a novel approach to develop a "comprehensive, integrated and coherent framework for all air legislation" that has been entitled "Clean Air for Europe," or CAFE

(European Commission 2001b, 47). This program sets out to focus particularly on particulate matter that has deleterious effects on human health and on ground-level ozone. Both are caused predominantly by traffic emissions and accumulate in urban areas to form a threat to human health. Further CAFE program objectives include

- a review of the implementation of air quality directives and effectiveness of air quality programs in the member states
- improving the monitoring of air quality and the provision of information to the public, based on indicators
- priorities for further actions, the review and updating of air quality thresholds and national emission ceilings, and the development of better systems for gathering information, modeling, and forecasting

In the first phase of the program, several reviews of existing air pollution policies have been carried out mostly by independent scientists from outside the European Commission. With regard to particulate matter, the program follows up on the first daughter directive to the Air Quality Framework Directive in 1999, which regulated these substances. It set the standard that 50 micrograms per cubic meter of particulate matter must not be exceeded on more than 35 days a calendar year effective as of 1 January 2005. Monitoring shows that this goal is being exceeded in many urban areas in Europe due to the nonimplementation of countering policy measures in the member states. Also the program's inclusion of even smaller particles that have been found to be similarly dangerous for human health will aggravate the situation of goal nonattainment (CAFE Working Group 2004).

With regard to the organizational structure, the CAFE program developed a detailed structure of working groups and organizational units to ensure the "strengthening of the scientific input to policy" as well as transparency and stakeholder involvement (European Commission 2001a, 12). The program also refers to the precautionary principle as the foundation for the policy formulation, but also that policy should nevertheless be founded on the latest and best-available scientific and technical information. Therefore, the program has been linked to the European Commission programs for research and development and to the EU's joint research centers. The central focal point for the program is the CAFE Secretariat located in Brussels with DG Environment, where the work is coordinated and the information collected and disseminated. Three additional working groups have been formed to assess relevant scientific and technical information and to feed it into the

CAFE policymaking processes—namely, the Working Group on Implementation, the Working Group on Particles, and the Working Group on Target Setting and Policy Assessment. All working groups consist of representatives from several member states, the commission, NGOs, industry, and the WHO to ensure a broad stakeholder involvement and a high level of transparency. In addition to the working groups, a steering group of stakeholders headed by the commission and supported by a Technical Analysis Group was installed to guide the CAFE process.

Scientific advice in the process is sought mostly from integrated models that give account of, among other things, abatement costs, possible damages, emission sources, and transport of pollutants. As in the CLRTAP case, the RAINS model constantly improved by IIASA is the prime modeling tool that also serves as a common reference point in this process. However, the other main source of scientific information and reference points is the WHO, which puts a strong emphasis on health concerns based on scientific research. For instance, the emission levels set in the CAFE program resulted from a risk assessment by the WHO. However, as Jørgen Wettestad and Andrew Farmer (2003) point out, many member states' growing emphasis on cost effectiveness conflicts with a rigorous protection of human health, and solutions need to be found by the European Commission and its bodies rather than through external advice.

The Overall Picture

Although the EU has achieved a high level of air quality policies, it has to struggle with numerous diverging national regulations that require further synchronization and harmonization to combat effectively the multitude of air pollutants and the damage they cause. Classical pollutants such as sulfur dioxide, lead, and nitrogen oxides have been reduced significantly over the past 30 years with the help of EU policies in coordination with the CLRTAP process and effective national policies based on scientific data and modeling results (see Selin and VanDeveer, chap. 3 in this volume). However, many other pollutants such as traffic-caused emissions of volatile organic compounds, particles, and certainly carbon dioxide have not been reduced to sufficient levels. Moreover, synergetic and accumulation effects continue to cause significant damage to human health, crops, and forests.

The EU's efforts to harmonize the various national approaches to air pollution policies and to combat transboundary air pollution effectively show an impressive track record of regulatory activity and have managed to put pressure on national governments. Due to the EU's supranational nature, it has also achieved the development of forceful legislation. However, actual implementation is left to the national governments, which often find ways to evade strong guidelines from Brussels. According to the Maastricht Treaty, environmental policy—including air pollution policy—falls within both the EU's and the member states' competences on the basis of the principle of subsidiarity. Isolated national initiatives in environmental policy are permitted only if they are urgent and required for public welfare—within a common control procedure. Transboundary air pollution is a prime case for a community concern, and the EU's policy measures have grown stricter and more encompassing. The changed voting system in environmental policy has helped the community on this path because it endows the commission with regulatory strength in coalition with several progressive countries such as the Scandinavian countries, the Netherlands, Germany, and the United Kingdom.

Interaction

By the explicit adoption of the precautionary approach, the EU acknowledged the fact that final certainties will probably never be found in many policy areas, in particular environmental policy. This principle demands regulatory action even in the absence of clear scientific proofs of causal chains. Expert judgment and well-reasoned likelihood estimations are now officially accepted as the foundation of advanced policy measures to prevent future damage. Granting scientific knowledge a significant role in policymaking, the precautionary principle gives science a voice even in situations where conventional certainty-based science is not available.

Scientific data support for community-level decision making has been provided mainly by two sources. First, the European Community draws on the data generated by the European monitoring network EMEP, which is the only Europe-wide monitoring system for air pollution (Selin and VanDeveer 2003). Established under the CLRTAP, the EMEP builds on domestic monitoring systems and collects and aggregates data from nearly all European countries. Second, like the CLRTAP process, the community also uses the RAINS model developed by IIASA as its reference model. Over the past 20 years, the model has constantly been improved on the basis of new and more disaggregated data and new research findings. As such, its role in the scientific advice of policymaking within the EU has been strengthened (Tuinstra 2007). Initially exhibiting a weak organization with regard to the science–policy interface in air pollution policies, the EU has increasingly become more transparent and inclusive. This design has helped to generate political effects in the formulation and implementation of its directives.

Participation

Over the years, the EU has developed a highly participatory process of decision making in air pollution policy. Although the first phase of its air pollution efforts was characterized by the inclusion of selected experts from member states, the process changed to include larger stakeholder groups, business groups, and different forms of scientific expertise. In addition, the EU policymaking processes within the Air Quality Framework Directive and the CAFE program are characterized by the inclusion of expertise from health policy and the WHO in particular. The WHO experts' participation thus ensured the inclusion of WHO's highly credible scientific knowledge on health problems caused by air pollution, but it was limited with regard to the actual decision making within the EU programs and their daughter directives (Wettestad and Farmer 2003). The broad participation of various stakeholders in developing EU air pollution policy is unique in international air pollution decision-making processes and has helped to bring about the exchange of scientific and nonscientific knowledge. It has not necessarily led to the stronger influence of scientific information and data per se, but it has helped to refocus the targets of air pollution policy—for example, on health issues vis-à-vis mere environmental damages.

Reflexive Mechanisms

EU air pollution policies have evolved tremendously over the past 30 years. The policy process is characterized by numerous checks and balances and feedback loops that can be viewed as reflexive mechanisms. The European Commission, Council, and Parliament review any legislation, with the intention to improve the legislation and to include more perspectives and knowledge backgrounds. In addition, the European Environment Agency and external experts are constantly reviewing EU air policies. The RAINS model and its applications within the CAFE program have also been reviewed within bilateral consultations between modelers and national political delegates. This reflection has allowed the reassessment of national outcomes, data sources, and implications of the model for individual countries (Tuinstra 2007).

These mechanisms have introduced reflexive elements into the implementation of the policy measures and have allowed for their continuous improvement. However, these efforts focus mainly on integrating non-scientific stakeholders and their knowledge, whereas scientific knowledge in the conventional sense is given less priority. Moreover, although reviews by the European Commission can result in sanctions and revisions of political processes, influences from scientific institutions, including the European Environment Agency, remain limited due to the lack of sanctioning powers and their generally lesser influence in the assessment and review processes.

Comparison and Conclusions

Both approaches studied in this chapter share many commonalities. The CLRTAP covers a larger geographical area than the EU, but through the enlargement process the EU is growing and thus reducing the CLRTAP's importance in the whole of Europe and North America. Nevertheless, the EU remains a party to the convention even though many of its standards are more ambitious than the consensus of the convention's parties allow.[22] The EU air pollution policy has drawn to a large extent on the experiences, expertise, and organizational structures established under the convention.

The two approaches have different means and coordination measures to address the transboundary nature of air pollution, and they have different takes on how to include scientific knowledge. With regard to the participating countries' powers, the EU has much more political leverage than the relatively weak CLRTAP. Whereas the EU can make decisions on the basis of a qualified majority, the CLRTAP process has to apply general UN rules of procedure, which demand consensus among all parties to the convention.[23] Moreover, the European Commission has sanctioning powers at its disposal that can be used against noncomplying member states. The CLRTAP process, by contrast, does not include a sanctioning mechanism. In addition, the convention has more members than the EU, which reduces the power of the central units vis-à-vis the large number of participating countries. On this account, the successes achieved under the CLRTAP are surprising and cannot be explained on the basis of regulatory power over its parties. The convention's success in arriving at international agreements and in reducing transboundary air pollution must rather be attributed to the organization of the process, the ambitious actors involved, and the effective inclusion of scientific knowledge in the policymaking process. By contrast, the science–policy interface in EU air pollution policies, due to its particular design, has proven less effective and less influenced by scientific actors in political decision-making processes.

A comparison of the interaction between scientists and the policymaking processes in the two approaches reveals only moderate differences. Whereas the CLRTAP process was characterized by close and direct ties and communication links between scientists and policymakers, the EU has attempted to include scientific expertise from outside through reviews, studies, and workshops. In the CLRTAP process, communication between modelers was founded on trust and mutual appreciation that even allowed the modelers to put pressure on the negotiators. Over the years, the European Commission has attempted to build up in-house expertise on the related topics, with some success. The CAFE program ended up with a similar structure of working groups and a coordinating secretariat.

What has become clear is that the hypothesis given earlier on in this chapter needs to be revised. It is not that more communication between scientific groups and decision makers helps to bring scientific knowledge into the decision-making process, but that the general conceptual framework helps to grant such an influence. Whereas the critical-loads approach relies heavily on scientific data and modeling results, the multieffects approach has given rise to questions regarding the scientific system's availability and capability to provide the required inputs. The EU's precautionary principle can be seen as an interesting in-between solution because it relies on scientific knowledge but endows policymakers with decision-making responsibilities in the case of remaining uncertainties and unknown problems. In both of these frameworks, scientific knowledge is valued differently with regard to the political decision-making processes.

Both processes differed significantly in their inclusion of the crucial policy field and the related experts from health policy. This concern was stronger in EU policymaking than in the CLRTAP process, in particular in the earlier phases until the Gothenburg Protocol in 1999. The EU Commission has carefully included WHO experts while at the same time ensuring enough independence in its decision making. This dividing line has granted credibility and legitimacy to high standards in air pollution policies and at the same time has limited the political influence of the scientific actors involved—namely, the WHO. By contrast, in the CLRTAP process scientists developed stronger agency and influence through their formally independent roles, the critical-loads concept, and policymakers' dependence on the modeling results. The initial hypothesis that more participation generally helps scientific actors to have a greater influence in decision making thus needs to be revised.

The design of reflexive mechanisms in the two policy processes shows some variation that might explain differences in policy effects. Whereas the reflexive mechanisms used within the CLRTAP process largely supported the scientific focus of the convention, EU processes are dominated by conventional policy-driven reflexive mechanisms with less scientific inputs. The latter mechanisms contribute to keeping the process under political control by EU decision makers vis-à-vis scientific expertise, in particular the WHO. Thereby, the design of the reflexive mechanisms within each domain has successfully directed the focus of the reflection and the resulting learning and change processes. In particular, the review mechanisms used in CLRTAP helped to increase scientific quality of the assessment processes and the facts and concepts that went into the related working groups. By contrast, the EU has taken significant care in ensuring fair participation among different stakeholder groups and the individual countries involved. This inclusion has led to more policy-relevant outcomes from the reviews, but it has helped relatively little to foster scientific credibility of the data and knowledge used in the processes.

What will Europe's air pollution policies look like in the future? On the basis of the current trends, three tendencies can be expected to continue. First, the analysis in this chapter leads to the conclusion that the CLRTAP process is more dominated by scientifically oriented actors than is the EU process with its more policy-focused approach. The latter can be expected to be more helpful in achieving a forceful implementation than the former. Second, the ongoing enlargement process will increase the EU's importance and partially crowd out the CLRTAP as the prime international agreement on the most pressing air pollution problems. Third, a global spread of the European experiences can be seen in Asia and some areas of Latin America, where similar air pollution problems are on the rise. Even though direct transfers of agreements and related bodies are difficult due to cultural differences, concepts and insights in the interaction between policymaking and scientific knowledge can generate fruitful policy approaches to combat air pollution in these regions effectively.

Acknowledgments

An earlier version of this article was presented at the Twelfth Asia–Europe Foundation University and benefited from the discussions with the international students and faculty members at the University Brunei Darussalam. I am particularly grateful to Göran Sundqvist, Rolf Lidskog, and three anonymous reviewers for their valuable comments.

Notes

1. Lee's concept is based on Price 1965.

2. For an overview of the different bodies and organizational connections under the convention, see the Web site http://www.unece.org/env/lrtap/welcome.html.

3. The evolution of the CLRTAP's political processes, its protocols, and their effectiveness are analyzed in Levy 1993; McCormick 1997, 1998; Swanson and Johnston 1999; Haas and McCabe 2001.

4. For a historical overview of the political processes under the convention, see Jagusiewicz 1999.

5. The EMEP had already been established in 1977 under the auspices of the UNECE, the World Meteorological Organization, and the UN Environment Program, but its funding remained insecure until the first protocol under CLRTAP had been signed.

6. These international cooperative programs are: the International Cooperative Program on Assessment and Monitoring of Air Pollution Effects on Forests; the International Cooperative Program on Assessment and Monitoring of Acidification of Rivers and Lakes; the International Program on Effects of Air Pollution on Materials, Including Historic and Cultural Monuments; the International Cooperative Program on Effects of Air Pollution on Crops; the International Cooperative Program on Integrated Monitoring of Air Pollution Effects on Ecosystems; and the International Cooperative Program for Mapping Critical Loads and Levels and Their Exceedances.

7. Keith Bull, secretary to the CLRTAP, interviewed by the author, Geneva, 21 March 2001.

8. Ibid.

9. For some examples of these maps, see http://www.rivm.nl/en/themasites/cce/data/critical_loads_databases_maps/index.html.

10. The model and its submodels are described in Hordijk 1988; Alcamo, Shaw, and Hordijk 1990. The integration of the model in the political processes is discussed in Tuinstra, Hordijk, and Amann 1999.

11. Also supported by Henning Wüster, economic affairs officer at UNECE, interviewed by the author, Geneva, 1 March 2001.

12. Leen Hordijk, Director of the Wageningen Institute for Environment and Climate Research, interviewed by the author, Cambridge, MA, 16 January 2001.

13. Further reviews of the convention processes might result in revisions or renegotiations of protocols.

14. Bull and Wüster interviews.

15. For a more critical view of the convention's success, see chapter 2 in this volume.

16. Bull interview.

17. Hordijk interview.

18. Ibid.

19. In the areas of fiscal policies, land use, and energy supply, unanimity was still required.

20. These daughter directives addressed the following issues: limit values for sulfur dioxide, nitrogen dioxide and nitrogen oxides, particulate matter, and lead in ambient air (Directive 1999/30/EC); limit values for benzene and carbon monoxide in ambient air (Directive 2000/69/EC); limit ground-level ozone (Directive 2002/3/EC); and limit arsenic, cadmium, mercury, nickel, and polycyclic aromatic hydrocarbons in ambient air (Directive 2004/107/EC).

21. In its eleventh annual report to the European Parliament, the European Commission (1994) cites 1,340 violations of the treaties in 1993. In this total, 383 of them can be ascribed to environmental policy.

22. For instance, the CAFE program overtly criticizes the Gothenburg Protocol under the CLRTAP: "Unfortunately the ceilings agreed by Parties to CLRTAP did not correspond to the level of ambition on which the technical work was based" (European Commission 2001a, 4).

23. As Wettestad argues in chapter 2 of this volume, parties to the convention rarely take recourse to their veto powers but generally refuse to sign the negotiated protocol.

References

Alcamo, Joseph, Roderick Shaw, and Leen Hordijk. 1990. *The RAINS Model of Acidification.* Science and Strategies in Europe. Dordrecht, Netherlands: Kluwer.

Biermann, Frank. 2001. Big Science, Small Impacts—in the South? The Influence of Global Environmental Assessments on Expert Communities in India. *Global Environmental Change* 11: 297–309.

Biermann, Frank. 2002. Institutions for Scientific Advice: Global Environmental Assessments and Their Influence in Developing Countries. *Global Governance* 8: 195–219.

Bull, Keith. 1992. An Introduction to Critical Loads. *Environmental Pollution* 77: 173–176.

Bull, Keith. 1995. Critical Loads—Possibilities and Constraints. *Water, Air, and Soil Pollution* 85: 201–212.

CAFE Working Group on Particulate Matter. 2004. *Second Position Paper on Particulate Matter*. Brussels: CAFE Secretariat.

Eckley, Noelle. 1999. *Drawing Lessons about Science–Policy Institutions: Persistent Organic Pollutants (POPs) under the LRTAP Convention.* ENRP

Discussion Paper E-99-11. Cambridge, MA: Kennedy School of Government, Harvard University.

European Commission (Commission of the European Communities). 1994. *Eleventh Annual Report to the European Parliament*. Brussels: European Commission.

European Commission (Commission of the European Communities). 2001a. *Communication from the Commission. The Clean Air for Europe (CAFE) Programme: Towards a Thematic Strategy for Air Quality*. COM (2001) 245 final. Brussels: European Commission.

European Commission (Commission of the European Communities). 2001b. *Environment 2010: Our Future, Our Choice—The Sixth Environment Action Programme*. COM (2001) 31 final. Brussels: European Commission.

European Commission. 1992. *Treaty on European Union*. Official journal C191. Brussels: European Commission.

Farrell, Alex, and Jill Jäger. 2006. *Assessments of Regional and Global Environmental Risks: Designing Processes for the Effective Use of Science in Decision Making*. Washington, DC: Resources for the Future.

Farrell, Alex, Stacy VanDeveer, and Jill Jäger. 2001. Environmental Assessments: Four Under-Appreciated Elements of Design. *Global Environmental Change* 11: 311–333.

Gieryn, Thomas F. 1996. Boundaries of Science. In *Handbook of Science and Technology Studies*, ed. Sheila Jasanoff, 393–443. Thousand Oaks, CA: Sage.

Guston, David H. 1999. Stabilizing the Boundary between US Politics and Science: The Role of the Office of Technology Transfer as Boundary Organization. *Social Studies of Science* 29: 1–25.

Haas, Peter M. 1992. Introduction: Epistemic Communities and International Policy Coordination. *International Organization* 46 (1): 1–35.

Haas, Peter, and David McCabe. 2001. Amplifiers or Dampeners: International Institutions and Social Learning in the Management of Global Environmental Risks. In *A Comparative History of Social Responses to Climate Change, Ozone Depletion, and Acid Rain*, vol. 1 of *Learning to Manage Global Environmental Risks*, by the Social Learning Group, 323–348. Cambridge, MA: MIT Press.

Hordijk, Leen. 1988. A Model Approach to Acid Rain. *Environment* 30 (2): 17–42.

Jagusiewicz, Andrzej. 1999. The History of the Convention on Long-Range Transboundary Air Pollution. *Pollution Atmosphérique* 164 (December): 13–21.

Jasanoff, Sheila. 1987. Contested Boundaries in Policy-Relevant Science. *Social Studies of Science* 17: 195–217.

Jasanoff, Sheila. 1990. *The Fifth Branch: Science Advisers as Policymakers*. Cambridge, MA: Harvard University Press.

Keohane, Robert O., Peter Haas, and Marc Levy, eds. 1993. *Institutions for the Earth: Sources of Effective International Environmental Protection*. Cambridge, MA: MIT Press.

Lee, Kai. 1993. *Compass and Gyroscope: Integrating Science and Politics for the Environment.* Washington, DC: Island Press.

Levy, Marc. 1993. European Acid Rain: The Power of Tote-Board Diplomacy. In *Institutions for the Earth: Sources of Effective International Environmental Protection*, ed. Peter M. Haas, Robert O. Keohane, and Marc A. Levy, 75–132. Cambridge, MA: MIT Press.

Lidskog, Rolf, and Göran Sundqvist. 2004. From Consensus to Credibility: Innovation. *European Journal of Social Science Research* 17 (3): 205–226.

McCormick, John. 1997. *Acid Earth: The Politics of Acid Pollution.* London: Earthscan.

McCormick, John. 1998. Acid Pollution: The International Community's Continuing Struggle. *Environment* 40 (3): 17–45.

Mitchell, Ronald B., William C. Clark, David W. Cash, and Nancy M. Dickson, eds. 2006. *Global Environmental Assessments: Information and Influence.* Cambridge, MA: MIT Press.

Nordic Council of Ministers. 2000. *Workshop on Future Needs for Regional Air Pollution Strategies.* Copenhagen: Nordic Council of Ministers.

O'Riordan, Timothy, and Andrew Jordan. 1995. The Precautionary Principle in Contemporary Environmental Politics. *Environmental Values* 4 (3): 191–212.

Patt, Anthony. 1998. *Analytic Frameworks and Politics: The Case of Acid Rain in Europe.* ENRP Discussion Paper E-98-20. Cambridge, MA: Kennedy School of Government, Harvard University.

Popper, Micha, and Raanan Lipshitz. 2000. Organizational Learning: Mechanisms, Culture, and Feasibility. *Management Learning* 31 (2): 181–196.

Price, Don Krasher. 1965. *The Scientific Estate.* New York: Oxford University Press.

Selin, Henrik, and Stacy D. VanDeveer. 2003. Mapping Institutional Linkages in European Air Pollution Politics. *Global Environmental Politics* 3 (3): 14–46.

Siebenhüner, Bernd. 2002. How Do Scientific Assessments Learn? *Environmental Science & Policy* 5 (5): 411–427.

Siebenhüner, Bernd. 2003. The Changing Role of Nation States in International Environmental Assessments: The Case of the IPCC. *Global Environmental Change* 13 (2): 113–123.

Siebenhüner, Bernd, Ralf Antes, and Hans-Ulrich Zabel. 1997. *Prevention through Process Policy and Product Policy—a Study of European and German Environmental Policy.* Discussion Paper of the Department of Economics no. 97/19. Halle, Germany: Martin Luther University of Halle-Wittenberg.

Swanson, Timothy, and Sam Johnston. 1999. *Global Environmental Problems and International Environmental Agreements.* Cheltenham, UK: Edward Elgar.

Tuinstra, Willemijn. 2007. Preparing for the European Thematic Strategy on Air Pollution: At the Interface between Science and Policy. *Environmental Science & Policy* 10 (5): 434–444.

Tuinstra, Willemijn, Leen Hordijk, and Markus Amann. 1999. Using Computer Models in International Negotiations: The Case of Acidification in Europe. *Environment* 41 (9): 33–42.

Tuinstra, Willemijn, Leen Hordijk, and Carolien Kroeze. 2006. Moving Boundaries in Transboundary Air Pollution Co-Production of Science and Policy under the Convention on Long-Range Transboundary Air Pollution. *Global Environmental Change* 16 (4): 349–363.

UNECE (United Nations Economic Commission for Europe). 1979. *Convention on Long-Range Transboundary Air Pollution*. Geneva: United Nations.

UNECE (United Nations Economic Commission for Europe). 1988. Protocol Concerning the Control of Nitrogen Oxides or Their Transboundary Fluxes. Geneva: United Nations.

UNECE (United Nations Economic Commission for Europe). 2007. *Review of the 1999 Gothenburg Protocol.* ECE/EB.AIR/WG.1/2007/14. Geneva: Executive Body for the Convention on Long-Range Transboundary Air Pollution.

UNECE (United Nations Economic Commission for Europe) Executive Body. 1999. *Report of the Seventeenth Session.* Geneva: United Nations, Executive Body of the Convention on LRTAP.

VanDeveer, Stacy D. 1998. *European Politics with a Scientific Face: Transition Countries, International Environmental Assessment, and Long-Range Transboundary Air Pollution.* Environment and Natural Resources Program Discussion Paper E-98-09. Cambridge, MA: Kennedy School of Government, Harvard University.

Wettestad, Jørgen, and Andrew Farmer. 2001. *The EU Air Quality Framework Directive: EU Pitcher; International Catcher?* Fridtjof Nansen Institute/Institute for European Environmental Policy discussion paper. Oslo: Fridtjof Nansen Institute.

Wettestad, Jørgen, and Andrew Farmer. 2003. *The EU Air Quality Framework Directive: Shaped and Saved by Interaction?* Fridtjof Nansen Institute/Institute for European Environmental Policy discussion paper. Oslo: Fridtjof Nansen Institute.

Young, Oran R. 2002. *The Institutional Dimensions of Environmental Change: Fit, Interplay, and Scale.* Cambridge, MA: MIT Press.

Young, Oran R., Leslie A. King, and Heike Schroeder, eds. 2008. *Institutions and Environmental Change: Principal Findings, Applications, and Research Frontiers*. Cambridge, MA: MIT Press.

II

Expertise and Learning

5

Organized Science, Usable Knowledge, and Multilateral Environmental Governance

Peter M. Haas and Casey Stevens

There is a widespread acceptance that effective multilateral environmental governance requires the involvement of environmental scientists (Gore 1996). Although speaking truth to power has long been a major theme in political science and policy studies (Wildavsky 1979), commentators are increasingly skeptical about whether modelers and scientists are capable of developing truth and whether power ever listens to them anyhow. Indeed, international relations scholars tend to be surprised by the occasions when it does. This chapter applies the political science literature to the related question of when power does listen to science, particularly with regard to the management of complex environmental issues.

The Convention on Long-Range Transboundary Air Pollution (CLRTAP) can assist in drawing larger lessons about the situations when science can speak to power and create effective multilateral environmental governance. This chapter places the analysis of CLRTAP in the wider empirical record of effective mobilization of usable knowledge. We begin by focusing on the development of sustainability science, which requires the substantial involvement of scientific expertise in policymaking. We then move to consider the typical arguments about the limits that knowledge places on power and the situations in which knowledge may actually contribute to power. Next, we look at the empirical examples of knowledge speaking to power in three sections: first, an overview of 30 different multilateral environmental agreements and their construction of procedural science–policy connections; next, the lessons from looking at the empirical record; and third, the limiting case of the Intergovernmental Panel on Climate Change (IPCC). The conclusion reflects on these empirical sections and on their implications for constructivism in international relations theory.

Knowledge, Sustainability Science, and the Social Construction of Consensus

Sustainable development is now one of the major mantras invoked in the area of international environmental governance. Sustainable development was popularized in the seminal 1987 World Commission on Environment and Development report *Our Common Future*. This new policy doctrine—or consensual wisdom within the international community of environmental policy analysts (and at times advocates)—rests on three key foundations (Ravetz 1986; National Research Council 1999; Clark 2001, 2007; Kates, Clark, Corell, et al. 2001): (1) a normative belief in systematic intervention in human activities to make those activities sustainable; (2) a procedural belief that policy should be participatory and transparent—in part to include new perspectives in the knowledge base brought to bear on understanding a particular problem as well as to promote buy-in and the inclusion of stakeholders in the subsequent application of these policies and thus to improve the prospects for more effective enforcement and compliance; (3) a substantive and more comprehensive belief that systematic approaches to planning and policy formulation are necessary, often through the technique of sustainability impact assessment and forms of interdisciplinary integrated modeling intended to bring together partial insights to sustainable development from different academic disciplines.

Although sustainability science has been percolating in the various laboratories of environmental analysts for nearly 20 years, there has been effective technology transfer from the laboratory to a government agency or international regime in few observable occasions and perhaps not in any. Even when scientists think they have developed truths for power, power appears disinterested at best and possibly even uninterested. Yet sustainable development has gradually encroached on environmental policymaking worldwide, both in international regimes and in national legislation and practices. Power listening to truth is rare, and science policy analysts and political scientists too often dismiss truth entirely.

Many arguments are harnessed to account for why science seldom influences policymaking, as indicated by the fact that science is seldom directly converted to policy. The co-production of science and social institutions is a leitmotif of the social studies of science literature (Jasanoff and Wynne 1998; Jasanoff 2004). Although the two activities do coevolve and may be mutually constitutive, in this chapter we seek to elaborate some of the causal mechanisms by which the two interact. For convenience, we keep scientific institutions and broader social institutions discrete for the purpose of closer analysis. The path from truth to power is circuitous at best. Let us cluster the reasons for the skepticism under a set of convenient categories (for a summary of arguments about science policy and an argument for a pragmatic procedural approach to generating scientific knowledge for public policy, see Ziman 2000).

1. Science isn't wisdom (or true). Science studies dismiss the prospects of objective knowledge about the world and stress the political dimensions to science and science policy. Scientific findings may in part reflect the bias of sponsors, but they may more deeply reflect the broader culture of the society from which they emerge and about which they may not be fully conscious (Litfin 1994).
2. Science is politically tainted and suspect. Organized modern science embodies implicit values of control, so that decisions made with scientific warrants may unconsciously reflect such hidden values. In addition, the distributional consequences of science-based advice are themselves political (Jasanoff and Society for Social Studies of Science 1995; Miller and Edwards 2001). Science is political in its consequences because some people and areas benefit and others suffer as a consequence of policy options that are supported by the application of scientific understanding. To the extent that those affected by the use of science in formulating policy are not consulted in its development and application, they potentially regard it as an illegitimate and exploitative set of discursive practices (Lidskog and Sundqvist 2002).
3. Power doesn't care about truth anyhow. Power doesn't want science; it wants a justification for preexisting political programs, which are driven principally by political anticipations of gain (Nelkin 1979; Miles 1998; Andresen, Walløe, and Rosendal 2005). In addition, science may not be sufficiently simple for policymakers. Harry Truman is supposed to have complained about economic advisers who would say "on the one hand" and "on the other hand," "Just give me a one-handed economist."

Before discussing situations in which power does listen to truth, let us first dismiss the standard rationalist account that major problems create the incentives for their resolution, and thus modern bureaucracies (or bureaucratic institutions) either develop effective responses almost automatically or are so powerfully constrained by the strategic interests of powerful member states or participants that institutional applications of knowledge are little different than the aggregate wishes of the more powerful members (for international institutions, see Koremenos, Lipson, and Snidal 2001; for domestic, see Moe 1984). We reject this proposition for two reasons.

One is that, particularly for sustainable development, the material incentives and the nature of the presumptive useful information are not in synch. Few organizations, if any, have the available mission and resources to be able to address the full integrative range of issues encompassed by sustainable development. Yet there is an empirical record of shifts in policies and institutions beyond what formal organizational assignments would predict (Haas and Haas 1995). The second reason can be addressed by way of a historical analogy. Scurvy was arguably the single most important limiting factor to the expansion of trade and geopolitical influence in the fifteenth and sixteenth centuries. Trade and exploration was significantly hampered because the mortality rate of sailors on long-distance expeditions was often in excess of 90 percent. Expeditions would return with far fewer ships not because of storms, but because too few crew members were left to be able to serve the entire fleet. And yet in the early 1600s the scurvy problem was solved by Captain James Lancaster, who suggested that citrus trees be brought along on ships, but the solution was forgotten for nearly 150 years, until Captain Cook rediscovered it in the 1760s (Milton 1999)! So much for rational societies responding effectively to important issues. We should not expect a fully formed sustainable development science to be developed and applied in response to the presumptive need for such a view. We shouldn't assume that all organizations are rational and will automatically recognize and adopt what prove to be the appropriate policy responses with the virtue of hindsight or reflect the material needs of their most powerful constituencies.

Usable Knowledge: When Knowledge Speaks to Power

We take a late-modern view of truth: the domain of science and its ability to confer truth is bounded, but we can talk of better and worse science (for an application of this view to the study of international institutions, see Haas and Haas 2002). In the context of this discussion, we can think of the domain of science that is likely to be adopted by decision makers as *usable knowledge* (the literature has multiple uses of the term: for earlier and slightly different usages, see Clark 1990 and Dimitrov 2003; Jasanoff 1990 calls a similar concept "serviceable truths"). Several different schools of thought exist about usable knowledge, although the core insights are quite similar and complementary. In short, usable knowledge encompasses a substantive core that makes it usable for policymakers and a procedural dimension that provides a mechanism for its transmission from the scientific community to the policy world, and it allows for agency when theorizing about broader patterns of social learning, policymaking, and international relations.

William Clark and Giandomenico Majone (1985) define the core ingredients to usable knowledge as adequacy, value, legitimacy, and effectiveness. Adequacy relates to including all the relevant knowledge or facts germane to the matter at hand. Value has to do with contributing to further understanding and meaningful policy. Legitimacy relates to the acceptance of the knowledge by others outside the community that developed it. Effectiveness relates to its ability to shape the agenda or advance the state of the debate and ultimately to improve the quality of the environment (see also Social Learning Group 2001). The substantive core of usable knowledge can be augmented in a number of ways.

Current research from comparative politics, international relations, policy studies, and democratic theory suggests that science remains influential if its expertise and claims are developed behind a politically insulated wall (Andresen 2000; Botcheva 2001; Social Learning Group 2001). Moreover, any efficiency gains by relying on one single source of policy advice are more than offset by the loss of legitimacy, analytic blinders imposed by relying on just one institutional source for usable knowledge, and the political doubts of bias that are raised by narrowing the source of information. Studies of international environmental assessments and science panels suggest the need for fluid bodies that can bring together multiple sources of information and are not beholden to one single funder or political sponsor (Siebenhüner 2002, 2003; Haas 2004b; Martello and Jasanoff 2004; Mitchell, Clark, Cash, et al. 2006). Studies of national-level environmental policy processes have convincingly argued against relying on individual institutions for research and policy advice because they may bias the information flow and control resources (Brown 1997; Skoie 2001). There is much national-level experience with establishing standing scientific panels, such as in the US National Academy of Sciences, the Environmental Protection Agency Science Advisory Board, and the now defunct Office of Technology Assessment, among others. Each panel is regarded as generating usable knowledge for the government and enjoys sufficient autonomy to identify research questions and to convene panels to develop reports. In CLRTAP, the European Monitoring and Evaluation Programme (EMEP), the working groups, and the Network on Air Pollution and Health (AIRNET) all serve to identify research questions and guide science with some degree of autonomy from the political process.

Based on these insights, the general factors that influence the likelihood that decision makers will apply scientific knowledge are credibility, legitimacy, and saliency (Schroeder, King, and Tay 2008). Credibility means that the knowledge claims are believed to be accurate: within a consensus theory of truth, this means that they are publically created through a deliberative process by people widely regarded as experts. Legitimacy means that knowledge is developed by people who have social authority and that it is accepted by people outside the community that developed it. In practice, the assignment of legitimacy often rests on peer review and scholarly reputation. Salience means that the information is timely and is organized on a politically meaningful timescale and scale of resolution.

In addition to the substantive core of usable knowledge, there is a focus on the processes of transmitting that usable knowledge from scientists to policymakers. A new consensus is emerging among social scientists who study the use of science in international regimes that a procedural or discursive model of the policy process is normatively superior and that the frequency of its application is growing, as against the former policy-analytic approach associated with cost-benefit analysis and analytic efforts to identify the best policy solution. Policy analysis is a process of exchange and mutual learning between policymakers and policy analysts. Policies themselves are experiments that participants monitor and about which they reflect in order to improve them over time (Asher 1986; Funtowicz and Ravetz 1991, 2001; Lee 1993; Lindner 2003; Siebenhüner, chap. 4 in this volume). International states are increasingly relying on this procedural approach in the development and application of truth to power. The Commission of the European Communities (European Commission) developed a set of guidelines and proposals for the collection and use of expertise by the commission that is very similar in orientation to expert groups assembled and utilized for specific, procedural tasks (European Commission 2001, 2002; Gornitzka and Sverdrup 2008). As a consequence, policymaking is a process rather than a fixed set of analytic techniques. Language may play a role (Fischer and Forester 1993; Hajer and Wagenaar 2003), but the constructivist argument pursued here stresses the political parameters within which debate occurs rather than the connotations generated by the language that is used.

Constructivists argue that under conditions of uncertainty—such as those associated with contemporary globalization and highly technical issues—it is impossible to ascertain ex ante sufficient information to follow the policy-analytic model (Checkel 1998; Hopf 1998; Kubalkova, Onuf, and Kowert 1998; Ruggie 1998; Christiansen, Jørgensen, and Wiener 1999; Guzzini 2000; Haas 2001b; Adler 2002). Alternatively, the key is to design policy-analytic processes from which actors learn about the world and about each other. This view is antirationalist in the sense that the process model entails path dependency and uncertainty assumptions that presume that outcomes and, indeed, preferences are often underspecified or indeterminate and thus that regimes and political interactions can best be appraised in terms of process rather than outcome. All outcomes will be suboptimal in some sense as compared to the ex post ideal outcome, but one can just hope that better outcomes may result from an open discursive process (Dryzek 1997).

Constructivist approaches to policy analysis suggest that science must be developed authoritatively and then delivered by responsible carriers to politicians. Doctors, scientists, and engineers remain the most esteemed professionals in Europe and thus command the greatest social legitimacy and deference when providing policy advice (Jasanoff and Wynne 1998; Drori, Meyer, Ramirez, et al. 2003; Drori and Meyer 2006). The transmission belt of like-minded scientists is called an "epistemic community" (Haas 2001a, 2007). The more autonomous and independent science is from policy, the greater its potential influence (Andresen 2000; Botcheva 2001; Haas 2001b). Consensus in isolation builds value and integrity, and then its consequences should be discussed publicly. Measures of autonomy and integrity include the selection and funding of scientists by intergovernmental organizations rather than by governments, the recruitment of scientists by merit on important panels, and reliance on individuals whose reputation and authority rest on their role as active researchers rather than on their role as policy advocates or science administrators. Accuracy can be achieved via peer review, interdisciplinary research, and independence from sponsoring sources. Sustainability scientists, themselves an epistemic community, increasingly argue for the need to include local knowledge along with the more formally technical understanding of traditional disciplinary elites associated with formulating sustainability policies (Haas and Haas 2002; Martello and Jasanoff 2004). However, the criteria for participation remain loosely defined, but perhaps no more so than the broad injunctions for multidisciplinary participation that do not clearly identify which disciplines need to be consulted for which types of questions.

International and transnational networks of scientists develop usable knowledge. That knowledge is heeded, to the extent that it is, after widely publicized shocks or crises. Although usable knowledge contributes to broader patterns of social learning, different communities often deliver and apply that knowledge. We are not talking of reflective learning by decision makers. Rather, we are talking about decision makers' recognition of the limits of their abilities to master new issues and the need to defer or delegate to authoritative actors with a reputation for expertise. In the aggregate, social learning and human betterment emerge (Levy, Cavender-Bares, Clark, et al. 2001) when the experts have been able to develop usable knowledge, and the decision makers feel compelled to apply it (Haas and Haas 2002).

Empirical Record of Usable Knowledge in Multilateral Environmental Governance

Over the past 30 years, there has been a dramatic adoption of international treaties and regimes to address many aspects of transboundary and global environmental threats (Tolba and Rummel-Bulska 1998; Haas 2001b; Miles, Underdal, Andresen, et al. 2002; Mitchell 2002; and see Mitchell's Web site at http://pages.uoregon.edu/rmitchel/data.html). When usable knowledge is successfully constructed and transmitted, it yields distinctive results: regimes are developed by a process of social learning; the regime rules reflect scientific consensus about environmental sustainability; and the regimes tend to be effective. Maurice Strong, secretary-general of the United Nations (UN) Conference on the Human Environment (in Stockholm in 1972) and of the UN Conference on Environment and Development (in Rio de Janeiro in 1992) and the first executive director of the UN Environment Program (UNEP), helped design the outlines of this process of social learning. Strong believed that "the process is the policy": that is, by generating an open political process in which states are exposed to consensual science, government officials may be persuaded to adopt more sustainable policies, and individual scientists may gain heightened political profiles at home, which may ultimately increase their effectiveness as well. In addition, Marc Levy and his colleagues explain that social learning "is characterized by acceptance of a core set of common values, mutual recognition of the legitimacy of each actor, and a high degree of trust" (Levy,

Cavender-Bares, Clark, et al. 2001, 110). International environmental negotiations develop through scientific discourse in which political debate and compromise reflect expert consensus on the behavior of ecosystems and their ability to sustain stress. The substance of regimes reflects scientific consensus about the most important environmental threats, and negotiated standards reflect consensus about the degree of environmental stress the target environment can sustain. For instance, the 1980 Land-Based Sources Protocol for the Mediterranean required more stringent emission controls on the industrialized countries than on the developing countries because the magnitude of degradation of the northern coast of the Mediterranean was much more severe than it was on the southern coast (Haas 1990).

The CLRTAP offers one case that demonstrates the permissive conditions for successful scientific guidance. Within CLRTAP, the EMEP serves as a body within the convention that has organized much of the science work on issues of air pollution. The EMEP and the various working groups under CLRTAP are composed of experts nominated by parties to the treaty. Most important, EMEP manages a series of centers in many member countries that monitor a wide variety of atmospheric conditions. Through discussions between EMEP and the working groups, a consensus has emerged regarding the amount of pollutants that need to be reduced below the critical-loads level. This dialogue between different groups of scientists with defined focuses, coupled with their ability to set their own agenda with guidance from the Conference of the Parties (COP) and the secretariat of the treaty, results in the construction of a politically usable scientific consensus. This consensus that emerged and the monitoring procedures that were established has resulted in a series of protocols leading to emission cuts and a system to supervise compliance. The construction of bodies that were largely insulated and respected by the members allowed the development of usable knowledge and the resultant social learning by the parties to the treaty (Tuinstra, Hordijk, and Amann 1999; Tuinstra 2008; Wettestad, chap. 2 in this volume). It is through these systems that CLRTAP was able to grow from a treaty that required little from members in terms of cutting emissions into a significant environmental regime that both lowered emissions from members and reduced the environmental pollution caused by transboundary air pollution. The chapter's final section investigates a limiting case where science was unable to influence policymakers.

Social learning regimes are substantively more comprehensive and attuned to the emergent sustainable development doctrine's sensibility and injunctions. Regimes built with usable knowledge appear to be more effective at inducing states to achieve their intended goals of improving environmental quality. For instance, the stratospheric ozone and European acid rain efforts are widely hailed as among the more successful and effective international environmental governance efforts of the contemporary era (Andresen 2000; Haas 2001b; Miles, Underdal, Andresen, et al. 2002). Social learning is aided through processes and organizational regime design that both keep the construction of science insulated and construct procedures for the connection of science to policy decisions. For example, most social-learning treaties have standing environmental monitoring and research committees to provide timely warnings of new problems, monitor achievements of regime goals, and educate politicians and policymakers on environmental issues.

Regimes vary widely in how they organize scientific participation. There are detailed case studies about many of these regimes (see, e.g., Farrell and Jäger 2005; Mitchell, Clark, Cash, et al. 2006). Regimes that delegate some responsibilities to organized science include many multilateral environmental treaties and organizations (for a partial list of some organizations that maintain some form of science panel, see appendix 5.1). Some institutions have effectively constructed connections between science and policymakers that assist the building of usable knowledge.

The maintenance and support of science bodies within multilateral environmental governance arrangements is absolutely vital for the construction of usable knowledge within the regime. Standing scientific panels allow the constant construction and transmission of accurate and timely information. Three types of standing-committee structure are typical in multilateral environmental regimes. The first type includes scientific groups that set their own schedules and research agendas. These groups meet as they see fit between meetings of policymakers and according to their own determined needs. The second type is made up of those groups whose meeting times are set by the COP. For example, most fishery agreements require their scientific bodies to meet a month or two months prior to the COP meeting. The UN's Joint Inspections Unit found that the Convention Concerning Desertification largely organized the science body meetings directly in line with COP meetings, thereby creating a polarized and political atmosphere in the meetings (Ortiz and Tang 2005). And the third type includes ad hoc science panels, which, especially if called for by member states, introduce a high level of political involvement in the reporting of science. In addition, long intervals between reports impede the timely involvement of science in policy

Table 5.1
Committee Organization in Multilateral Environmental Regimes

Standing Committee: Sets Its Own Schedule	Standing Committee: Schedule Set by COP	Ad Hoc Committee
Stratospheric Ozone	Climate Change	CCMPRCBS
CLRTAP	Ramsar Convention on Wetlands	IATTC
CBD	Bonn Convention on Migratory Species	ICCAT
CCAMLR	IWC	IPHC
CITES	IMO	NASCO
MEDPOL	CCSBT	South Pacific
North Sea	IOTC	Persian Gulf (pre-1990)
Stockholm POPs	UNCCD	PERSGA
Polar Bears	IOTC	
	NAFO	
	SEAFO	
	Baltic Sea	
	Basel Convention on Hazardous Wastes	
	Rotterdam PIC	
	Southeast Pacific	

Note: Abbreviations are spelled out in the appendix.
Source: UNEP 1998, 2003; Fritz 2000; Haas 2004b.

discussions. Table 5.1 shows the distribution among the various types of science committee organization in multilateral environmental agreements. Standing committees that set their own agenda and schedule appear to be the most politically insulated, whereas ad hoc science panels appear to be the most vulnerable to political involvement.

In addition to the timing of scientific bodies' meetings, the process of selecting scientists to serve on various bodies varies, with different effects on the legitimacy of the science created. The COP and the states involved in political negotiations sometimes directly select authors and experts. For some Multilateral Environmental Agreements' (MEAs) science bodies, the selection of experts is begun by the secretariat (sometimes in addition to selection by the parties) or even by secretariats of other international organizations in order to insulate the process (Fritz 2000). Table 5.2 illustrates the different mechanisms for selecting scientific

Table 5.2
Selection Procedures for Scientific Bodies

Selected by States/COPs	Selected by Secretariat	Selected by Other International Organizations
IPCC (climate change)	Ozone	STAP GEF
CLRTAP	Baltic Sea	Polar Bears
CBD (biodiversity)	MEDPOL (Mediterranean Sea)	IMO
CCAMLR (Antarctica)	South Pacific	
Ramsar (wetlands)	Southeast Pacific	
Bonn (migratory species)	North Sea	
CITES (endangered species)		
IWC (whaling)		
IOTC (fishery)		
Basel (hazardous wastes)		
Stockholm (POPs)		
Rotterdam (PICs)		
Persian Gulf (pre-1990)		
PERSGA (Red Sea)		

Note: Abbreviations are spelled out in the appendix.
Sources: UNEP 1998; Fritz 2000; Haas 2004b.

bodies. The most insulated involves selection of authors from secretariats or even scientific bodies of other intergovernmental organizations. Most MEAs, however, select the authors and peer reviewers of scientific reports using either the secretariat or the COP.

The actual design of a scientific body in multilateral environmental governance reflects one of three different design types (see table 5.3). The most popular institutional design (50 percent of regimes) is an open-access design that allows each member country to appoint a representative to a scientific body. Some of these open-membership bodies require the representatives from states to meet minimal and often vague expert qualifications in order to serve (for example, the Bonn Convention on Hazardous Wastes). This design does not typically create proper insulation between the scientific bodies and the policymakers. It is popular in many regimes and institutions dealing with biodiversity and fishery management. A second type of design is the small expert group. For example, the Ramsar Convention on Wetlands uses a small expert group of 12 members nominated and accepted by the COP and distributed

Table 5.3
Different Forms of Scientific and Expert Bodies in MEAs

Open Membership Body	Small Expert Panel	Roster of Experts
CBD (biodiversity)	CLRTAP	STAP GEF
CCAMLR (Antarctica)	Ozone	IPCC (climate change)
Bonn (migratory species)	Ramsar (wetlands)	CBD (biodiversity)
CCMPRCBS (fishery)	CITES (endangered species)	UNCCD (desertification)
IMO (fishery)	IWC (whaling)	Basel (hazardous wastes)
CCSBT (fishery)	IPHC (fishery)	Stockholm POPs
IATTC (fishery)	Stockholm POPs	
ICCAT (fishery)	Rotterdam PIC	
IOTC (fishery)	MEDPOL (Mediterranean)	
NAFO (fishery)	Polar Bears	
NASCO (fishery)	Persian Gulf (pre-1990)	
SEAFO (fishery)	Baltic Sea	
Basel (hazardous wastes)	Southeast Pacific	
UNCCD (desertification)	South Pacific	
PERSGA (Red Sea)		

Note: Abbreviations are spelled out in the appendix.
Sources: Fritz 2000; Haas 2004; UNEP 1998, 2003.

geographically based on the convention's membership. The trade-off in these panels is that although they are efficient, the complex environmental–social linkages and large amount of information necessary often prevents the development of adequate knowledge. A third and relatively recent design is a roster of experts that can either stand alone or be coupled to either of the previous forms. Jan-Steffan Fritz identifies these rosters as being increasingly important in filling the need for specialists to provide cutting-edge information but within broader contexts (2000, 30). Although this form has been most extensively used in the Scientific and Technical Advisory Panel (STAP) of the Global Environment Facility (GEF) (along with a small expert panel), other organizations are moving toward creating a roster of experts and some unique ways to use them (Fritz 2000). These rosters are most applicable in monitoring functions—for example, in the Convention on Biological Diversity (CBD) biosafety protocol (Hill and Sendashonga 2006). The STAP of the GEF uses the roster of experts for a variety of different functions, from providing peer review of reports to serving as regional science hubs to providing funding oversight. The sheer number of scientists and the evaluation system developed in the STAP ensures insulation between the science and the policymakers.

Despite all these constructions that aim to facilitate the creation of usable knowledge, social learning takes time. Comprehensive treaties are slower to negotiate than others are because they require persuasion and consensus rather than mere compromise. From a policy perspective, though, comprehensive regimes are likely to be superior in their ability to protect the environment in a cost-effective and politically acceptable manner. Moreover, treaties developed with help from the scientific community typically enter into force more rapidly than without it, presumably because of the weight that scientists' involvement carries in the ratification process.

A small number of international institutions has supported the development and transmission of usable knowledge. Epistemic communities often work in conjunction with broader policy networks, functional bureaucrats, transnational scientific organizations, nongovernmental organizations, and international civil servants. UNEP has played a powerful role in environmental protection the past 30 years. With a staff of less than 200 professionals and a budget now on the order of $150 million a year, UNEP has led global environmental monitoring efforts, catalyzed environmental protection activities in other UN bodies, served as the environmental conscience of the UN system, and sponsored the conclusion of dozens of international environmental treaties. The UN Economic Commission for Europe and the International Institute for Applied Systems Analysis helped develop and circulate usable knowledge for the effective management of European acid rain. Since 1991, the GEF has been a major supporter of the transmission of usable knowledge.

Lessons for Generating and Mobilizing Usable Knowledge

Given the 30-year experience with addressing transboundary and global environmental threats, what lessons are available about developing and mobilizing usable knowledge for multilateral environmental governance? The lessons listed here seem evident, based on comparative case studies of multilateral environmental regimes with and without usable knowledge and based on process-tracing studies of efforts to institutionalize usable knowledge (Haas 1991, 2000a, 2000b, 2004a, 2004b; Victor, Raustiala, and Skolnikoff 1998; Weiss and Jacobson 1998; Young 1999; Andresen 2000; Reinicke, Deng, Witte, et al. 2000; Miles, Underdal, Andresen, et al. 2002; Kanie and Haas 2004; Oberthür and Gehring

2006). Thus, these conditions are found in effective regimes and are absent in ineffective regimes. Effective regimes

1. Create standing international interdisciplinary scientific panels or committees to address specific topics.
2. Create and rely on separate subcommittees responsible for different functions of governance, such as basic research, environmental monitoring, policy analysis, and policy verification and evaluation.
3. Carefully survey the population of scientists to identify a core group that shares values and causal beliefs. For instance, in the Mediterranean a UNEP consultant spent nine months visiting national laboratories to inventory national capabilities and to build the scientific network personally.
4. Ensure that network and international panels have interdisciplinary representation, including social sciences. Ensure that Individuals, are highly regarded in their own disciplines and are able to talk to experts from other disciplines.
5. Recruit carefully for national and regional institutions. Base judgments on professional credentials and networking ability.
6. Avoid relying on one national institution to provide or sponsor research and training.
7. Provide professional outlets for members through conferences and publications in refereed professional journals. By this means, elevate the domestic profile of individual scientists in the community of experts, who may then be recruited to fill positions in national administrations.
8. Promote scientific discussions on topics that are likely to lead to consensus—that is, ripe research topics. Critical loads and the Regional Acidification Information System are excellent examples of ripe research topics (Sundqvist, chap. 7 in this volume).
9. Avoid government assignment of scientists to international meetings.
10. Try to make use of joint international panels for environmental risk assessment rather than relying on national assessments. Avoid capture by one scientific discipline or school of expert analysis (Wettestad, chap. 2 in this volume).
11. Assure the timely submission of scientific advice in advance of meetings. Submit reports in a timely way according to the legislative cycle in the major countries.

12. Arrange for focused interactions between scientists and policymakers to discuss the technical substance of the issues. For instance, in CLRTAP the International Institute of Applied Systems Analysis arranged for two-day sessions to familiarize policymakers with acid rain transfer and deposition models developed by scientists.

13. Maintain momentum within the community by continuing to have projects and research opportunities so members don't drift away, which avoids having to reconstitute the community each time a new problem emerges.

14. Seek funding for studies from multiple sources in order to avoid budgetary shocks if money is withheld from a principal funder. Thus, economically inefficient redundancy is politically warranted.

15. Formally institutionalize a roster of experts within the organization so that it can operate more effectively. The risk is that without institutionalizing and embedding the roster of experts the group will float freely and largely fail to affect policymaking.

16. Construct models so that effects are calculated at meaningful political scales—that is, corresponding to significant political divisions that are relevant in developing policy applications. For instance, at the international level this means that models should explicitly identify effects by country and even, if possible, by domestic districts (at least in countries without proportional representation). Climate-change models have so far yielded effects only at a scale of resolution sufficient to demonstrate to countries able to vote in the UN General Assembly that they are likely to suffer, and thus they keep the issue on the international agenda.

17. Train or recruit scientists who are capable of communicating with the media, politicians, and popular audiences.

18. Try to recruit networks from as broad a national basis as possible because governments are more likely to rely on experts who share their nationality than on experts who are not.

Broader considerations of the proper institutional design of science policy entails timing. When consensus is achieved before an issue reaches the agenda and policy discussions have begun, scientists can merely be introduced as experts, following these lessons. However, at times it is necessary to develop scientific consensus and to advance policy debates simultaneously. For these issues, such as in the Mediterranean and ozone regimes, the parallel development of science and policy must be kept insulated from ongoing policy debates, with the two streams united only when consensus has been achieved. In other cases, where consensus remains elusive and policy debates have already attained their own momentum, as in climate change and biodiversity, it may be best if the two activities are kept as separate as possible.

Overall, there has been extensive social learning in governance of Mediterranean pollution (Program for the Assessment and Control of Pollution in the Mediterranean Region, MEDPOL), European acid rain (CLRTAP), and stratospheric ozone. Thus, for example, epistemic communities were vital in the creation of MEDPOL and then for the degree of social learning that occurred within the organization. As a consequence, independent panels helped to develop standards for achieving reduction of land-based pollutants (Haas 1990). With respect to MEDPOL, CLRTAP, and stratospheric ozone, epistemic communities have been institutionalized generally in the regimes and played key roles in legal negotiations. Social learning has occurred in the creation of many environmental regimes (e.g., the International Whaling Convention [IWC] and the North Sea), but these regimes have mostly stalled, and little social learning has happened within them since they were established. In the IWC, the scientific committee developed a revised management procedure to allow whaling of nonendangered species. Years later this revised management scheme has still not been adopted, and Japan, Iceland, and Norway's continued hunting of nonendangered species threatens the legitimacy of the regime (Andresen 2002). There has been moderate social learning in some biodiversity and chemical conventions (Rotterdam Convention on Prior Informed Consent, Bonn Convention on Migratory Species, Ramsar, and Convention on International Trade of Endangered Species [CITES]). For example, the establishment of scientific committees in the 1987 CITES allowed a continued scientific appraisal of the regime's effectiveness and the expansion or reduction of the list of endangered species. The scientific committees have been able to update and monitor endangered species status effectively, but these functions have translated minimally into changing the organization's policy significantly. There has been little to no social learning in a number of regimes regarding fisheries, biodiversity (namely, the CBD), desertification, and climate change (see relevant chapters in Social Learning Group 2001; Miles, Underdal, Andresen, et al. 2002). Table 5.4 gives a breakdown of various MEAs.

The insulation of the science body within the MEA is key to the social learning that can occur within the organization (see appendix 5.1 for a

Table 5.4
Degrees of Social Learning in MEAs

No Significant Social Learning	Some Social Learning within Regime	Social Learning
Climate Change	Bonn (migratory species)	MEDPOL (Mediterranean)
CCMPRCBS (fishery)	Ramsar (wetlands)	CLRTAP (air pollution)
IMO (fishery)	Rotterdam PIC	Ozone
CCSBT (fishery)	CITES (endangered species)	CCAMLR (Antarctica)
IATTC (fishery)	IWC (whaling)	Polar Bears
ICCAT (fishery)	North Sea	Stockholm POPs
IOTC (fishery)	Baltic	
IPHC (fishery)	Persian Gulf (pre-1990)	
NAFO (fishery)	South Pacific	
NASCO (fishery)	Southeast Pacific	
SEAFO (fishery)		
UNCCD (desertification)		
Basel (hazardous wastes)		
PERSGA (Red Sea)		

Note: Abbreviations are spelled out in the appendix.
Sources: See Young 1999; Social Learning Group 2001; Miles, Underdal, Andresen, et al. 2002.

combination of the information given in all the tables). For example, in MEDPOL, CLRTAP, and the ozone regime there is an insulated and separate science panel. Most cases in which there are small groups of experts that are selected by the secretariat and that can set their own agenda lead to some degree of social learning (this is true in the ozone, MEDPOL, CLRTAP, Rotterdam, and Baltic Sea regimes). For example, with regard to CLRTAP, Bernd Siebenhüner (chap. 4 in this volume) discusses various *reflexive mechanisms*, including bringing international experts into the panels of the CLRTAP negotiations and external reviews as forms of insulation to increase the legitimacy of the science. There are several cases in which the use of some of these procedures has produced varied results in terms of social learning. The Bonn Convention on Migratory Species is the only institution with an open advisory system (each nation appoints one member to the science committee) that has demonstrated any social learning. Bonn, however, has several other safeguards and mechanisms that insulate science from political control. Fisheries management in general has failed to insulate science bodies from direct or indirect political control. The design of the science committees in fisheries reflects an institutional pathology that limits the ability of science to speak to power. Insulation is vital procedurally if usable knowledge is to lead to social learning; however, many organizations that aim toward deliberation and transparency are still limited in their level of social learning. Thus, the mere existence of a roster of experts or a small expert group does not guarantee that social learning will occur. Issues of comprehensiveness and adequacy are vital if the regime is to maintain social learning.

Climate Change, Usable Knowledge, and the Limits to Social Learning

The climate-change regime, the focus of the majority of global environmental diplomatic efforts during the 1990s, has also developed through a deliberate effort by diplomats to base policy on strong scientific foundations through the IPCC). The IPCC is of interest because it highlights the way in which states may choose to shape the science advisory process. A closer look at the interplay of science and power in the IPCC reveals empirically how this dynamic interaction operates in this key contemporary issue and establishes analytically the political limits to autonomous science and social learning. The knowledge base developed so far has not proven particularly usable largely because the construction of knowledge in the IPCC is politically suspect and because the results are not presented in a way that connects to individual experiences (Haas 2008).

The IPCC is one of the most concerted efforts by the international community to harness usable knowledge for addressing transboundary and global environmental threats (Agrawala 1998a, 1998b; Skodvin 2000a, 2000b; Siebenhüner 2002, 2003). The IPCC was established in 1988 as an international science policy advisory body for global warming, but the general belief is that it was formed politically so that governments would be able to reassert control over the science process in an issue that was accelerating on the policy agenda more rapidly than most leaders in the North were comfortable with. The IPCC was the consequence of a UN General Assembly initiative for a climate-change regime and did not rely on UNEP, as had most previous international environmental regime initiatives. The IPCC consists of three working groups, with members chosen by governments subject to the scientific reputation of the candidates. Working Group One addresses questions of atmospheric science; Working Group Two assesses social and economic impacts and adaptation measures; and Working Group Three focuses on mitigation alternatives. Each working group is administered by a bureau composed of the

IPCC Bureau members, working group cochairs and vice chairs, and a technical advisory unit drawn from the country of the work group's chair. To date, the IPCC has produced 4 three-volume assessment reports (1990 with 1992 revisions, 1995, 2001, and 2007), each with a summary for policymakers as well as various ad hoc special reports and technical papers. The working groups report to a plenary composed of government delegates, who review the summary reports on a line-by line-basis and approve the working group assessments and special reports in more blocklike fashion. All reports other than the summary reports rely on extensive peer reviews, and the material presented must come from peer-reviewed journals or be in the process of publication in a peer-reviewed journal. The summary reports, which command the greatest publicity and hence public attention, are written by the working group leaders along with the lead authors and specially invited experts (Skodvin 2000b; Swart, Mitchell, Moita, et al. 2002). The bureaus are responsible for drafting an initial table of contents and topics for each chapter. The IPCC political plenary then approves this agenda. The scientific committees draft reports, and then government representatives in the plenary approve them subject to careful scrutiny.

Thus, governments have sought to exercise control over the scientific process while also allowing for some degree of scientific latitude in order to generate accurate advice, even if the process is designed in such a way that the advice is unlikely to be particularly salient. Governments appoint the IPCC chair, and in 2002 the United States vetoed the appointment of the climatologist Robert Watson, a well-regarded American candidate, in favor of Rajendra Pachauri, an Indian engineer, based on the belief that Watson was too independent of the US administration. The United States criticized the *Third Assessment Report* (2001) for containing language that the United States claimed was stronger than had been approved by the plenary, although the authors were able to prove subsequently that their draft was consistent with the IPCC's rules of procedure (Miller and Edwards 2001; Siebenhüner 2002). Diplomats in the plenary evaluate the summaries for policymakers on a line-by-line basis. All individuals are nominated and chosen by governments, although there is little evidence of direct government manipulation in recruitment or the inclusion of material. The procedures are carefully designed to maximize the seeming procedural scientific legitimacy and accuracy of the work by stressing peer review and reliance on peer-reviewed material.

The degree to which the IPCC is capable of generating usable knowledge is largely politically circumscribed. The state of scientific understanding of the key global systems that affect global warming remains relatively immature. It is unclear to what extent funding and research choices have been shaped by political factors, but there is no strong evidence to assert that the state of knowledge about the phenomenon is directly biased or controlled by political influences. The saliency of the IPCC science remains limited. The estimates of global warming and their effects are crude, and global carbon models are unable to account fully for the global carbon cycle.

The IPCC does not enjoy a high degree of legitimacy in the eyes of many consumers of science policy (Biermann 2002). It suffers from the appearance of governmental control because governments appoint the scientists and vote on the reports. The distribution of scientific participants comes overwhelming from the North, despite the bureaus' best efforts to increase participation from developing countries since the release of the *First Assessment Report* (Biermann 2002). Table 5.5 shows the crude national breakdown of scientific involvement in the Fourth Assessment Report by the scientific participants' country of origin. The IPCC is limited in its legitimacy through its seeming lack of equity in participation. Although the IPCC has helped pay for the participation of scientists from developing countries, it faces the deeper structural science

Table 5.5
IPCC Fourth Assessment Report: Distribution of Nationality of Scientific Participants

Nationality	WG1 Authors (*n* = 625)	WG2 Authors and Expert Reviewers (*n* = 1,503)	WG3 Authors (*n* = 272)	Synthesis Lead Authors (*n* = 39)
Industrialized Countries	88%	75%	68%	62%
Developing Countries	12%	25%	32%	38%
Technical Support Unit	United States	United Kingdom	Netherlands	

Note: Industrialized countries are considered those that are members of the Organization of Economic Cooperation and Development (OECD), which includes those countries at high levels of industrialization and development. Developing countries are those that are not members of OECD. See http://www.oecd.org/home/0,2987,en_2649_201185_1_1_1_1_1,00.html.
Source: See http://www.ipcc.ch, retrieved 10 December 2007.

policy issue that the overwhelming majority of climate-change research is conducted in the North by northern scientists.

The IPCC's saliency is particularly poor. For one thing, it has been unable to develop policy advice that resonates domestically in any of the major countries. To date, its scientific work has narrowed the range of likely warming that will occur in 2100 and generated scenarios of what the global environmental consequences may be of such effects. However the scenarios are sufficiently crude that they do not engage any significant political interest in any of the member countries, other than reinforcing the prior knowledge that Egypt, Bangladesh, and the small island countries may well be submerged. Thus, the political effect of the new knowledge is adequate only for the aggrieved parties to continue to keep the issue on the General Assembly agenda, where they have sufficient votes, but not to influence policy outcomes directly. More salient science would present data at a level of resolution that corresponded to domestic political divisions (i.e., states or congressional districts in the United States and parliamentary districts in parliamentary systems). In addition, in attempting to be politically neutral, the IPCC aims to provide "policy-relevant, not policy-prescriptive" information, which results in information that is too broad to provide even basic policy assistance (Hasselman and Barker 2008, 220).

The publication of the assessment reports did not fit in well with the negotiating cycle of the Kyoto Protocol and efforts to follow up the 1992 UN Framework Convention on Climate Change. A few of the special reports—such as the 2000 *Report on Land Use, Land Use Changes, and Forestry*—corresponded to COP meetings and thus were able to provide timely scientific policy advice to the actual negotiators. The reports have more often not coincided closely with the negotiation process, so they have had little impact other than periodically reminding the mass public of the potential severity of global warming. IPCC scientists have not fared well at communicating their findings to the public or in engaging in either elite or public education. This failure can be contrasted with the timely issuing of reports by other organizations to precede major political conferences on the issue. For example, both the *Scientific Assessment of Ozone Depletion* in the 1990s and the *World Water Report* in the 2000s preceded major negotiating cycles internationally. However, many reports from scientists are timed on cycles (for example, every two years) that seem to have little or nothing to do with the political negotiations. The IPCC seems to have altered this pattern slightly with the Fourth Assessment Report, which was released in early 2007 following a momentous year that saw Al Gore's *An Inconvenient Truth* receiving every possible award for a movie, book, and book on tape (with an opera in the planning stages), the IPCC's receiving the 2007 Nobel Peace Prize, and then major international climate-change negotiations. However, this slight alteration in timing failed to result in effective negotiations among the COP at the Copenhagen UN Framework Convention on Climate Change (COP 15) in 2009, where, in the measured words of one analyst, "the conference was a failure whose magnitude exceeded our worst fears, and the resulting Copenhagen Accord was a desperate attempt to mask that failure" (Dimitrov 2010, 18).

In short, the IPCC is designed to keep science on a tight leash by controlling the selection and autonomy of individual scientists engaged in the assessment process. As a consequence, the degree of usable knowledge generated by the IPCC has been limited. And IPCC scientists not surprisingly have been unable to exercise sufficient discretion to develop more politically tractable advice. In turn, negotiated treaties within the climate-change regime have not reflected a strong degree of scientific basis despite the ongoing IPCC efforts. The 1992 UN Framework Convention on Climate Change and the 1997 Kyoto Protocol reflect broad rhetorical commitments or political compromises about target emissions but do not point to any scientifically grounded targets for specific atmospheric concentrations of greenhouse gases. Few governments outside the European Union have proven willing to make national decisions based on this collective enterprise, and European Union greenhouse gas targets reflect political and energy policy realities rather than scientific justifications (Haas 2008).

The limits of the IPCC's ability to speak truth to power can be explained by principal-agent theory (Moe 1984; Koremenos, Lipson, and Snidal 2001; see also Guston 1998 and 2003 for an effort to capture science policy issues more generally within a principal-agent framework). According to this theory, the principals strategically design the institution (the agent) on which they rely for information and advice in order to minimize surprises from the agent and to ensure that the agent will not provide advice that runs counter to the principals' ex ante interests and preferences. The principals in the IPCC case are the governments of the industrialized countries. In 1988, they were tired of being lambasted by UNEP in multilateral environmental negotiations and were concerned that uncontrolled scientific panels might give rise to policies that they did not deem warranted. The Villach meetings from 1985 to 1988 had forcefully put the issue of global warming on the international agenda, and the 1988

Toronto conference convened by nonstate actors had called for 20 percent reductions in greenhouse gas reductions, thus alerting foreign ministry officials of the political threats posed by nonstate actors' potential control of the climate-change agenda. The recent lesson of the Ozone Trends Panel's powerful influence on the rapid conclusion of the Montreal Ozone Protocol was also fresh in the minds of foreign ministries. Moreover, the short-term costs of addressing the problem for the industrialized countries (or, put another way, the net discounted value of efforts for mitigation adaptation) appeared prohibitive so that they wished to reign in any independent political pressure that would be generated from an organized scientific involvement in collective discussions on climate change.

The agents in this regard are the scientific community, who are recruited and organized through the IPCC in a way that assures governments that they will be able to exercise maximum control over individual scientists and will remain able to shape the political agenda for climate-change negotiations. Through the reliance on extensive peer review, some degree of control was admittedly lost, but at the end of the day governments appoint the individuals, and foreign-ministry officials approve reports on a line-by-line basis. Thus, the principals establish and enforce the parameters within which the agents have to act, thus ensuring cautious outcomes by the agents.

But the IPCC is an extreme case in terms of the clarity with which the powerful industrialized countries recognized their material interests in the matter and in terms of the widespread transformations in industrial and individual behaviors that effective preventive policies were presumed to entail. In addition, the concurrent negotiations and science advisory process made governments extremely sensitive to the strategic implications that scientific assessments may have for national policies. In most other transboundary and global environmental issues, the scientific consensus preceded active political discussions so that governments were less acutely aware of the implications of their choice to delegate rule making to scientists. Most other environmental and sustainable development issues do not share the characteristics of the IPCC case, so there is greater potential for governments to create more legitimate and porous scientific processes to guide their policy work.

Conclusion

Thirty years of speaking truth to power has yielded tangible beneficial outcomes. Power listens sometimes. There now exist far more extensive legal commitments on environmental protection than ever existed before, and most areas of human activity in most countries of the world are now regulated as a consequence. In the aggregate, major changes have occurred internationally in environmental governance. Many multilateral environmental regimes are more comprehensive than before science was systematically applied to multilateral environmental diplomacy; stronger national environmental protection commitments have been undertaken in compliance with international obligations; and green markets for cleaner technologies have grown as a consequence. Markets estimated at more than $650 billion a year now exist for green technologies ("Global Market Overview" 2006). These markets were created in response to the regulatory obligations resulting from international treaties negotiated based on the application of new scientific understandings of the behavior of ecosystems (Garcia-Johnson 2000; Haufler 2001). The protection of ecological integrity has become elevated to the position of a state interest on a par with the pursuit of national wealth and power. Sustainable development is now the doctrine by which states seek to reconcile (or optimize) the pursuit of those three objectives.

Knowledge can speak volumes to power. Much current work suggests that science remains influential if its expertise and claims are developed behind a politically insulated wall. Epistemic communities are the transmission belts by which new knowledge is developed and transmitted to decision makers. Knowledge must also possess the substantive characteristics of usable knowledge: credibility, legitimacy, and saliency. And successful regimes demonstrate practical administrative lessons about generating and mobilizing epistemic communities and usable knowledge.

Real limits to learning persist. National administrative design often inhibits learning and the diffusion of environmental lessons. Most learning remains highly localized: lessons are limited to changes in national policy with regard to one particular environmental medium—for instance, air pollution or marine pollution. This limitation is due to administrative design in most countries. Epistemic-community members' expertise is highly specialized, so their influence and involvement in agency choices are limited to a specific regime and environmental issue area. Moreover, environmental responsibilities within governments vary by environmental issue because different agencies are responsible for coordinating national activities in different areas. Thus, epistemic-community members' influence in the short term is highly restricted. Few governments have the domestic potential for these lessons to spread within the government because the functional agencies have little influence over other agencies.

The prospects for improving national sensitivity to epistemic-community influence rest on recruitment and retention patterns within national administrations. Environmental agencies often lack the ability to spread their influence within government administrations because most of them are so specialized and their staffs hold their positions for only limited periods, so there is limited potential for individuals to educate each other.

Scientific consensus is often poorly represented to decision makers. Eloquent and articulate specialists from within scientific disciplines are needed to communicate across technical and cultural divides to policy analysts and politicians (Koehn and Rosenau 2002). CLRTAP shows that the gradual development of scientific consensus can allow policy consensus to emerge. In contrast, the IPCC has made it clear that it is much harder to mobilize knowledge to address politicized and high stakes issues when knowledge is being developed at the same time as policy debates are occurring.

Although constructivists are in general reluctant to rely on the rationalist assumptions of theories such as the principal-agent theory, in the particular issue of global warming the material incentive to the principals are so clear that the principal-agent theory's rationality assumptions are warranted. It is only in cases with greater uncertainty about material national interests and national effects from environmental harm that more constructivist approaches are warranted, although some dimensions of the global-warming issue are open to fruitful constructivist analysis (Miller and Edwards 2001; Martello and Jasanoff 2004). In conclusion, institutionalized knowledge has contributed to more effective multilateral environmental management. Further work on this topic can advance the understanding of the interplay between knowledge and power by studying when states are willing to create autonomous institutions and defer to their authority; of the interplay between different ideational factors in shaping collective action; and of the interplay between various nonstate actors at and across multiple scales of multilevel global governance (Annan 1998; Camilleri, Malhotra, and Tehranian 2000; Kanie and Haas 2004).

Acknowledgments

We are very indebted to Rolf Lidskog and Göran Sundqvist for offering insightful and helpful editorial comments on earlier drafts of this chapter.

The chapter expands and updates the analysis in Peter M. Haas, "When Does Power Listen to Truth? A Constructivist Approach to the Policy Process," *European Journal of Public Policy* 11 (4) (2004): 569–592.

Appendix 5.1: Collection of Information from Tables

egime Name	Abbreviation	Committee Organization	Expert Selection	Type of Science Committee	Degree of Social Learning
Vienna Convention for the Protection of the Ozone Layer (and Protocols)	Stratospheric Ozone	Standing; Sets Agenda	Secretariat	Group of Experts	Social Learning
UN Framework Convention on Climate Change	Climate Change	Standing	COP	Group of Experts	Little/ None
Convention on Long-Range Transboundary Air Pollution	CLRTAP	Standing; Sets Agenda		Group of Experts	Social Learning
Convention on Biological Diversity	CBD	Standing; Sets Agenda	COP	Open	Little/ None
Convention for the Conservation on Marine Living Resources	CCAMLR	Standing; Sets Agenda	COP	Open	Social Learning
Ramsar Convention on Wetlands	Ramsar	Standing	COP	Group of Experts	Some
Bonn Convention on Migratory Species	Bonn	Standing	COP	Open	Some

Appendix 5.1: (continued)

Regime Name	Abbreviation	Committee Organization	Expert Selection	Type of Science Committee	Degree of Social Learning
8 Convention on International Trade of Endangered Species	CITES	Standing; Sets Agenda	COP	Group of Experts	Some
9 International Whaling Convention	IWC	Standing	COP	Group of Experts	Some
10 International Maritime Organization	IMO	Standing	Other Intergovernmental Organizations	Open	Little/ None
11 Convention on the Conservation and Management of the Pollock Resources in the Central Bering Sea	CCMPRCBS	Ad Hoc			Little/ None
12 Convention for the Conservation of Southern Bluefin Tuna	CCSBT	Standing		Open	Little/ None
13 Inter-American Tropical Tuna Commission	IATTC	Ad Hoc		Open	Little/ None
14 International Commission for the Conservation of Atlantic Tuna	ICCAT	Ad Hoc		Open	Little/ None
15 Indian Ocean Tuna Commission	IOTC	Standing	COP	Open	Little/ None
16 International Pacific Halibut Commission	IPHC	Ad Hoc		Group of Experts	Little/ None

ppendix 5.1: (continued)

egime Name	Abbreviation	Committee Organization	Expert Selection	Type of Science Committee	Degree of Social Learning
7 Northwest Atlantic Fishery Organization	NAFO	Standing		Open	Little/ None
8 North Atlantic Salmon Conservation Organization	NASCO	Ad Hoc		Open	Little/ None
9 Southeast Atlantic Fishery Organization	SEAFO	Standing		Open	Little/ None
0 Program for the Assessment and Control of Pollution in the Mediterranean Region	MEDPOL	Standing; Sets Agenda	Secretariat	Group of Experts	Social Learning
1 Convention for the Protection of the Marine Environment of the North-East Atlantic	North Sea	Standing; Sets Agenda	Secretariat		Some
2 Convention on the Protection of the Marine Environment of the Baltic Sea Area	Baltic Sea	Standing	Secretariat	Group of Experts	Some
3 Convention for the Protection of the Natural Resources and Environment of the South Pacific Region (and Protocols)	South Pacific	Ad Hoc	Secretariat	Group of Experts	Some

Appendix 5.1: (continued)

Regime Name	Abbreviation	Committee Organization	Expert Selection	Type of Science Committee	Degree of Social Learning
24 Convention for the Protection of the Marine Environment and Coastal Zones of the South-East Pacific (Lima Convention)	Southeast Pacific	Standing	Secretariat	Group of Experts	Some
25 Regional Organization for the Protection of the Marine Environment (Persian Gulf, pre-1990)	ROPME	Ad Hoc	COP	Group of Experts	Some
26 Program for the Environment of the Red Sea and Gulf of Aden	PERSGA	Ad Hoc	COP	Open	Little/ None
27 Basel Convention on the Control of Transboundary Movements of Hazardous Wastes and Their Disposal	Basel Convention	Standing	COP	Open	Little/ None
28 Stockholm Convention on Persistent Organic Pollutants	Stockholm POPs	Standing; Sets Agenda	COP	Group of Experts	Social Learning
29 Rotterdam Convention on Prior Informed Consent	Rotterdam PIC	Standing	COP	Group of Experts	Some

ppendix 5.1: (continued)

egime Name	Abbreviation	Committee Organization	Expert Selection	Type of Science Committee	Degree of Social Learning
0 UN Convention concerning Desertification	UNCCD	Standing	COP	Open	Little/ None
1 Agreement on Conservation of Polar Bears	Polar Bears	Standing; Sets Agenda	Other Intergovern-mental Organizations	Group of Experts	Social Learning

References

Adler, Emanuel. 2002. *Changing Identities: The Road to Peace.* Toronto: University of Toronto.

Agrawala, Shardul. 1998a. Context and Early Origins of the Intergovernmental Panel on Climate Change. *Climatic Change* 39:605–620.

Agrawala, Shardul. 1998b. Structural and Process History of the Intergovernmental Panel on Climate Change. *Climatic Change* 39:621–642.

Andresen, Steinar. 2000. *Science and Politics in International Environmental Regimes: Between Integrity and Involvement.* Issues in Environmental Politics. New York: Manchester University Press.

Andresen, Steinar. 2002. The International Whaling Commission (IWC): More Failure than Success? In *Environmental Regime Effectiveness: Confronting Theory with Evidence*, ed. Edward M. Miles, Arild Underdal, Steinar Andresen, Jørgen Wettestad, Jon Birger Skjaerseth, and Elaine M. Carlin, 379–404. Cambridge, MA: MIT Press.

Andresen, Steinar, Lars Walløe, and Kristin Rosendal. 2005. The Precautionary Principle: Knowledge Counts, but Power Decides? In *Biodiversity and the Precautionary Principle: Risk and Uncertainty in Conservation and Sustainable Use*, ed. Barnie Dicson and Rosie Conney, 39–54. London: Earthscan.

Annan, Kofi. 1998. The Quiet Revolution. *Global Governance* 4 (2): 123–138.

Asher, William. 1986. The Evolution of the Policy Sciences. *Journal of Public Analysis and Management* 5 (2): 365–389.

Biermann, Frank. 2002. Institutions for Scientific Advice. *Global Governance* 8 (2): 195–219.

Botcheva, Liliana. 2001. Expertise and International Governance. *Global Governance* 7 (2):197–224.

Brown, G. 1997. Environmental Science under Siege in the U.S. Congress. *Environment* 39 (2): 12–20, 29–31.

Camilleri, Joseph A., Kamal Malhotra, and Majid Tehranian. 2000. *Reimagining the Future*. Bundoora, Victoria: La Trobe University.

Checkel, Jeffrey T. 1998. The Constructivist Turn in International Relations Theory. *World Politics* 50 (2): 324–348.

Christiansen, Thomas, Knud Erik Jørgensen, and Antje Wiener. 1999. The Social Construction of Europe. *Journal of European Public Policy* 6 (4): 528–544.

Clark, William C., ed. 1990. *Usable Knowledge for Managing Global Climate Change*. Stockholm: Stockholm Environmental Institute.

Clark, William C. 2001. Knowledge Systems for Sustainable Development. *Environment* 43 (8): 1.

Clark, William C. 2007. Sustainability Science: A Room of Its Own. *Proceedings of the National Academy of Sciences of the United States of America* 104 (6): 1737–1738.

Clark, William C., and Giandomenico Majone. 1985. The Critical Appraisal of Scientific Inquiries with Policy Implications. *Science, Technology, & Human Values* 10 (3): 6–19.

Dimitrov, Radoslav. 2003. Knowledge, Power, and Interests in Environmental Regime Formation. *International Studies Quarterly* 47 (1): 123–151.

Dimitrov, Radoslav. 2010. Inside Copenhagen: The State of Climate Governance. *Global Environmental Politics* 10 (2): 18–24.

Drori, Gili S., and John W. Meyer. 2006. Global Scientization: An Environment for Expanded Organization. In *Globalization and Organization: World Society and Organization*, ed. Gili S. Drori, John W. Meyer, and Hokyu Hwang, 50–68. New York: Oxford University Press.

Drori, Gili S., John W. Meyer, Francisco O. Ramirez, and Evan Schofer. 2003. *Science in the Modern World Polity: Institutionalization and Globalization*. Stanford, CA: Stanford University Press.

Dryzek, John S. 1997. *The Politics of the Earth: Environmental Discourses*. Oxford, UK: Oxford University Press.

European Commission (Commission of the European Communities). 2001. *Report of Working Group on Broadening and Enriching the Public Debate on European Matters: Democratising Expertise and Establishing Scientific Reference Systems*. Brussels: European Commission.

European Commission (Commission of the European Communities). 2002. *Improving the Knowledge Base for Better Policies*. Luxembourg: Communication from the Commission on the Collection and Use of Expertise by the Commission, Principles and Guidelines.

Farrell, Alexander E., and Jill Jäger, eds. 2005. *Assessments of Regional and Global Environmental Risks: Designing Processes for the Effective Use of Science in Decisionmaking*. Washington, DC: Resources for the Future.

Fischer, Frank, and John Forester. 1993. *The Argumentative Turn in Policy Analysis and Planning*. Durham, NC: Duke University Press.

Fritz, Jan-Steffan. 2000. *The Second Report on International Scientific Advisory Processes on the Environment and Sustainable Development*. Paris: United Nations Environmental Program.

Funtowicz, Silvio O., and Jerome R. Ravetz. 1991. A New Scientific Methodology for Global Environmental Issues. In *Ecological Economics*, ed. Robert Costanza, 137–152. New York: Columbia University Press.

Funtowicz, Silvio O., and Jerome R. Ravetz. 2001. Global Risk, Uncertainty, and Ignorance. In *Global Environmental Risk*, ed. Jeanne X. Kasperson and Roger Kasperson, 173–194. London: Earthscan.

Garcia-Johnson, Ronie. 2000. *Exporting Environmentalism: U.S. Multinational Chemical Corporations in Brazil and Mexico*. Cambridge, MA: MIT Press.

Global Market Overview. 2006. *Environmental Business Journal*. 19 (5–6): 1–3.

Gore, Al. 1996. The Technology Challenge: What Is the Role of Science in American Society? Paper read at American Association for the Advancement of Science conference, Baltimore.

Gornitzka, Åse, and Ulf Sverdrup. 2008. Who Consults? The Configuration of Expert Groups in the European Union. *West European Politics* 31 (4): 725–750.

Guston, David H. 1998. Principal-Agent Theory and the Structure of Science Policy. *Science & Public Policy* 23 (4): 229–240.

Guston, David H. 2003. Principal-Agent Theory and the Structure of Science Policy, Revisited. *Science & Public Policy* 30 (5): 347–357.

Guzzini, Stefan. 2000. A Reconstruction of Constructivism in International Relations. *European Journal of International Relations* 6 (2): 147–182.

Haas, Peter M. 1990. *Saving the Mediterranean: The Politics of International Environmental Cooperation, the Political Economy of International Change*. New York: Columbia University Press.

Haas, Peter M. 1991. Save the Seas: UNEP's Regional Seas Programme and the Coordination of Regional Pollution Control Efforts. In *Ocean Yearbook*, vol. IX, ed. Elisabeth Mann Borgese, Norton Ginsburg, and Joseph R. Morgan. Chicago: University of Chicago Press.

Haas, Peter M. 2000a. Letter to the Editor. *Marine Policy* 24 (6): 499–500.

Haas, Peter M. 2000b. Prospects for Effective Marine Governance in the NW Pacific Region. *Marine Policy* 24 (6): 341–348.

Haas, Peter M. 2001a. Environment: Pollution. In *Managing Global Issues*, ed. P. J. Simmons and Chandal de Jonge Oudraat. Washington, DC: Carnegie Endowment for International Peace.

Haas, Peter M. 2001b. Epistemic Communities and Policy Knowledge. In *International Encyclopedia of Social and Behavioral Sciences*, ed. N. J. Smelser, James Wright, and P. B. Baltes, 11578–11586. New York: Elsevier.

Haas, Peter M. 2004a. Science Policy for Multilateral Environmental Governance. In *Emerging Forces in Environmental Governance*, ed. Norichika Kanie and Peter M. Haas, 115–136. Tokyo: United Nations University Press.

Haas, Peter M. 2004b. When Does Power Listen to Truth? A Constructivist Approach to the Policy Process. *Journal of European Public Policy* 11 (4): 569–592.

Haas, Peter M. 2007. Epistemic Communities. In *The Oxford Handbook of International Environmental Law*, ed. Daniel Bodansky, Jutta Brunnee, and Ellen Hey, 791–806. New York: Oxford University Press.

Haas, Peter M. 2008. Climate Change Governance after Bali. *Global Environmental Politics* 8 (3): 1–7.

Haas, Peter M., and Ernst B. Haas. 1995. Learning to Learn: Improving International Governance. *Global Governance* 1 (1): 255–285.

Haas, Peter M., and Ernst B. Haas. 2002. Pragmatic Constructivism and the Study of International Relations. *Millennium: Journal of International Studies* 31 (3): 573–601.

Hajer, Maarten A., and Hendrik Wagenaar, eds. 2003. *Deliberative Policy Analysis: Understanding Governance in the Network Society*. Cambridge, UK: Cambridge University Press.

Hasselman, Klaus, and Terry Barker. 2008. The Stern Review and the IPCC Fourth Assessment Report: Implications for Interaction between Policymakers and Climate Experts. An Editorial Essay. *Climatic Change* 89:219–229.

Haufler, Virginia. 2001. *A Public Role for the Private Sector: Industry Self-Regulation in a Global Economy*. Washington, DC: Carnegie Endowment for International Peace.

Hill, Ryan, and Cyrie Sendashonga. 2006. Conservation Biology, Genetically Modified Organisms, and the Biosafety Protocol. *Conservation Biology* 20 (6): 1620–1625.

Hopf, Ted. 1998. The Promise of Constructivism in International Relations Theory. *International Security* 23 (1): 171–200.

Jasanoff, Sheila. 1990. *The Fifth Branch: Science Advisers as Policymakers*. Cambridge, MA: Harvard University Press.

Jasanoff, Sheila. 2004. *States of Knowledge: The Co-Production of Science and Social Order*. New York: Routledge.

Jasanoff, Sheila, and the Society for Social Studies of Science. 1995. *Handbook of Science and Technology Studies*. Thousand Oaks, CA: Sage.

Jasanoff, Sheila, and Brian Wynne. 1998. Science and Decisionmaking. In *Human Choice and Climate Change*, ed. Steve Rayner and Elizabeth L. Malone, 1–87. Columbus, OH: Batelle Press.

Kanie, Norichika, and Peter M. Haas, eds. 2004. *Emerging Forces in Environmental Governance*. Tokyo: United Nations University Press.

Kates, Robert W., William C. Clark, Robert Corell, J. Michael Hall, Carlo C. Jaeger, Ian Lowe, James J. McCarthy, et al. 2001. Sustainability Science. *Science* 292 (5517): 641–642.

Koehn, Peter H., and James N. Rosenau. 2002. Transnational Competence in an Emergent Epoch. *International Studies Perspectives* 3 (2): 105–127.

Koremenos, Barbara, Charles Lipson, and Duncan Snidal. 2001. Rational Designs: Explaining the Form of International Institutions. *International Organization* 55 (4): 1–32.

Kubalkova, Vendulka, Nicholas Onuf, and Paul Kowert, eds. 1998. *International Relations in a Constructed World*. Armonk, NY: M. E. Sharpe.

Lee, Kai N. 1993. *Compass and Gyroscope: Integrating Sciences and Politics in the Environment*. Washington, DC: Island Press.

Levy, Marc A., Jeannine Cavender-Bares, William C. Clark, Gerda Dinkelman, Elena Nikitina, Ruud Pleune, and Heather Smith. 2001. Goal and Strategy Formulation in the Management of Global Environmental Risks. In *Learning to Manage Global Environmental Risks: A Comparative History of Social Response to Climate Change, Ozone Depletion, and Acid Rain*, ed. Social Learning Group. Cambridge, MA: MIT Press.

Lidskog, Rolf, and Göran Sundqvist. 2002. The Role of Science in Environmental Regimes: The Case of CLRTAP. *European Journal of International Relations* 8 (1): 77–101.

Lindner, Johannes. 2003. Institutional Stability and Change: Two Sides of the Same Coin. *Journal of European Public Policy* 10 (6): 912–935.

Litfin, Karen. 1994. *Ozone Discourses: Science and Politics in Global Environmental Cooperation*. New York: Columbia University Press.

Martello, Marybeth Long, and Sheila Jasanoff. 2004. Introduction: Globalization and Environmental Governance. In *Earthly Politics: Local and Global in Environmental Governance*, ed. Sheila Jasanoff and Marybeth Long Martello, 1–30. Cambridge, MA: MIT Press.

Miles, Edward L. 1998. Personal Reflections on an Unfinished Journey through Global Environmental Problems of Long Timescale. *Policy Sciences* 31 (1): 1–33.

Miles, Edward L, Arild Underdal, Steinar Andresen, Jørgen Wettestad, Jon Birger Skjaerseth, and Elaine M. Carlin. 2002. *Environmental Regime Effectiveness: Confronting Theory with Evidence*. Cambridge, MA: MIT Press.

Miller, Clark A., and Paul N. Edwards. 2001. *Changing the Atmosphere: Expert Knowledge and Environmental Governance*. Cambridge, MA: MIT Press.

Milton, Giles. 1999. *Nathaniel's Nutmeg*. New York: Farrar, Straus and Giroux.

Mitchell, Ron. 2002. International Environment. In *Handbook of International Relations*, ed. Thomas Risse, Beth Simmons, and Walter Carlsnaes, 500–516. New York: Sage.

Mitchell, Ronald B., William C. Clark, David W. Cash, and Nancy M. Dickson, eds. 2006. *Global Environmental Assessments: Information and Influence*. Cambridge, MA: MIT Press.

Moe, Terry M. 1984. The New Economics of Organization. *American Journal of Political Science* 28: 739–777.

National Research Council. 1999. *Our Common Journey: A Transition toward Sustainability*. Washington, DC: National Academy Press.

Nelkin, Dorothy, ed. 1979. *Controversy: Politics of Technical Decisions*. London: Sage.

Oberthür, Sebastian, and Thomas Gehring, eds. 2006. *Institutional Interaction in Global Environmental Governance*. Cambridge, MA: MIT Press.

Ortiz, Even Fontaine, and Guangting Tang. 2005. *Review of the Management, Administration, and Activities of the Secretariat of the United Nations Convention to Combat Desertification (UNCCD)*. Geneva: United Nations Joint Inspection Unit.

Ravetz, Jerome R. 1986. Usable Knowledge, Usable Ignorance: Incomplete Science with Policy Implications. In *Sustainable Development of the Biosphere*, ed. William C. Clark and R. E. Munn, 415–432. New York: Cambridge University Press.

Reinicke, Wolfgang H., Francis Deng, Jan Martin Witte, and the International Development Research Centre (Canada). 2000. *Critical Choices: The United Nations, Networks, and the Future of Global Governance*. Ottawa: International Development Research Centre.

Ruggie, John Gerald. 1998. What Makes the World Hang Together? Neo-Utilitarianism and the Social Constructivist Challenge. *International Organization* 52 (4): 855–885.

Schroeder, Heike, Leslie A. King, and Simon Tay. 2008. Contributing to the Science–Policy Interface: Policy Relevance of Findings on the Institutional Dimensions of Global Environmental Change. In *Institutions and Environmental Change: Principal Findings, Applications, and Research Frontiers*, ed. Oran Young, Leslie A. King, and Heike Schroeder, 261–275. Cambridge, MA: MIT Press.

Siebenhüner, Bernd. 2002. How Do Scientific Assessments Learn? *Environmental Science & Policy* 5 (5): 411–427.

Siebenhüner, Bernd. 2003. The Changing Role of Nation States in International Environmental Assessments. *Global Environmental Change* 13: 113–123.

Skodvin, Tora. 2000a. The Intergovernmental Panel on Climate Change. In *Science and Politics in International Environmental Regimes*, ed. Steinar Andresen, Tora Skodvin, Arjun Underdal, and Jutta Wettestad, 146–180. Manchester, UK: Manchester University Press.

Skodvin, Tora. 2000b. *Structures and Agents in the Scientific Diplomacy of Climate Change*. Dordrecht, Netherlands: Kluwer.

Skoie, Hans. 2001. *The Research Councils in the Nordic Countries—Developments and Some Challenges.* Oslo: Norsk Institutt for Studier av Forskning og Utdanning.

Social Learning Group, ed. 2001. *Learning to Manage Global Environmental Risks: A Comparative History of Social Response to Climate Change, Ozone Depletion, and Acid Rain.* Cambridge, MA: MIT Press.

Swart, Rob, John Mitchell, Tsuneyuki Moita, and Sarah Raper. 2002. Stabilisation Scenarios for Climate Impact Assessment. *Global Environmental Change* 12 (3): 155–165.

Tolba, Mostafa, and Iwona Rummel-Bulska. 1998. *Global Environmental Diplomacy: Negotiating Environmental Agreements for the World, 1973–1992.* Cambridge, MA: MIT Press.

Tuinstra, Willemijn. 2008. European Air Pollution Assessments: Co-Production of Science and Policy. *International Environmental Agreement: Politics, Law, and Economics* 8 (1): 35–49.

Tuinstra, Willemjin, Leen Hordijk, and Markus Amann. 1999. Using Computer Models in International Negotiations. *Environment* 41 (9): 33–42.

UNEP (United Nations Environment Program). 1998. *Report on International Scientific Advisory Processes on the Environment and Sustainable Development.* Paris: UNEP.

UNEP (United Nations Environment Program). 2003. *Global Marine Assessments: A Survey of Global and Regional Marine Environmental Assessments and Related Scientific Activities.* Paris: UNEP.

Victor, David G., Kal Raustiala, and Eugene B. Skolnikoff. 1998. *The Implementation and Effectiveness of International Environmental Commitments: Theory and Practice.* Cambridge, MA: International Institute for Applied Systems Analysis and MIT Press.

Weiss, Edith Brown, and Harold Karan Jacobson. 1998. *Engaging Countries: Strengthening Compliance with International Accords.* Cambridge, MA: MIT Press.

Wildavsky, Aaron B. 1979. *Speaking Truth to Power: The Art and Craft of Policy Analysis.* Boston: Little, Brown.

Young, Oran R. 1999. *The Effectiveness of International Environmental Regimes: Causal Connections and Behavioral Mechanisms.* Cambridge, MA: MIT Press.

Ziman, John. 2000. *Real Science: What It Is, and What It Means.* Cambridge, UK: Cambridge University Press.

6

Scientists Learn Not Only Science but Also Diplomacy: Learning Processes in the European Transboundary Air Pollution Regime

Atsushi Ishii

International environmental governance is so complex that its entirety cannot be properly envisaged from the beginning. Governance is therefore inherently a learning process and at the same time is one of the best arenas for studying such processes. Learning generally means the recognition and acceptance of information that can change the actors' motives and behavior. A rich collection of literature explores how various actors have learned in governance processes and the factors that explain such learning. However, the scientific community, which has occupied a very important role in the political process of governance, remains an underresearched actor despite the fact that scientists, who usually live in the "Republic of Science" (Polanyi 1962), must overcome enormously complex interactions between science and policy through learning. This chapter focuses on how advisory scientists to the Convention on Long-Range Transboundary Air Pollution (CLRTAP) regime have collectively learned in the process of scientific assessment for the 1994 Protocol on Further Reduction of Sulfur Emissions (Second Sulfur Protocol, SSP).

Why is the analysis of learning processes in the CLRTAP assessments important for the regime? Three potential contributions come to mind. First, such analysis can shed light on one of the few missing pieces to produce a more comprehensive explanation and nuanced understanding of the regime's scientific assessments: the internal dynamics of scientific assessment and its outputs and outcomes in terms of learning. Although the existing literature does indeed focus very often on the regime's scientific assessment, it is focusing mostly on the assessment as a given "objective" input to the governance process or on surrogate variables such as advisory scientific communities that undertake the assessments (Zito 2001), assessment processes (communication) (Tuinstra 2006), discourses (Sundqvist, Letell, and Lidskog 2002), and institutions for science–policy interaction (Wettestad 2000). What is lacking, in other words, is a treatment of the process, institutions, actors, scientific knowledge, and the production of scientific knowledge as an integral part of the regime's scientific assessment and an analysis of these elements in a systematic way in the context of CLRTAP governance.

Second, an analysis of the CLRTAP learning process can contribute to the transition management of European transboundary air pollution policies (see Lidskog and Sundqvist, chap. 1 in this volume). Such transition management should include enhancing the transparency of the science–policy complex developed thus far. Göran Sundqvist (2003) argues also that such transparency should be complemented with a profound understanding of the internal dynamics of scientific assessments and specifically the strategies employed by the advisory scientists to achieve credibility for their advice. An intensive focus on the learning process of CLRTAP assessments can make a major contribution to achieving these suggestions by opening up the assessment process and explaining the learning process, including how the scientists learned strategies to enhance the credibility of their advice.

The third contribution, which is related to the second, is that such an analysis may elicit strategies for designing proper learning processes for scientific assessment. There is an obvious and urgent need for the regime to design such learning processes, given the challenge of overcoming the growing scientific and political complexities of transboundary air pollution (cf. Lidskog and Sundqvist, chap. 1 in this volume). Analyzing past CLRTAP assessments can systematically explain the factors that encouraged or discouraged collective learning and thus led to the outcomes of the assessments and can identify which learning strategies worked better or worse (Siebenhüner 2002b). Analyzing past cases in the CLRTAP process can generally contribute to accelerating learning in environmental assessments (Siebenhüner 2002a), but it also has the advantage of situating them in a context relatively more similar to the current context of the European transboundary air pollution governance than to the context of other governance issues. Therefore, the lessons from the past can at least be insightful and possibly applicable to the current transition period for the European transboundary air pollution policy.

To discuss these contributions fully, a proper conceptual framework of learning and its factors is essential. According to Edward Parson and William Clark (1995), every conceptualization of learning needs to answer the following questions: Who or what learns? What is to be learned? What counts as learning? Why bother asking? Seminal work by Bernd Siebenhüner (2002a, 2002b) takes up the challenge of answering these questions by using an institutionalist approach.[1] However, the resulting conceptual framework cannot avoid a fundamental criticism of institutionalism, which is that it does not give scientific knowledge a distinct role in the learning process. Therefore, this framework cannot provide the necessary foundation for pursuing the contributions made in this chapter and for achieving the general task of accelerating learning processes in scientific assessments. This criticism does not mean, however, that the institutional approach is useless. On the contrary, it provides a very useful starting point for developing an alternative conceptual framework.

In this chapter, I first briefly summarize the institutionalist approach to learning and critically review it to specify how the alternative conceptual framework for learning assessments should be developed.[2] I then establish the alternative framework based on the critical review and apply it to the cases of the critical-loads (CLs) concept and the integrated assessment model Regional Air Pollution INformation and Simulation (RAINS) utilized in the negotiations of the SSP, thereby explaining the learning assessment and its stimulating or discouraging factors. I conclude with some notes on the extent of the alternative framework's usefulness and with suggestions for transition management in the European transboundary air pollution policy that can be derived from the case study.

Developing an Alternative Conceptual Framework

The Institutionalist Approach

The institutionalist approach to learning assessments has thus far been interested in explaining how assessments improve procedures and enhance effectiveness of scientific advice to the policy process through collective learning. The research questions are: "How did the assessments at hand learn over the years and in the different phases of the assessment process? How could the learning process be characterized? Which factors did influence learning in assessments? Which conclusions can be drawn to improve learning processes in assessments in general?" (Siebenhüner 2002a, 411–412). This approach identifies lessons and promising learning strategies applicable to future scientific assessments. Here, I focus on some features of the institutionalist approach categorized mainly by the questions Parson and Clark (1995) pose.

First, based on the fields of social learning and organizational learning, the institutionalist approach defines learning by scientific assessments as "a process of long-lasting change in the behavior or the general ability to behave in a certain way that is founded on changes in knowledge" (Siebenhüner 2002a, 412). This definition is rather general in that it can be applied not only to scientific assessments, but also to other institutions and actors. However, it is less general in terms of the prescribed direction of what learning should deliver: change in the behavior in a certain way.

This prescribed direction is not left as an open question. A second feature of this approach is that it identifies learning endeavors only when learning procedural knowledge results in enhancing three criteria that characterize the learning assessment: saliency, credibility, and legitimacy (Siebenhüner 2002a, 412). These criteria are in line with the work of the Harvard University's Global Environment Assessment (GEA) project and are considered the necessary conditions to design influential scientific assessments in international negotiations (e.g., Mitchell, Clark, Cash, et al. 2006).

The field of organizational learning holds that to choose the appropriate theoretical and analytical approach, it is crucial to distinguish individual from collective learning (Siebenhüner 2002a). A third feature of the institutionalist approach is that it justifies learning by scientific assessments as a collective action by arguing that, given the complexity of the mandate of scientific assessments, division of labor among the assessment's participants allows for the possibility of collective learning, which goes beyond the simple aggregation of individual learning. This justification is also espoused in the alternative conceptual framework explained later.

Fourth, the institutionalist approach sets procedural knowledge as the cognitive element to be learned. Procedural knowledge "is concerned with how the assessment is designed referring to questions like who participates in it, which decision-making procedures are adopted, on which scale and scope of the problem at hand it will focus, and how uncertainty is dealt with" (Siebenhüner 2002a, 412). The counterpart of procedural knowledge is substantive knowledge, which refers to knowledge generated by research about, for example, cause-and-effect relationships, adverse impacts, and abatement strategies (Siebenhüner 2002a, 412). If we apply these categories to the case of the RAINS model, procedural knowledge encompasses the fact that the RAINS model is used in the decision-making process, and substantive knowledge encompasses the actual acidification impacts and abatement strategies suggested by the model.

A fifth feature concerns the typology of learning modes, which answers the question "Learn to what extent?" The modes are distinguished according to the direct effects of learning on scientific assessments. The institutionalist approach employs the topology that Chris Argyris and Donald Schön (1978; 1996) developed in the field of organizational learning: single-loop and double-loop learning. The former consists of error correction only, with the organization's objective and underlying beliefs held constant. The latter "occurs when error is detected and corrected in ways that involve the modification of an organization's underlying norms, policies and objectives" (Argyris and Schön 1978, 3). In short, this typology distinguishes between incremental error correction and more complex learning.

A Critical Review of the Institutionalist Approach to Learning

Cross-fertilization of the institutionalist approach with science and technology studies (STS) approach provides a critical perspective of the former. Whereas "co-production" (Jasanoff 2004; Lidskog and Sundqvist, chaps. 1 and 12 in this volume) emphasizes primarily the interaction between science and policy (and the broader society), the STS approach is also concerned with what kind of science such interaction constructs. The STS literature has established that the production of scientific knowledge exhibits diverse modes with diverging practices, norms, and behavior contingent on their situated context. With respect to these modes, academic (or research) sciences offer a good contrasting case to scientific assessments in the CLRTAP regime and in global governance in general.

Academic sciences are done by citizens of the "Republic of Science" and have four social norms, often called "Mertonian norms": communism (knowledge open to public), universalism (evaluation based on universal criteria), disinterestedness (not pursued for personal gain), and organized skepticism (knowledge must be critically scrutinized by empirical and logical criteria—through, for example, peer review) (Ziman 2000). Robert Merton also suggested empiricism as one of the technical norms that is obeyed by academic sciences, accompanying and consistent with these social norms (Schuster 1995). Empiricism is "a family of traditions in philosophy of science, which argue that scientific truths grow out of, and are properly generalized from, appropriate empirical observations" (Law 2004, 16). Closely related to empiricism is positivism, which says that "scientific truths are rigorous sets of logical relations or laws that describe the relations between (rigorous) empirical descriptions" (Law 2004, 16). These kinds of (at least conceptually) coherent sets of social and technical norms correspond to the concept of paradigm, which provides a defining framework of what science should and should not do in a given context. Though I am aware of the heated debates about this concept, throughout this chapter I generally refer to the social and technical norms that define science as a "paradigm" (e.g., Smokler 1983).

As Sheila Jasanoff (1990) proposed and Karin Bäckstrand (2001) has argued, the most relevant mode for the science–policy interaction involved in the CLRTAP regime is "regulatory science." This mode becomes operational in the context of producing policy advice in domestic settings and involves heterogeneous actors that include at least policymakers and advisory scientists as well as, more broadly, industry. Whereas scientists pursuing academic sciences are held accountable only to the "Republic of Science," regulatory scientists must be socially accountable to policymakers, courts, and the media through regulatory peer review, judicial review, and legislative oversight. Although Jasanoff (1990) does not explicitly name citizens, they can also be included in the list of "accountability checkers" and thereby can render regulatory science more reflexive to multiple views. The quality-control system of regulatory science accordingly becomes more complicated in that the traditional peer-review system is supplemented by additional socially dependent criteria such as policy usefulness, fairness, and communicability. The overall "science policy paradigm" (Jasanoff 1990, passim) in regulatory science thus emerges from complex interactions of these attributes and is heavily context dependent.

This brief discussion of the modes of science has important implications for analyzing learning assessments. First, it reconfirms that the production of scientific knowledge pursued by learning assessments cannot be explored in any way unless the norms, institutions, procedures, and research involved in such production are treated in an integrated way. The implication is that the output and effectiveness of learning assessments should be regarded as a consequence of learning all aspects of scientific assessments, including procedural and substantive knowledge.

Second, it demonstrates that for a learning assessment to be effective in the governance process, scientific assessment may have to develop an enabling paradigm and thereafter constantly adjust to reflect on the context of policy application in which it is situated. The advisory scientists obviously cannot accomplish such development and adjustment of an appropriate paradigm beforehand. The case studies of regulatory science show how regulatory scientists learned to construct their work socially and to adjust the prevailing paradigm over time to maintain their cognitive authority in the deconstructive science–policy interaction while taking into account the internal dynamics of the scientific community, external social factors, and their interactions (Jasanoff 1990, 1992). Therefore, it is very important for analysts of learning assessments to distinguish between those learning modes that achieve new paradigm development or adjustment of the existing paradigm (paradigm shift) through learning and those that hold the dominant paradigm of knowledge production constant. By employing this kind of typology, the analysis of learning assessments can provide answers to questions such as: Which factors lead to a paradigm shift? What kind of paradigm shift leads to enhanced or diminished political acceptance? When is it appropriate to avoid paradigm shift to enhance political acceptance of learning assessments?

Identified Refinements

First, setting only procedural knowledge as what is to be learned by assessments should be replaced with focusing on all the knowledge and experience relevant to the learning assessment. In addition, it is extremely difficult to discern the results of learning procedural knowledge from learning substantive knowledge; in other words, the causal relationship between procedural knowledge, which is set to be what learning assessments learn, and the results of such learning may suffer indeterminacy. The actual case study on the CLRTAP assessments using the institutionalist approach (Siebenhüner 2002b) is self-revealing on this point. In that study, the narrative shows clearly that the adoption and utilization of the CLs concept and the RAINS model resulted not from just error correction, but from more complex learning. Although I believe this conclusion to be correct, it implicitly argues that what the CLRTAP assessment learned is a mixture of procedural and substantial knowledge[3] relevant to CLs and RAINS. This argument contradicts the case study's own conceptual framework, though, which sets procedural knowledge as the only cognitive element to be learned by scientific assessments. This contradiction may have been inevitable because it is quite inconceivable that the assessment participants adopted novel concepts such as the CLs concept only from learning the procedural elements related to CLs. Moreover, because analyzing learning assessments has just begun, in order to maximize the theoretical and practical potential of this analytical challenge we should not prescribe what kind of knowledge should be acquired through them (e.g., Clark, Jäger, and van Eijndhoven 2001).

Second, prescribing the direction of the learning process using the GEA criteria should be avoided. The GEA criteria, though cited extensively, have not gained consensus among scholars, and there are other criteria derived by various approaches (e.g., Jasanoff 1996). Even if we assume that the GEA criteria are correct, there is a clear possibility that learning may result in diminished political acceptance of the assessment's policy advice, which I call "detrimental learning." One can easily imagine that a person who learned too many things may complicate the situation rather than arrive at the right solution. Some experts in the CLRTAP assessments are concerned about the relative, if not excessive, complexity of the scientific assessment for the Protocol to Abate Acidification, Eutrophication, and Ground-Level Ozone (Gothenburg Protocol) compared to assessment for the SSP (Lidskog and Sundqvist, chap. 1 in this volume). In fact, the former was less politically accepted than the latter in terms of the level of the recommended reduction targets calculated by the RAINS model (Ishii 2001). Perhaps the former assessment learned too much so that it became incomprehensible and therefore resulted in lower political acceptance. More careful analysis is certainly needed to make this explanation compelling. The point here for analyzing learning assessment is that a prescription of the direction of learning makes it impossible to explore very important and policy-relevant research questions such as the detection of factors that stimulate (or discourage) *detrimental learning*. Therefore, the prescription incorporated in the institutionalist approach should be abandoned. Instead, we should carefully reconsider the complex causal relationship between the learning processes of scientific assessments and various types of effectiveness to the political process, which is outside the scope of this chapter.

Third, as argued in the discussion of the STS literature, the typology of learning modes should be changed to distinguish between learning assessments that achieve paradigm shift through learning and those that hold the dominant paradigm of knowledge production constant. The institutionalist approach cannot detect this distinction because it distinguishes only between error correction and more complex learning.

Establishing the Alternative Conceptual Framework

I basically share the research questions outlined for the institutionalist approach. Hence, my intention in establishing the alternative framework is not to replace the institutionalist agenda entirely, but to make its framework more suitable to answering important research questions on its agenda.

The definitions of scientific assessments and their learning endeavor have to be determined to establish the alternative approach. For the former, I employ a modified version of GEA's definition (GEA 1997, 53): the entire social process by which expert knowledge related to a policy problem is produced, organized, evaluated, integrated, and documented to inform policy or decision making. The word *produced* is added to reflect on the preceding critical review, which makes it clear that production should not be ignored in explaining scientific assessments.

Although espousing Siebenhüner's justification of analyzing learning as a collective behavior (2002a, 413), I adopt a definition of learning that does not prescribe the direction of learning: learning in scientific assessments is a deliberate attempt to adjust the operation of scientific assessments in response to relevant experience and knowledge. The emphasis is on the deliberative nature of learning, which means that any changes in response to experience and knowledge have to be intentional, and, conversely, that any unintended changes do not count as learning. From an empirical point of view, the participants of the learning assessments must recognize experience and knowledge and explicitly change the operation of the learning assessments in response to those cognitive elements, which makes it possible to establish causal pathways between the cognitive elements and the resulting changes. It is important to note that this definition does not exclude *detrimental learning*. The definition automatically answers the "what to learn" question: relevant experience and knowledge.

I now turn to learning modes. I have argued that the institutionalist typology of learning is not appropriate for the study of learning assessments. Because no other typologies have been proposed so far in the literature, it may be useful to return to the basics: Gregory Bateson's (1972) general logical categorization of learning. Bateson argues that theories of learning must establish a typology of learning according to the theory of logical types to avoid logical contradictions and thereby to develop robust theories of learning. Bateson specifically proposes the following typology as being applicable to all kinds of learning theories. Learning I is error correction within a set of alternatives. Learning II is change in the process of Learning I, which is change in the set of alternatives from which choices are made. Learning III is change in the process of Learning II, which is change in the system of sets of alternatives, from which choices are made. According to Bateson (1972), logical contradiction in learning means, for example, applying any theory or factors of Learning I to Learning II. More examples of such logical contradictions may include treating Learning I and II as having the same level of learning or misidentifying change in the set of alternatives resulting from learning as mere error correction in the same set of alternatives (in other words, misidentifying Learning II as Learning I). If strict logical typing of learning is not applied, one can never be sure about the applicability of any theory or factors because any theory or factors about Learning I should strictly be applied to and validated by the same level of learning. Given the general typology of learning, the question now is how to adapt it to the context of the study of learning assessments.

I begin with the simplest type of learning, Learning I—namely, error correction within the same set of alternatives. This type of error correction in the context of scientific assessments may include adjusting parameters of a numerical model or the number of participants in the assessment because it is a choice from the same set of the alternative parameters or number of participants. In other words, Learning I occurs when the scientific or institutional methodology of scientific assessments is adapted while other things remain equal. Learning II must change the set of alternatives from which the methodology is chosen. A crucial feature of scientific assessments that change the set of alternative methodologies would be the assessment's objective. For instance, the choice between investigating water acidification and investigating soil acidification as the objective demands different sets of alternative methodologies. Of course, there is some overlap between those methodologies, but the point here is that a *different set of methodologies* corresponds to each objective. Learning III requires changing the system of sets of alternative objectives. The brief discussion of the STS literature has implications for this discussion as well. It suggests that the system of sets of alternative objectives can be changed by paradigm shift because the shift changes the underlying definition of science and thereby the system of sets of alternatives. For example, a shift from academic science to regulatory science changes the sets of overall objectives and methodologies from those consistent with academic norms to those oriented toward policy.

In sum, Learning I occurs when scientific assessments engage in error correction by changing the methodology while the objective and paradigm remain the same. Learning II does not change the overall paradigm but changes the assessment's objective and accordingly the methodology. Learning III requires a fundamental shift in the dominant paradigm, followed by an often drastic change in the overall objective and methodology of the learning assessment. For more clarity and direct reference to the traits of the learning modes, I name Learning I, II, and III "adaptive learning," "reformative learning," and "paradigmatic learning," respectively.

For the factors that stimulate or discourage learning, I employ those used in the institutional approach with some additions (see Siebenhüner 2002a for factors not explained here). The dynamics of production of scientific knowledge has strong path-dependent elements once they are institutionalized (Haas 2007). Moreover, it is easy to imagine that the learning endeavors of scientific assessments can be affected by collective perception of the performance of past operations as a failure or a success (Brown and Kenney 2006). Therefore, to the alternative framework I add "path-dependent factors" such as evaluative perceptions of the past operations of the scientific assessment itself, including the science-policy interaction. Table 6.1 summarizes the differences between the institutionalist approach and my alternative approach, focusing on learning assessment.

As I demonstrate in the next section, the main methodology to operationalize this framework is process tracing of the scientific assessments in question. One should first identify concretely the changes in a scientific assessment that resulted from collectively learning relevant knowledge and experience as well as the stimulating/discouraging factors incorporated in the framework. Next, one should assess the resulting changes in terms of the extent of learning based on the framework's typology. However, for clarity's sake, the presentation structure of the case study in the next section does not correspond to this order of action.

The Case Study: Learning Assessments for the Second Sulfur Protocol

The focus of this case study is the twin pillars of the scientific assessment for the SSP: the CLs approach and the integrated assessment model RAINS. The empirical data comprise interviews and written communications with participating scientists and review of primary and secondary materials (both natural and social sciences).

To provide the basis for analyzing how scientists collectively learned in this case, I begin by examining the relevant scientific activities in the pre–case study period to identify the prevailing paradigm, objectives, and methodologies because learning modes focus on changes in those elements of learning assessments and therefore can be examined only by comparing them to past operations. I then apply the alternative framework to the scientific assessments of the CLs concept and the RAINS

Table 6.1
Comparison between the Institutionalist Approach and the Alternative Conceptual Framework

	Institutionalist Approach	Alternative Conceptual Framework
Who learns?	Collective learning by participants in scientific assessments	
What is to be learned?	Only procedural knowledge	Relevant knowledge and experience (no separation between procedural and substantive knowledge)
What counts as learning? (What is the definition of learning?)	A process of long-lasting change in the behavior or the general ability to behave in a certain way (that is, enhancing saliency, legitimacy, or credibility of the scientific assessment) that is founded on changes in knowledge	A deliberate attempt to adjust the operation of scientific assessments in response to past experience and relevant knowledge
Learn to what extent?	Distinguishing between error correction (single-loop learning) and more complex learning (double-loop learning)	Adaptive learning (error correction) Reformative learning (changes in methodology and objective under constant paradigm) Paradigmatic learning (changes in methodology, objective, and paradigm)
What are the stimulating/ discouraging factors considered in the framework?	*Structural factors (the factors with respect to formal structure in learning assessments)* • Storage of knowledge • Hierarchy/leadership • Communication structures • Reflective mechanisms *Personal factors (the factors with respect to the individuals involved in learning)* • Individual capabilities • Dissatisfaction and conflicts *Cultural factors (the factors with respect to human relations)* • Values, norms, and beliefs • Informal communication networks *Contextual factors (factors external to scientific assessments)* • Political pressure • New scientific findings • Media coverage • Other assessments	
		Path-dependent factors • Evaluative perception of the past operations of the scientific assessment itself, including the science–policy interaction

Note: The institutionalist approach is discussed in Siebenhüner 2002a, 2002b.

model and explain what the scientific assessment learned, the learning modes, and the factors stimulating or preventing learning.

The Relevant Scientific Activities in the Pre–Case Study Period (1970s to Early 1980s)

Because there is no space to summarize all the scientific activities with respect to the SSP in this period, I focus mainly on large research projects, which were generally regarded as of international importance in the European context: the Organisation for Economic Co-operation and Development (OECD) Programme on Long-Range Transport of Air Pollutants (1972–1977, with eleven participating European countries) and the Norwegian research project Acid Precipitation—Effects on Forest and Fish, or the so-called SNSF project (1972–1980). The Norwegian Institute for Air Research was the body responsible for coordination of both projects, so the two projects were closely connected (Roll-Hansen 2002).

The two projects were obviously not pursuing academic science. Instead, as their objectives indicate, they were planned and designed to answer policy-relevant research questions. The OECD program had the objective of determining "the relative importance of local and distant sources of sulfur compounds in terms of their contribution to the air pollution over a region" (OECD 1977, 3), and the SNSF project had the objective of establishing "as precisely as possible the effects of acid precipitation on forest and freshwater fish" and investigating "the effects of air pollutants on soil, vegetation and water" (Overrein, Seip, and Tollan 1981, 3). However, they were far from being policy prescriptive and made no explicit policy recommendations.

The two projects were accordingly organized to incorporate and maintain this policy-relevant focus. The established organizational setting can be characterized as securing relatively close relationships between scientists and policymakers to the extent that policymakers were involved as supervisors and "accountability checkers" of the scientific work but never directly participated in research activities. For the SNSF project, it was found that due to this close contact "the [Norwegian] Ministry of Environment greatly influenced the development and content of the project and that this [influence] was somewhat disadvantageous" (Roll-Hansen and Hestmark 1990, quoted in Seip 2001, 4). The point is that at that time a clear separation with close contacts between science and policy was regarded as a tool for maintaining scientific integrity (Seip 2001).

The mandate of policy relevance also affected the scientific methodologies employed in these projects. The entire scientific endeavor can be characterized as interdisciplinary[4] in the sense that the two projects set research objectives that formed a common goal for involved disciplines and necessitated development of comprehensive theories of long-range transboundary atmospheric transport of pollutants and their effects by integrating relevant disciplines: for the OECD program, meteorology, atmospheric chemistry, and physics; and for the SNSF project, atmospheric chemistry, limnology, and physiology. For instance, the OECD program developed several types of atmospheric dispersion models that integrated meteorological and emission data to calculate atmospheric concentration and ground deposition of air pollutants. These calculations were then compared to measured data of the same kinds. The SNSF project developed conceptual models to investigate causal relationships between acid precipitation and freshwater acidification. As a result of these interdisciplinary studies, there was close agreement among the participating scientists that long-range transboundary atmospheric transport of air pollutants was very likely and that it may have caused significant decline in fish populations in southern Scandinavia. However, no agreement was reached on the quantification of such effects (Rosenqvist and Seip 1986, cited in Seip 2001).

However, the two projects' dominant paradigm in this period is one of the few attributes they have in common with academic sciences: empiricism and positivism. This commonality can be detected by the prevailing method of comparing the calculated data and measured "ground truth" and constructing models to connect the "ground truths" in a rigorous and logical way; it is also evident in the fact that the success of the scientific assessment was measured against the demonstrated reproducibility of the "real" world in light of the policy-relevant objectives. Interdisciplinary scientific methods were inevitably chosen as a way to accomplish the policy-relevant objectives because traditional disciplinary studies were not suited to this daunting task.

In sum, although the given objective in these projects was to answer policy-relevant research questions, the chosen scientific methodology was interdisciplinary research based on a combination of empiricism and positivism as the dominant paradigm (hereafter, I refer only to positivism for brevity). It is well known that the political acceptability of scientific advice regarding transboundary air pollution in this period was limited, but it was also positive to the extent that more and more countries had come to recognize that transboundary air pollution might be a significant international problem (e.g., Wettestad 2000).

Examining the Learning Process

The Critical Loads

The emergence of the CLs concept can be conceived as an intersection of at least three environmental discourses that had been emerging at the time. First, given the definition of CLs as a critical threshold of adverse ecosystem effects, the concept can be conceived as an attempt to operationalize the notion of an ecosystem's "carrying capacity" embedded in the sustainable development discourse. Second, because the definition inherently entails science to operationalize it, it can also be conceived in line with the overall discourse of ecological modernization (Bäckstrand 2001). This discourse regards science and technology as powerful tools to manage environmental problems. The third discourse is on the precautionary principle. Because some of the uncertainties surrounding the CLs concept and the possibility of ignorance were irreducible, the CLs concept was embedded in and thereby justified in part by the precautionary discourse (e.g., Bull 1995). These discourses altogether have certain affinities and synergies with each other, and it is very likely that they provided an enabling context for the CLs concept to be operationalized.

The direct origin of the CLs approach can be traced back to the idea of Swedish scientists who had close channels of communication with policymakers (Bäckstrand and Selin 2000). The initiative to develop the CLs approach and operationalize it in the CLRTAP regime was basically motivated by the need to accommodate the criticism that the flat-rate reduction protocols adopted to date (in both the Protocol on the Reduction of Sulfur Emissions [First Sulfur Protocol] and the Protocol Concerning the Control of Nitrogen Oxides [Nitrogen Oxides Protocol]) were politically biased (the reduction targets were achievable without any additional national reduction policies), and arbitrary. In 1986, a core group of Scandinavian scientists was formed to hold a workshop in Oslo under the auspices of the Nordic Council of Ministers. An additional 30 experts participated in the Oslo workshop as peer reviewers. The overall conclusion of the workshop was positive, and a scientific definition of "critical loads for sulphur and nitrogen" was adopted: "the highest load that will not cause chemical changes leading to long-term harmful effects on the most sensitive ecosystems" (Nilsson 1986, 4). The results of the various national acidification research projects (such as the SNSF project) and the International Institute of Applied Systems Analysis's (IIASA) RAINS model were used and referenced in the resulting workshop report. The latter especially indicates the early linkage of CLs development with the RAINS model. Despite the positive outcome, it was observed that "the concept had to be evaluated further in order to get more reliable figures and to cover a larger variety in ecosystems" (Nilsson and Grennfelt 1988b, 4).

Expansion of this scientific community and more political involvement came soon when the CLRTAP's Working Group on Effects decided to hold a workshop on CLs during its fifth session in June 1986. This move was somewhat natural to respond to the criticisms raised against the First Sulfur Protocol agreed to in 1985, including the lack of scientific basis of its 30 percent flat-rate reduction target. This workshop was held in Skokloster, Sweden, in March 1988 and was attended by the United Nations Economic Commission for Europe (UNECE) secretariat and experts from 16 countries. The workshop participants experienced problems in defining CLs and therefore decided to use the definition agreed upon by the UNECE Working Group on Nitrogen Oxides in February 1988 in a deliberation in which both negotiators and scientists participated (Nilsson and Grennfelt 1988a). This was the birth of the famous definition: "[a] quantitative estimate of an exposure to one or more pollutants, below which significant harmful effects on specified sensitive elements of the environment do not occur according to present knowledge" (Nilsson and Grennfelt 1988c, 9). The difference between this definition and the one given in the previous paragraph lies in the shift in focus from the previous "most sensitive ecosystems" to "significant harmful effects," along with the added wording "according to present knowledge." This shift is interpreted elsewhere as rendering a "more politically qualified concept" (Tickle 1995, 82). To put it differently, this shift and the added wording broadened the political applicability and strengthened the robustness of the scientific assessments of the CLs concept; the previous definition cannot be applied unless the most sensitive ecosystems are identified and is thus vulnerable to criticisms of the surrounding uncertainties because it does not acknowledge them.

The participants in the Skokloster workshop also "agreed on the usefulness of the concept of critical load as well as on best estimates for critical loads to various exosystems [*sic*]" and recognized that "[t]he concept of critical load is only in its infancy. The underlying scientific knowledge is in many cases weak and the importance of different processes involved is not always quantified" (Nilsson and Grennfelt 1988b, 4). More specifically, for forest soils and groundwater, CLs values were proposed according to a classification scheme based on characteristics of soil material. For surface waters, CLs for sulfur deposition were estimated based on available information and proposed with the recognition that "[c]ritical loading estimates must be made specific to the area of application" (Nilsson and Grennfelt 1988c, 22).[5]

As for the basic methodology for calculating CLs, the Steady-State Mass Balance (SSMB) method was adopted for soils, surface waters, and ground water (see Sverdrup, de Vries, and Henriksen 1990). The idea behind the method can be traced to a background document prepared for the Skokloster workshop in which the CLs values, derived from a steady-state soil chemistry model, were compared to empirical data, which confirmed that the two gave nearly the same results (Sverdrup and Warfvinge 1988). The SSMB ignores time-dependent variations in chemical interactions relevant to acid rain effects (Hettelingh, Downing, and de Smet 1991b). As many "outside" scientists became increasingly involved in the CLs debate, they criticized these kinds of oversimplifications of representing ecosystems but largely joined the advocating scientists to form consensus when they were challenged to suggest alternative approaches and realized that simple science can influence policy.[6] The dynamic modeling method, which takes time-dependent variations into account, was not chosen as the standard method for two reasons: first, it was too complex in terms of applicability to a European scale, in which not all countries have the necessary data needed for it, and in terms of comprehensibility to the policymakers; second, it could not be mapped and applied to policy.[7] There were other battles, "for instance, between the soil experts and water experts who proposed different methods. Eventually groups accepted that there was more than one way to look at the problem and do the calculations."[8] As a consequence, the steady-state water chemistry method was suggested for application to surface waters also (Hettelingh, Downing, and de Smet 1991b). An empirical method based on a Skokloster workshop recommendation was also suggested for application when no sufficient data are available for the methods using models.

After adoption of the SSMB, some additional modifications were implemented. The most significant is the division between CLs for sulfur and other acidifying substances. This division was required based on the request from the CLRTAP's Working Group on Strategies, the main negotiation arena for protocols, to produce a map showing the ecosystem impact of sulfur deposition only because this working group was responsible for developing a draft protocol that would address sulfur deposition exclusively (Hettelingh, Downing, and de Smet 1991b). At the session of the Coordination Centre for Effects (CCE) in 1991, "the sulphur fraction was . . . proposed as the ratio of the present load of sulphur to the present load of sulphur, nitrogen and ammonia based on 1988 EMEP [European Monitoring and Evaluation Programme] emission data" (Hettelingh, de Vries, Schöpp, et al. 1991a, 41). Although in hindsight it was observed that this modification is "difficult to justify scientifically, it was felt necessary in practice if a critical loads approach was to be used to estimate emission reduction of S [sulfur]" (Bull 1995, 205).

Another modification was made in the aftermath of the initial calculation of CLs. Standardized methods calculated negative CLs values for 150 lakes in Scandinavia, 128 of which were located in Norway (Henriksen, Kämäri, Posch, et al. 1993, cited in Downing, Hettelingh, and de Smet 1993), and some of those became the influential grid cell that led to problematic scenario analysis (requiring 100 percent reduction of sulfur emissions to certain countries to satisfy the environmental objective based on CLs). When the CLs value exceeds the natural background deposition of the same area, the adopted solution is to set the former the same as the latter based on the idea that CLs should be subject only to anthropogenic emissions reduction to incorporate the diplomatic context where natural sources of acidifying substances are totally out of scope. As for the validation of the CLs approach, a comparison between a CL map developed by the Stockholm Environment Institute (SEI) and a CCE map was implemented. It was concluded that "the SEI and CCE critical load maps, even though produced using different methodologies, do result in broadly similar abatement strategies. This should promote confidence in the use of critical load maps" (Kuylenstierna, Gough, Chadwick, et al. 1991, 75).

The RAINS Model

The development of the RAINS model commenced in 1983 at the IIASA. At the beginning, the RAINS team adopted the following modeling guidelines based on internal discussions (Alcamo, Shaw, and Hordijk 1990, 23–24):

The model should be codesigned by analysts, experts, and potential users.

The model should be of modular construction.

Submodels should be as simple as possible and yet be based, where possible, on more detailed data or models.

To facilitate its use, the model should have interactive inputs and clear graphical outputs.

What the RAINS team tried to do with this set of guidelines was to improve the RAINS model dynamically through interactive learning among potential users, mainly the policymakers in charge of the acid rain issue, other experts, and the team members themselves. In other words, this interaction made it clear that the RAINS team did not consider learning from external actors as a "residual," but rather incorporated it into the modeling process as an inherent component from the outset to win policymakers' acceptance of the RAINS model.

The first guideline means opening communication channels among the modelers, outside experts, and policymakers (or their advisors) to enable the modelers to learn policymakers' needs as well as submitting the RAINS model to peer review and addressing the peer experts' concerns. However, communication is not enough. Academic modeling exercises usually end up with a model that policymakers can never comprehend. The third and fourth guidelines indicate the facilitation of comprehensibility for policymakers by emphasizing simplicity and visibility in the model structure and results. A clear example is the way the modelers dealt with the EMEP atmospheric transfer model (ATM);[9] because the ATM is too complex and too demanding of time and data to be a communication tool, the modelers instead incorporated the submodel consisting of the source–receptor matrix (the so-called "blame matrix" because it clearly visualizes the polluting and victim countries) into the RAINS model (Alcamo, Kauppi, Posch, et al. 1984). As analysts put it, "This represents a great simplification of the detailed work on atmospheric chemistry and meteorological modeling to construct a heuristic that, in the eyes of the experts at EMEP, provided an adequate representation of pollutant import and export in Europe" (Gough, Castells, and Funtowicz 1998, 24). The model can become more detailed if model users and scientific experts feel that more detail is required (Alcamo, Hordijk, Kämäri, et al. 1985, 49). The last requirement for the model to reflect on what was learned from outside actors is that it should be flexible enough to incorporate the things learned. This reflection is done by following the second guideline; a model with a modular structure (for example, dividing the model structure into parts that handle emissions, technology costs, atmospheric processes, and environmental impacts of acid rain) enables the modelers to make changes or to add new perspectives more easily than does a model without a modular structure because modelers do not have to examine the whole model, only the relevant modules, to ensure the model's consistency and integrity, which can be compromised because of the model change.

The RAINS team began with a very simple model having atmospheric transfer coefficients borrowed from the EMEP's ATM. In addition to avoiding complex and demanding computations, this decision was made because the spatial coverage of the EMEP model covers the negotiating countries of the CLRTAP, and scientists and policymakers recognized the EMEP as having sufficient scientific quality with an international status (Hordijk 1991, 600). The pollutant initially focused on was sulfur because the team felt that there was consensus on the primary importance of sulfur in transboundary air pollution (Alcamo, Kauppi, Posch, et al. 1984).

To implement the first guideline, a series of review meetings took place beginning in the mid-1980s. After two rounds of review meetings, the RAINS team concluded from participants' written and verbal comments that "(1) the modular and flexible design of RAINS makes it possible to easily update the model, as additional expert opinion and data become available; and (2) RAINS links many different parts of the acidification problem in Europe in a comprehensible and usable manner to both scientists and non-scientists" (Alcamo, Hordijk, Kämäri, et al. 1985, 61). According to the then project leader Leen Hordijk, the initial changes were mainly fourfold.[10] First, the fixed available energy scenarios were made flexible so that the user could generate the scenario for use. Second and third, control cost and optimization modules were added to the model, which enabled the model to perform optimization aimed at the least costly abatement strategies and other forms of optimization. Fourth, more attention was paid to model uncertainties, not by incorporating stochastic variables or uncertainties into the model, but by making supplementary (and basically post hoc) analysis of the model outputs and maintaining the model's decisive character.[11] This last change came in part from Hordijk's personal experience: "I found a tendency [among policymakers] to use the uncertainty ranges as excuses for non-action, and as vehicles for questioning the credibility of the work. Later, I realized that making uncertainty explicit was very important."[12] This change also constituted one of the reasons to abandon cost–benefit analysis as explained below.

In the later stage, the environmental impact modules underwent significant changes. The first impact module was concerned with forest soil acidification, and the coverage of ecosystems later expanded to lake surface water and direct forest impacts. However, as it became clear that "the link between forest damage and acid rain was not as strong as believed in the 80s,"[13] and the uncertainties in soil acidification was so high that RAINS could not use this module (Skeffington 1990, cited in Hordijk 1991), the RAINS team dropped the forest soil module and turned to the CLs concept in 1991 (Hordijk 1991, 601). The subjected pollutant was concurrently largely expanded from the initial focus of sulfur to nitrogen and ammonia to reflect on review meeting participants' opinion that "it was not until the inclusion of NO_x that RAINS was considered to be good enough for policy use" (Hordijk 1991, 601).

What advice was not included in the RAINS model? First, a suggestion for a cost–benefit analysis was rejected. This decision was made in part because of Hordijk's experience:

> Several times [Hordijk] had conducted [cost–benefit analysis], and had been careful to warn policymakers about the great uncertainties inherent in the benefits estimations, uncertainties often extending over an order of magnitude. Repeatedly, he had observed policy-makers use just one number, taken from within that range of uncertainty, as a single point estimate to justify their preferred policy. Hordijk had been personally criticized, and had found the selective use of data by policy-makers troubling and counterproductive. He decided that similar problems could occur within LRTAP, undercutting the legitimacy of the RAINS model. (Patt 1999, 119)

The additional reason for rejecting cost–benefit analysis was the Soviet Union's resistance to it because its orientation toward market economy was inconsistent with the Soviet system.

The second piece of advice rejected was to use more detailed ATMs than those used in the EMEP.[14] RAINS was occasionally criticized for using the EMEP's simple (in terms of process characterization and the low resolution) ATM, so it is understandable that more detailed ATMs were suggested. However, the suggested alternative did not cover the UNECE region,[15] so it was rejected on the grounds that "[t]his [spatial coverage] is a necessity for a model that should be used in a negotiation setting" (Hordijk 1991, 600).

Inclusion of energy models was the third suggestion rejected even though it came from the CLRTAP Executive Body (Hordijk 1991, 601–602). The lack of energy models was in fact considered a limitation of the RAINS model because without them it is impossible to investigate policy measures that abate pollutants through structural changes in the energy sector, such as energy conservation and market-based abatement measures. This suggestion was rejected because the technical task of including a country-scale energy model for all participating countries and linking them to RAINS was considered enormous and difficult (Hordijk 1991, 602).[16] However, some supplementary studies were made for consideration of the SSP.[17]

Because RAINS was not the only model built for use in the international negotiations, the RAINS team learned from the competition among those models as well. The other candidates were the Abatement Strategies Assessment Model developed by the Imperial College London and the Coordinated Abatement Strategy Model developed by the Stockholm Environment Institute, which were broadly similar in design and scope but had slightly different optimization procedures (Castells and Ravetz 2001, 414–415). The competition was also one of the driving forces to make RAINS more relevant and acceptable to the policymakers.[18] In addition, "[the competition] . . . stressed the need for a very clear explanation of what we were doing, how we were using EMEP, how the cost functions had been derived and how the institutional position of IIASA [as an international institute] could be better used. Moreover, . . . [it] forced the IIASA team to explain the differences in model outcome between RAINS, [the Abatement Strategies Assessment Model and the Coordinated Abatement Strategy Model]."[19] In the actual negotiations, the RAINS model was finally chosen as the guiding model, and other models were used for checking (or relativizing) its runs and outputs.

Learning Mode

To what extent did the scientific assessments for the SSP learn to develop the CLs approach and the RAINS model? As shown, the advisory scientists abandoned their positivistic paradigm and shifted to a more diplomacy-oriented paradigm that would hold them accountable to country parties, which is a prerequisite in diplomatic settings and makes scientists adhere to the overall norm of usefulness in the diplomatic context. A contrasting point is that whereas in the pre–case study period the over-involvement of policymakers and political influence was regarded as an obstacle to maintaining scientific integrity, such involvement was contrarily regarded as a necessary condition and not an afterthought to secure the assessment's integrity in the diplomatic context.

Specifically in the CLs case, the definition of CLs was changed as a result of learning, going from the early form where only most sensitive ecosystems were focused on to a more politically robust form in the context of CLRTAP diplomacy. The data were calculated according to an agreed-upon methodology, but at the same time the participating countries were given choices of the type of ecosystem and the relevant parameters, and no data were used without an "official stamp" on them, which countries could interpret as a fulfillment of the advisory scientists' obligation to respect the countries' sovereign rights. A separation between the acidifying effects of nitrogen and sulfur was even introduced through supplementary research, although it was scientifically indefensible, in order to make the CLs approach useful in the SSP diplomatic context, in which only sulfur was subject to negotiations and nitrogen was out of the scope.

This paradigm shift is easier to detect in the RAINS case because it is explicitly stated in the RAINS guidelines. The inclusion of optimization modes, incorporation of the simple "blame matrix," the choice of spatial scale, and resistance to both non-EMEP ATMs and complex energy models were pursued to secure accountability to policymakers and usefulness in the SSP diplomacy and would have never been realized if the positivistic norm of the pre–case study period had dominated.

The objective and the dominant methodology were accordingly shifted to materialize the paradigm shift. The objective was set to guide the diplomatic negotiations in terms of the least-cost abatement strategy to achieve CLs all over Europe stipulated in the 1988 Nitrogen Oxides Protocol, which is a more diplomacy-oriented and policy-prescriptive objective compared to the pre–case study period. As for methodology, transdisciplinary research methods were introduced to meet the objective. They were transdisciplinary in the sense that scientists and nonscientists (policymakers) had been integrated into the assessment process to produce integrated knowledge about the necessary political choices of reduction targets and their costs to achieve certain environmental targets and, conversely, about the environmental consequences of political choices of reduction targets and measures. Therefore, the case study exhibits *paradigmatic learning*: a shift in the dominant paradigm, objective, and methodology of scientific assessments compared to the pre–case study period.

Stimulating/Discouraging Factors

One of the stimulating factors is certainly the continuity of participating organizations and scientists, which enabled *storage of knowledge and experience* within the learning assessment (structural factor) (Siebenhüner 2002b). For example, the Norwegian Institute for Air Research was the organization responsible for coordinating the OECD program and the SNSF project and was one of the EMEP's Meteorological Coordinating Centers as well. This continuity in participating institutions and actors also enabled the scientists to build on the positive outcome—though that outcome was limited in terms of political acceptability—of the interdisciplinary research of the OECD program and the SNSF project to develop and implement the CLs approach and the RAINS model, which is a path-dependent factor. Interdisciplinary research and transdisciplinary research were not so common at that time as they are now, and they have some common elements, such as transcending disciplinary boundaries, setting common goals, and thereby producing integrated knowledge (Tress, Tress, and Fry 2006). From a counterfactual point of view, it would have been difficult for scientists in the 1980s to become smoothly engaged in transdisciplinary research in the context of diplomacy and achieve successful results without having any positive perception of a scientific endeavor in a more or less similar context.

The case study also confirms that *reflective mechanisms* (structural factor) such as international workshops, critical review by policymakers, and "outside" scientists were powerful stimulating factors in the context of the SSP (see also Siebenhüner, chap. 4 in this volume). In the CLs case, such reflective mechanisms made it possible for the advisory scientists to make the CLs concept more diplomatically acceptable. They resulted in a more politically robust definition of CLs, modifications of the CLs approach through separation of sulfur and nitrogen, and broader peer support for the CLs approach by convincing "outside" scientists through the learning process that "oversimplification" of the CLs calculation method was necessary to make it useful in the SSP context. In the case of the RAINS model, the review workshop led to several changes in the RAINS model: the fixed available-energy scenarios were made flexible; a control cost and optimization modules were added; and more attention was paid to model uncertainties. Moreover, competition among similar integrated assessment models also proved to be a useful reflective mechanism through which the RAINS team learned the importance of explaining to policymakers the model itself and its differences from the competing models.

The reflective mechanisms also helped to build *trust* (cultural factor of values, norms, and beliefs) among the participants in the science–policy complex. Through the RAINS review workshop, it was learned that participants regarded RAINS as a model that is comprehensible and useful for both scientists and nonscientists, which means that the RAINS team convinced the participants that it is able to cooperate with scientists and nonscientists in a constructive way, and it thereby gained *trust* from the participants. Further factors include the norm prohibiting transboundary air pollution declared by the 1972 Stockholm Conference and basically shared among all the participants of the science–policy complex (cultural factor) (Siebenhüner 2002b) as well as the criticisms of the First Sulfur Protocol and the Nitrogen Oxides Protocol regarding the lack of scientific basis for their reduction targets (contextual factor of political pressure).

What about discouraging factors? As observed earlier, the criticism by the "outside" scientists and the conflict between soil and water scientists in the CLs case might have become discouraging factors, but they did not. Instead, the advisory scientists learned to manage such criticisms and conflicts. Some scientist continued to criticize the CLs approach as being oversimplified, arbitrary, and not scrutinized by the usual scientific peer-review process, and they argued for an ecological approach that encompasses living organisms in soils and waters as an alternative to the CLs approach (Skeffington 1995). However, such criticism was never given serious attention in the learning assessments, perhaps in part because it was not valid in the diplomacy-oriented paradigm in which the learning assessment operated.

Conclusion

This case study demonstrates the usefulness of the alternative framework by showing that a focus on procedural and substantive knowledge, a paradigm shift acquired through learning, and path-dependent factors are important when explaining learning assessments. For the first two, the cross-fertilization between STS and institutionalist approaches enabled the alternative framework to emphasize their importance. It may now be established that, in addition to the existing literature, the shift in the dominant paradigm of the learning assessment from positivism to a diplomacy-oriented one can be one of the candidate factors that explains the high political acceptability of the integrated assessments using the CLs and the RAINS model in the SSP negotiations. However, the political effectiveness of learning assessments is complex and requires careful examination of causal pathways between learning processes and such effectiveness.

In the chapter introduction, I argued that analysis of learning assessments can contribute to the democratization of expertise in the European transboundary air pollution governance by explaining the learning process, including how scientists learn what kind of strategies enhance the credibility of their advice, and that the analysis can suggest useful ways to enhance the learning process needed for the democratization process. I conclude with some implications of the case study to these ends.

Again, the case study showed that the main strategy advisory scientists employed to enhance credibility of their advice was paradigmatic learning. The resulting diplomacy-oriented paradigm, however, is substantially different from an ordinary understanding and image of science and may therefore be very difficult for an outsider to understand. This difference becomes particularly important when it comes to the democratization of expertise because the diplomacy-oriented paradigm introduces newcomers to the science–policy complex who generally have no idea of this diplomacy-oriented paradigm. The necessary communication for the democratization process can function properly only when the context and the corresponding paradigm are well understood and shared. When they are not, miscommunication may result among the experienced personnel and the newcomers, and the situation may turn out to be detrimental to the democratization process. Therefore, it is of the utmost importance to explain the paradigmatic learning and the resulting paradigm to the newcomers.

What kind of learning strategy does the case study offer? As shown, reflective mechanisms are promising learning strategies to be used in undertaking the democratization process (see also Siebenhüner, chap. 4 in this volume). Based on a proper understanding of the contexts and paradigms in past science–policy interaction, a good first step might be to ask the experienced participants and newcomers to consider jointly what context should be taken into consideration and what kind of corresponding paradigm should be developed through deliberate and interactive workshops. Unless those elements have been decided, it is very difficult to determine an effective objective and methodology. In addition, these kinds of workshop will certainly contribute to developing trust among the experienced participants and the newcomers, which was also an important stimulating factor in the case study discussed here.

Although the case study does not reveal any necessary changes in the alternative framework, needless to say the framework should be reevaluated and updated by further case studies. Exploring the learning process is itself a learning process.

Acknowledgments

An earlier version of this chapter was presented at the 2007 Annual Convention of the Japan Association of International Relations, Fukuoka,

27 October 2007, and at the Forty-ninth Annual Convention of the International Studies Association, San Francisco, 26 March 2008. I am grateful to Arne Henriksen, Leen Hordijk, and Keith Bull for providing me with very useful information. I also thank the editors of this book, Göran Sundqvist and Rolf Lidskog; three anonymous reviewers; as well as Yasuko Kameyama and Norichika Kanie for their very valuable comments. This work was funded in part by Grants-in-Aid for Scientific Research (ref. no. 17310025 and 20310025) of the Japan Society for the Promotion of Science.

Notes

1. Most of the later works by Siebenhüner (for example, Siebenhüner 2006, 2008) are not very different from earlier works (Siebenhüner 2002a, 2002b). Siebenhüner 2002a and 2002b represent the most explicitly tailored and comprehensive approach to analyzing international scientific assessments and therefore the most relevant for this chapter.

2. I use the term *learning assessment(s)* as shorthand to refer to "scientific assessments of learning."

3. Siebenhüner provides the following examples of procedural and substantive knowledge, respectively, to be learned in the introduction of the CLs: "Previous pollution abatement strategies concentrated on overall emission reductions at all sources largely irrespective of the actual damages of specific emissions in certain regions" (2002b, 422); and "In the critical-loads approach, the different emission reduction requirements could be deduced for different regions. This allows for more cost-effectiveness in abatement strategies" (2002b, 422).

4. I use the definition of the terms *interdisciplinary* and *transdisciplinary* summarized in Tress, Tress, and Fry 2006.

5. This argument was made with reference to Henriksen and Brakke 1988.

6. Keith Bull, secretary to the CLRTAP, e-mail message to author, 5 March 2007.

7. Ibid.

8. Ibid.

9. For detailed description of the EMEP model, see Eliassen and Saltbones 1983.

10. Leen Hordijk, then director of the International Institute for Applied Systems Analysis, e-mail message to author, 21 February 2007.

11. For RAINS versions one to five, uncertainty analysis was done only of a limited number of modules. For RAINS version six, which was actually used in the SSP negotiations, no uncertainty analysis was done (Hordijk and Kroeze 1997, 412).

12. Hordijk, e-mail message to author, 21 February 2007.
13. Ibid.
14. Ibid.
15. Ibid.
16. Ibid.
17. For example, see Rentz, Haasis, Jattke, et al. 1994.
18. Hordijk, e-mail message to author, 21 February 2007.
19. Ibid.

References

Alcamo, Joseph, Leen Hordijk, Juha Kämäri, Pekka Kauppi, Maximilian Posch, and Eliodoro Runca. 1985. Integrated Analysis of Acidification in Europe. *Journal of Environmental Management* 21 (1): 47–61.

Alcamo, Joseph, Pekka Kauppi, Maximilian Posch, and Eliodoro Runca. 1984. *Acid Rain in Europe: A Framework to Assist Decision Making.* IIASA Working Paper WP-84-32. Laxenburg, Austria: International Institute for Applied Systems Analysis.

Alcamo, Joseph, Roderick Shaw, and Leen Hordijk. 1990. *The RAINS Model of Acidification: Science and Strategies in Europe.* Dordrecht, Netherlands: Kluwer Academic.

Argyris, Chris, and Donald A. Schön. 1978. *Organisational Learning: A Theory of Action Perspective.* Reading, MA: Addison-Wesley.

Argyris, Chris, and Donald A. Schön. 1996. *Organisational Learning II: Theory, Method, and Practice.* Reading, MA: Addison-Wesley.

Bateson, Gregory. 1972. *Steps to an Ecology of Mind.* New York: Ballantine Books.

Bäckstrand, Karin. 2001. *What Can Nature Withstand? Science, Politics, and Discourses in Transboundary Air Pollution Diplomacy.* Lund, Sweden: Lund University Press.

Bäckstrand, Karin, and Henrik Selin. 2000. Sweden—a Pioneer of Acidification Abatement. In *International Environmental Agreements and Domestic Politics: The Case of Acid Rain*, ed. Arild Underdal and Kenneth Hanf, 87–108. Aldershot, UK: Ashgate.

Brown, M. Leann, and Michael Kenney. 2006. Organization Learning: Theoretical and Methodological Considerations. In *Organizational Learning in the Global Context,* ed. M. Leann Brown, Michael Kenney, and Michael Zarkin, 1–17. Aldershot, UK: Ashgate.

Bull, Keith. 1995. Critical Loads: Possibilities and Constraints. *Water, Air, and Soil Pollution* 85 (1): 201–212.

Castells, Nuria, and Jeremy Ravetz. 2001. Science and Policy in International Environmental Agreements: Lessons from the European Experience on

Transboundary Air Pollution. *International Environmental Agreement: Politics, Law, and Economics* 1 (4): 405–425.

Clark, William C., Jill Jäger, and Josee van Eijndhoven. 2001. Managing Global Environmental Change: An Introduction to the Volume. In *A Comparative History of Social Responses to Climate Change, Ozone Depletion, and Acid Rain*, vol. 1 of *Learning to Manage Global Environmental Risks*, ed. Social Learning Group, 3–19. Cambridge, MA: MIT Press.

Downing, Robert J., Jean-Paul Hettelingh, and Peter A. M. de Smet, eds. 1993. *Calculation and Mapping of Critical Loads in Europe: Status Report 1993*. Bilthoven, Netherlands: Coordination Center for Effects, National Institute of Public Health and Environmental Protection.

Eliassen, Anton, and Jørgen Saltbones. 1983. Modelling of Long-Range Transport of Sulphur over Europe: A Two-Year Model Run and Some Model Experiments. *Atmospheric Environment* 17 (8): 1457–1473.

GEA (Global Environmental Assessment Project). 1997. *A Critical Evaluation of Global Environmental Assessments: The Climate Experience*. Calverton, MD: CARE.

Gough, Clair, Nuria Castells, and Silvio Funtowicz. 1998. Integrated Assessment: An Emerging Methodology for Complex Issues. *Environmental Modeling and Assessment* 3 (1–2): 19–29.

Haas, Peter M. 2007. Epistemic Communities. In *The Oxford Handbook of International Environmental Law*, ed. Daniel Bodansky, Jutta Brunnée, and Ellen Hey, 791–806. New York: Oxford University Press.

Henriksen, Arne, and David F. Brakke. 1988. Sulphate Deposition to Surface Waters: Estimating Critical Loads for Norway and the Eastern United States. *Environmental Science & Technology* 22 (1): 8–14.

Henriksen, Arne M. Forsius, Juha Kämäri, Maximilian Posch, and Anders Wilander. 1993. *Exceedance of Critical Loads for Lakes in Finland, Norway, and Sweden: Reduction Requirements for Nitrogen and Sulfur Deposition*. Acid Rain Research Report 32/1993. Oslo: Norwegian Institute for Water Research.

Hettelingh, Jean-Paul, Wim de Vries, Wolfgang Schöpp, Robert J. Downing, and Peter A. M. de Smet. 1991a. Methods and Data. In *Mapping Critical Loads for Europe*, ed. Jean-Paul Hettelingh, Robert W. Downing, and Peter A. M. de Smet, 31–48. Coordination Centre for Effects Technical Report no. 1. Bilthoven, Netherlands: Coordination Center for Effects, National Institute of Public Health and Environmental Protection.

Hettelingh, Jean-Paul, Robert W. Downing, and Peter A. M. de Smet, eds. 1991b. *Mapping Critical Loads for Europe*. Coordination Centre for Effects Technical Report no. 1. Bilthoven, Netherlands: Coordination Center for Effects, National Institute of Public Health and Environmental Protection.

Hordijk, Leen. 1991. Use of the RAINS model in Acid Rain Negotiations in Europe. *Environmental Science & Technology* 25 (4): 596–603.

Hordijk, Leen, and Carolien Kroeze. 1997. Integrated Assessment Models for Acid Rain. *European Journal of Operational Research* 102 (3): 405–417.

Ishii, Atsushi. 2001. Merging into the EU Acidification Strategy: Evaluating the 1999 Gothenburg Protocol to Abate Acidification, Eutrophication, and Ground-Level Ozone. *Review of European Community & International Environmental Law* 10 (2): 210–226.

Jasanoff, Sheila. 1990. *The Fifth Branch: Science Advisers as Policymakers*. Cambridge, MA: Harvard University Press.

Jasanoff, Sheila. 1992. Politics and the Renegotiation of Expertise at EPA. *Osiris* 7 (1): 194–217.

Jasanoff, Sheila. 1996. Science and Norms in Global Environmental Regimes. In *Earthly Goods: Environmental Change and Social Justice*, ed. Fen O. Hampson and Judith Reppy, 173–197. Ithaca, NY: Cornell University Press.

Jasanoff, Sheila, ed. 2004. *States of Knowledge: The Co-Production of Science and Social Order*. London: Routledge.

Kuylenstierna, Johan C. J., Clair A. Gough, Michael J. Chadwick, Jean-Paul Hettelingh, Robert J. Downing, and Peter A. M. de Smet. 1991. Comparison of the Stockholm Environment Institute Critical Loads Map and the CCE Critical Loads Map Based on European and National Data. In *Mapping Critical Loads for Europe*, ed. Jean-Paul Hettelingh, Robert W. Downing, and Peter A. M. de Smet, 58–75. Coordination Centre for Effects Technical Report no. 1. Bilthoven, Netherlands: Coordination Center for Effects, National Institute of Public Health and Environmental Protection.

Law, John. 2004. *After Method: Mess in Social Science Research*. Oxford, UK: Routledge.

Mitchell, Ronald B., William C. Clark, David. W. Cash, and Nancy M. Dickson, eds. 2006. *Global Environmental Assessments: Information and Influence*. Cambridge, MA: MIT Press.

Nilsson, Jan. 1986. Introduction. In *Critical Loads for Nitrogen and Sulphur*, ed. Jan Nilsson, 4–5. Copenhagen: Nordic Council of Ministers.

Nilsson, Jan, and Peringe Grennfelt, eds. 1988a. *Critical Loads for Sulphur and Nitrogen*. Copenhagen: Nordic Council of Ministers.

Nilsson, Jan, and Peringe Grennfelt. 1988b. Foreword. In *Critical Loads for Sulphur and Nitrogen*, ed. Jan Nilsson and Peringe Grennfelt, 4. Copenhagen: Nordic Council of Ministers.

Nilsson, Jan, and Peringe Grennfelt. 1988c. Workshop report. In *Critical Loads for Sulphur and Nitrogen*, ed. Jan Nilsson and Peringe Grennfelt, 7–32. Copenhagen: Nordic Council of Ministers.

OECD (Organization for Economic Cooperation and Development). 1977. *The OECD Programme on Long Range Transport of Air Pollutants: Summary Report*. Paris: Organization for Economic Cooperation and Development.

Overrein, Lars N., Hans M. Seip, and Arne Tollan. 1981. *Acid Precipitation—Effects on Forest and Fish: Final Report of the SNSF-Project 1972–1980*. Oslo: Norwegian Council for Scientific and Industrial Research, Agricultural Research Council of Norway, and Norwegian Ministry of Environment.

Parson, Edward A., and William C. Clark. 1995. Sustainable Development as Social Learning: Theoretical Perspectives and Practical Challenges for the Design of a Research Program. In *Barriers & Bridges to the Renewal of Ecosystems and Institutions*, ed. Lance H. Gunderson, Crawford S. Holling, and Stephen S. Light, 428–460. New York: Columbia University Press.

Patt, Anthony. 1999. Separating Analysis from Politics: Acid Rain in Europe. *Policy Studies Review* 16 (3–4): 104–137.

Polanyi, Michael. 1962. The Republic of Science: Its Political and Economic Theory. *Minerva* 1 (1): 54–74.

Rentz, Otto, Hans-Dietrich Haasis, Andreas Jattke, Peter Russ, Martin Wietschel, and Markus Amann. 1994. Influence of Energy-Supply Structure on Emission-Reduction Costs. *Energy* 19 (6): 641–651.

Roll-Hansen, Nils. 2002. *Ideological Obstacles to Scientific Advice in Politics? The Case of "Forest Death" from "Acid Rain."* Report no. 48, Makt-og demokratiutredningens rapportserie. Available at http://www.sv.uio.no/mutr/publikasjoner/rapp2002/Rapport48.html (accessed 20 July 2009).

Roll-Hansen, Nils, and Geir Hestmark. 1990. *Miljøforskning mellom vitenskap og politikk* (Environmental Research between Science and Policy). Oslo: Norwegian Research Council for Science and Humanities.

Rosenqvist, Ivan T., and Hans M. Seip. 1986. Reduser svovelutslippene. Forsurning av vassdrag—hvor stor er uenigheten? (Reduce Sulfur Emissions: Acidification of Surface Waters—How Large Is the Disagreement?). *Kjemi* 3.

Schuster, John A. 1995. *The Scientific Revolution: An Introduction to the History and Philosophy of Science*. Available at http://descartes-agonistes.com/index.php?option=com_docman&task=cat_view&gid=37&Itemid=53 (accessed 16 August 2010).

Seip, Hans M. 2001. Acid Rain and Climate Change: Do These Environmental Problems Have Anything in Common? *Cicerone* 6.

Siebenhüner, Bernd. 2002a. How Do Scientific Assessments Learn? Part 1. Conceptual Framework and Case Study of the IPCC. *Environmental Science & Policy* 5 (5): 411–420.

Siebenhüner, Bernd. 2002b. How Do Scientific Assessments Learn? Part 2. Case Study of the LRTAP Assessments and Comparative Conclusions. *Environmental Science & Policy* 5 (5): 421–427.

Siebenhüner, Bernd. 2006. Can Assessments Learn, and If So, How? A Study of the IPCC. In *Assessments of Regional and Global Environmental Risks: Designing Processes for the Effective Use of Science in Decisionmaking*, ed. Alexander E. Farrell and Jill Jäger, 166–186. Washington, DC: Resources for the Future Press.

Siebenhüner, Bernd. 2008. Learning in International Organizations in Global Environmental Governance. *Global Environmental Politics* 8 (4): 92–116.

Skeffington, Richard A. 1990. Commentary at the NAPAP Review Meeting, Hilton Head Island, SC.

Skeffington, Richard A. 1995. Critical Loads and Energy Policy. In *Proceedings of a Conference Held at the Environmental Change Research Centre, University College London, London, 1993*, 91–97. London: Environmental Change Research Centre, University College London.

Smokler, Howard. 1983. Institutional Rationality: The Complex Norms of Science. *Synthese* 57 (2): 129–138.

Sundqvist, Göran. 2003. Recovery in the Acid Rain Story: Transparency and Credibility in Science-Based Environmental Regulation. *Journal of Environmental Policy and Planning* 5 (1): 57–79.

Sundqvist, Göran, Martin Letell, and Rolf Lidskog. 2002. Science and Policy in Air Pollution Abatement Strategies. *Environmental Science & Policy* 5 (2): 147–156.

Sverdrup, Harald, Wim de Vries, and Arne Henriksen. 1990. *Mapping Critical Loads: A Guidance Manual to Criteria, Calculation, Data Collection, and Mapping*. Miljørapport 1990:14. Copenhagen: Nordic Council of Ministers.

Sverdrup, Harald, and Per Warfvinge. 1988. Assessment of Critical Loads of Acid Deposition on Forest Soils. In *Critical Loads for Sulphur and Nitrogen*, ed. Jan Nilsson and Peringe Grennfelt, 81–129. Miljørapport 1988: 15. Copenhagen: Nordic Council. of Ministers.

Tickle, Andrew K. 1995. Critical Loads and NGO Policy. In *Proceedings of a Conference Held at the Environmental Change Research Centre, University College London, London, 1993*, 81–89. London: Environmental Change Research Centre, University College London.

Tress, Barbel, Gunther Tress, and Gary Fry. 2006. Defining Concepts and the Process of Knowledge Production in Integrative Research. In *From Landscape Research to Landscape Planning: Aspects of Integration, Education, and Application*, ed. Barbel Tress, Gunther Tress, Gary Fry, and Paul Opdam, 13–26. Dordrecht, Netherlands: Springer.

Tuinstra, Willemijn. 2006. Reducing Air Pollution in Europe: A Study of Boundaries between Science and Policy. Ph.D. diss., Wageningen University.

Wettestad, Jørgen. 2000. The ECE Convention on Long-Range Transboundary Air Pollution: From Common Cuts to Critical Loads. In *Science and Politics in International Environmental Regimes: Between Integrity and Involvement*, ed. Steinar Andresen, Tora Skodvin, Arild Underdal, and Jørgen Wettestad, 95–121. Manchester, UK: Manchester University Press.

Ziman, John. 2000. *Real Science: What It Is, and What It Means*. Cambridge, MA: Cambridge University Press.

Zito, Anthony. 2001. Epistemic Communities, Collective Entrepreneurship, and European Integration. *Journal of European Public Policy* 8 (4): 585–603.

7

Fewer Boundaries and Less Certainty: The Role of Experts in European Air Policy

Göran Sundqvist

Scholars in the field of science and technology studies (STS) often repeat the message that the relationship between science and society is unstable. Spurred by regulatory failures in dealing with problems such as "mad cow disease" (bovine spongiform encephalopathy) and public criticism of genetically modified organisms, the European Commission has prioritized expert credibility as an important issue in need of political attention. In preparing for the *White Paper on Governance* published in 2001, the commission set up a working group on "democratizing expertise" whose main objective was to propose ideas on how to restore the credibility of expertise (EC 2001). In 2005, the commission established an expert group consisting of STS scholars to address the widely recognized problem of public unease with science (Wynne and Felt 2007). These events and activities obviously indicate that the relationship between science and society must be strengthened; they also suggest that STS insights can contribute to this strengthening.

According to STS scholars, the problem is that the public receives conflicting advice from experts about what to think and how to act. The solution to this problem, however, is not unanimous advice, but a better understanding of how to handle disagreeing experts. The problem consists of naive expectations as to what expert advice can in fact deliver—expectations arising from the widely held but misleading view that scientific advice is certain. This false aura of certainty surrounding the role of the scientific expert is the main problem and the source of unrealistic expectations (Collins and Pinch 1993, 1998, 2005; Shapin 1995; Latour 1998; Hilgartner 2000; Wynne 2001; Jasanoff 2003; Irwin 2006; cf. Callon, Lascoumes, and Barthe 2009). When conflicting advice is understood from the perspective of certainty, problems arise. The solutions to the problem, according to STS, are for scientific experts to stop presenting their work as about certainty and for the public to lower its expectations and to understand that science, like all other human enterprises, is uncertain and fallible. The only way to strengthen expert credibility is to foster a realistic understanding of what science is and what it can provide to policymakers.

Doubts can be raised, though, as to whether this view is applicable to all policy areas in that there seem to be some instances of stable and trusting cooperation between scientific experts, policymakers, and the general public. One of the most obvious examples of such cooperation is European air policy. For almost four decades, scientific experts from all over Europe have cooperated in developing a scientific understanding of the problems related to long-range transboundary air pollution. Policymakers, with strong public support, have based their negotiations on this foundation, which has led to seven binding protocols on emission-reduction commitments.

Expert conflicts as well as exaggerated expectations on the part of the public have been part of this story, but the result has not been instability or decreased expert credibility. How can this result be explained? In what ways has the interaction between science and policymaking been organized and perceived by involved actors and by the general public in this case? Can this example provide a clue to how more generally to restore the credibility of expertise and to foster robust relationships between science and society? This chapter is devoted to answering these questions. The overall aim is to investigate how key STS findings can help us understand expert advice in European air policy connected to the Convention on Long-Range Transboundary Air Pollution (CLRTAP) process and thereby critically assess these findings as well as the work of air policy experts. An additional aim is to review international relations (IR) research, which often treats science as a neutral, external input to the policy process, and to compare this research with STS findings regarding how to understand the interaction between science and policy in regulating transboundary air pollution.

The next section describes CLRTAP and its success, emphasizing in particular the science–policy interaction that is often seen as an important explanation of this success. The aim is to define the characteristics of the purported success as described by involved actors and IR researchers. In the third section, the STS description of a general crisis in the relationship between science, policy, and the public is spelled out in more detail. As already mentioned, the STS portrayal of a general crisis differs from how European air policy work is usually assessed. In the fourth section, the regulation of transboundary air pollution and associated communications are juxtaposed to the STS key findings by examining four formative periods in CLRTAP's history. Here, the analysis centers on how controversies, uncertainties, and instability are dealt with and, above all, how scientific experts communicate to an audience of policymakers, politicians, and the general public. The final section compares how IR and STS scholars as well as involved experts understand the scientist–society interplay in European air policy, drawing conclusions about important findings for understanding and designing a robust (i.e., stable and trusted) interplay. In such an endeavor, key STS findings as well as air policy experience are important building blocks, though of different kinds.

Successful Interplay between Science and Policy in Transboundary Air Pollution

European air policy is widely described as a success. Involved actors and policy analysts seem to agree that CLRTAP is successful and has "become an outstanding example of fruitful international cooperation" (Dovland, Ballaman, and Thompson 2004, 3; cf. Tuinstra, Hordijk, and Kroeze 2006). But what is actually being seen as a success, and what serves to explain this success? One obvious answer is the establishment of CLRTAP, one of the first international treaties for environmental protection, and the ratification of eight protocols without serious conflicts or delays. Also impressive are the emission cuts made since the signing of the convention. European emissions of the most harmful acidifying compounds, sulfur dioxides (the target of a protocol agreed on in 1985), have been reduced by more than 75 percent, from 53 million tonnes in 1980 to 12.5 million tonnes in 2005 (Elvingson and Ågren 2007).[1]

What explains the combination of political will and relevant emission cuts is scientific sophistication. The protocols signed in the name of the convention have evolved from flat-rate emission cuts of one pollutant—the same for every party—to geographically differentiated abatement strategies referring to several specific ecosystem-effect criteria, based on the critical-loads concept and cost-effectiveness. There seems to be agreement that CLRTAP has become increasingly science based as it has developed because of a close relationship between scientific experts and policy negotiators. In short, what is generally recognized as the convention's real success is that the policy is science based: "Many scientists representing different subjects have been involved, contributing to a solid foundation on which policymakers have based their negotiations on emission reduction commitments" (Dovland, Ballaman, and Thompson 2004, 1). This means that abatement strategies that are good for both the environment (i.e., based on critical loads) and society (i.e., based on cost-effectiveness) have been designed.

The concept of critical loads is often emphasized when citing the success of CLRTAP and its science-based approach. IR policy analysts have described the importance of this concept as "a revolution in the management of transborder pollutants" (Levy 1993, 100) and as a clever way to establish science-based policy. "For the first time the basis of negotiations of reduction targets was formed by scientifically founded environment standards rather than arbitrarily chosen proposals" (Gehring 1994, 192). With the introduction of the critical-loads concept, abatement strategies are considered to rest on a sound and consensual scientific basis.

IR scholars often take the scientific knowledge underlying the formation of environmental regimes for granted, assuming that knowledge "represents a 'neutral' input on behalf of the resource or environment in question" (Andresen and Skjærseth 2007, 186) and viewing it as something external that can make an important contribution to policy negotiations—that is, science capable of "speaking truth to power" (Haas and Stevens, chap. 5 in this volume). This assumption has also been made in the analyses of CLRTAP, in which IR researchers often highlight the importance of scientific certainty, stressing a linear development from science to policy (Sundqvist, Letell, and Lidskog 2002).

However, the experts involved do not single out scientific certainty and unidirectional influence—going from science to policy—as most important. On the contrary, they value the flexible use of science in the policy process as well as the negotiable character of the science–policy interaction; or, more exactly, they value the flexible way the convention uses science to design policy that can be accepted by society as a whole. Science, used as a flexible tool, can be a catalyst for political change, it is argued. The convention has been able to expand its work to include new issues and chemical compounds and to focus on various kinds of effects to garner support from all involved parties (Letell 2002, 1–3; cf. Lidskog and Pleijel, chap. 8 in this volume). A book published on the occasion of the convention's twenty-fifth anniversary and authored by key involved people emphasizes flexibility as a key determinant of the convention's success:

> The Convention has shown a remarkable ability to adapt to different circumstances and to shift operative focus. While the attention in the first years lay with building a sound scientific knowledge base, in the second phase lasting through the 1990s the focus shifted to development and negotiation of gradually more advanced and comprehensive protocols, while at the same time ensuring that the scientific knowledge was kept up to date. . . . This has provided adaptability to new demands and circumstances and introduced an element of dynamism in policy development. (Dovland, Ballaman, and Thompson 2004, 2, 4)

A key to this flexibility and dynamism is the close relationship between scientific experts and policymakers fostered under CLRTAP. This description of CLRTAP's success runs counter to IR scholars' claims, however, as well as to STS scholars' assertion of a crisis in the science–policy interplay. It is therefore of great interest to explore in greater depth CLRTAP's assessed success. However, the STS diagnosis must be presented in more detail first.

Crisis in the Relationship between Science, Policy, and the Public: The STS Diagnosis

In recent decades, governments and their scientific experts have failed to convince the public of the benefit of technologies such as nuclear power and gene-based biotechnology. For example, the public has strongly opposed the disposal of nuclear waste and the introduction of genetically modified organisms (Sundqvist 2002, chap. 2; Blowers and Sundqvist 2010). Various groups have resisted these technologies, raising concerns about "neglected" risks. Technocratic experts have been criticized for one-sided advocacy that claims that the risks are under control and that the technologies should be accepted. What we have been witnessing is nothing less than a technocratic failure in which experts try to explain technical details to people who question the necessity of the technology as such. This clash is an important factor underlying the European Commission's initiative of a "democratized expertise" in response to a regulatory approach that leads to public distrust in expertise (EC 2001, 19).

This technocratic failure has led to the conclusion that communication between scientific experts and the public should be developed from new angles—in other words, that more intense communication is needed between science and society. Part of this effort is the founding of the academic field called "public understanding of science" (PUS). The conventional view of the public in PUS, however, is based on the old technocratic and science-centered frame called "the deficit model" (Wynne

1993, 322; cf. Leach, Scoones, and Wynne 2005). Public knowledge is thus compared with and judged in relation to scientific knowledge. It is assumed that science is central to people's everyday lives and that it is possible to measure the extent to which people live up to a scientific understanding in their beliefs and practices. From such a perspective, scientific understanding is seen as sufficient, lay understanding as deficient (Gross 1994; Miller 2001).

Contrary to a technocratic perspective, STS does not emphasize the deficiency of lay understanding but tries to reconcile the state of unease between experts and laypeople by acknowledging laypeople as participants equal in value to experts but commanding different viewpoints and types of knowledge (Irwin and Wynne 1996, 7; Wynne 1993, 321). If the limitations of science are concealed, scientific experts will exaggerate the scope and power of scientific knowledge in the public domain and ignore cultural factors and lay knowledge (Wynne 2001, 460).

Three key findings in the STS literature indicate how to understand problems of communication between scientific experts and laypeople. Bruno Latour appositely captures the STS understanding of science in his distinction between science and research, with science characterized as certainty and research as uncertainty: "Science is supposed to be cold, straight, and detached; research is warm, involving, and risky. Science puts an end to the vagaries of human disputes; research creates controversies" (1998, 208; cf. Latour 1987). According to STS, science in this sense is a myth, something we find in textbooks, but not in laboratories. Scientists never do science, just research. If we wish to retain the word *science*, it should be understood as a craft that is fallible, clumsy, untidy and that does wrong and makes mistakes, and it should above all be seen as a human activity (Irwin 1995, 2; Collins and Pinch 2005, vii–viii). Thus, the first STS key finding states that *science is an uncertain human enterprise* and is not about certainty, as the dominant public view holds.

The second key finding is that *scientists know very well that science is something uncertain, but in public they present their results as certain.* Backstage, when speaking to colleagues, scientists have no problem admitting that science is uncertain. In fact, uncertainties are what they speak about most when talking to each other. The question is therefore why this uncertainty has never become part of a widespread public understanding of science. The answer is the dominance of the orthodox idea of science that scientists must live up to when publicly presenting their findings: science is about certainty, so uncertainties cannot be part of good science. Therefore, the more correct view of science as uncertain (of which scientists are well aware) has never been disseminated to a public audience (Collins and Pinch 1993, 2).

Harry Collins has noted the paradox that laypeople often draw firm conclusions about subjects that are topics of dispute among the most competent scientists (1988, 725). According to Collins, this paradox has nothing to do with laypeople's competence but concerns how scientific experts present their results in public. What explains the paradox is that scientists speak with two voices, sounding certain in public but voicing uncertainty to other scientists. The former gives people the impression that science is more certain than it is. When experts relegate uncertainties to the backstage, to the realm of their profession, laypeople viewing experts in public cannot do otherwise than draw firm conclusions that the scientific results presented are certain (cf. Hilgartner 2000, 12).

However, it is hard work to keep the uncertainties hidden backstage. Thus, the third STS key finding says that *uncertainties are sometimes shifted from backstage to center stage*, becoming visible to the public. This shift leads to what Harry Collins and Trevor Pinch have called "flip-flop" thinking about science. Science becomes black or white; its presentation of itself as the automatic way to certain knowledge (the flip) leads to disappointed expectations (the flop) (2005, 222–223). Brian Wynne has portrayed the mission of STS as saving scientific culture from itself (2001, 472). To increase public legitimacy, science must become more self-reflective about its focus on certainty as part of its identity (Wynne 1993, 322).

The STS approach does not mean that science will become of less value when its uncertainty is acknowledged openly. When adapted to realistic expectations, science is still the best way to understand an uncertain world (Collins and Evans 2007, 2; cf. Callon, Lascoumes, and Barthe 2009). Scientists are competent and skilled workers in special domains of expertise; they are like gardeners in the domain of gardening, for example, except that no one expects a gardener to be infallible. In short, there is no flip-flop thinking about gardening, just an understanding of capable and less capable gardeners (Collins and Pinch 1998, 155).

STS offers a more nuanced understanding of what science is while advising experts to focus less on certainty and more on expertise. Collins and Pinch conclude: "Scientists should promise less; they might then be better able to keep their promises. Let us admire them as craftspeople: the foremost experts in the ways of the natural world" (1993, 142). Long-term credibility cannot be achieved by hiding uncertainties backstage, but by being less confident about the results of science and acknowledging its limitations. Science as uncertain knowledge must be disseminated as a center-stage performance. Everyone would gain from this dissemination. But how true is this diagnosis? In the next section, the three STS findings are evaluated in light of 40 years of expert efforts to understand and regulate transboundary air pollution.

Analyzing Air Policy Experts from the STS Perspective

Experience from the air policy process seems to contradict the STS view of how scientific experts present their expert knowledge in public. Scientific experts involved in air pollution regulation do not seem to have had any strong focus on certainty in their communication with policymakers and the public. Instead, they have focused more on the negotiability and adaptability of scientific knowledge than on certainty. Interestingly, although the field has never been in crisis, the experts have been embroiled in controversies, despite which policymakers, the media, and the public have kept their high expectations that scientific experts can deliver knowledge to act on. By examining these experts' behavior during some of the most formative and controversial phases of establishing and developing transboundary air pollution policy, we can empirically explore the STS hypothesis that scientific experts publicly shroud their results in a false aura of certainty and that this is the main reason for the loss of expert credibility in the public sphere. This case can also offer lessons for other policy areas as well as for the field of STS.

I discuss four cases of the development of transboundary air pollution regulation in Europe: (1) the Swedish scientist Svante Odén's efforts in 1967 to introduce the acidification of precipitation as a regional and transboundary problem due to the long-range transport of sulfur dioxides; (2) expert conflicts over the explanations of forest dieback in the early 1980s; (3) expert efforts in the late 1980s and early 1990s to introduce the critical-loads concept as a regulatory mechanism for protocols and the use of the Regional Acidification Information System (RAINS) model as an instrument to facilitate political negotiations; and (4) World Health Organization (WHO) health expert assessments of particulate matter as a risk to public health in the early 2000s.

I have chosen these cases due to their importance in relation to CLRTAP and the establishment of some of its key protocols: the first for formulating the convention as such (signed in 1979); the second for formulating the Protocol on the Reduction of Sulfur Emissions or Their

Transboundary Fluxes by at Least 30 Percent (First Sulfur Protocol, signed in 1985); the third for formulating the Protocol on Further Reduction of Sulfur Emissions (Second Sulfur Protocol, signed in 1994); and the fourth for the ongoing efforts to revise the most recent multipollutant/multieffect protocol (signed in 1999).

My focus here is on how experts present their findings. I analyze existing conflicts and uncertainties as well as how policymakers and the public are served in terms of being provided a realistic picture of what science can deliver.

Svante Odén and the Transboundary Nature of the Acidification of Precipitation

On 24 October 1967, the Swedish scientist Svante Odén published an article entitled "The Acidification of Precipitation" in the leading Swedish daily *Dagens Nyheter.* By integrating findings from various scientific fields with practical observations, he presented a picture of what he called a new large-scale environmental problem. Interestingly, he chose to publish his findings first in a daily paper a week before the same ideas were published as part of the final report of a Swedish government commission on natural resources and environmental research, for which Odén was one of the consulted experts. A year later he individually authored a more comprehensive treatment of the subject in a Swedish bulletin (Odén 1968). Not until 1976, almost 10 years later, did he publish his findings in an international scientific journal. New scientific findings were in this case presented first to the Swedish public and by the scientist himself. What were the new findings about, and how were they presented?

Odén (1967) argued that through the combustion of fossil fuels, enormous quantities of sulfur are pumped into the atmosphere: "For Sweden alone 0.5 million tonnes are emitted every year. Over the whole of Europe the quantity probably is about 20 million tonnes. . . . The total amount of acid precipitated over Sweden is calculated to be 1 million tonnes of concentrated sulfuric acid, or about 20 kg per hectare and year. . . . There is a more or less permanent mass of polluted air over Central Europe, including central and southern Sweden." What he saw was a "problem of large-scale environmental change" (my translation).

This "mass of polluted air" meant the acidification of air and precipitation and the spread of European pollution to Scandinavia that constituted "a latent threat to important conditions in our environment."

Odén discussed the concrete effects of acidification and mentioned that fish were dying in lakes in southwestern Sweden and that the loss of nutrients due to the filtering of acidic water through the soil might make it difficult to maintain forest growth. The great amount of sulfur in the atmosphere was also leading to a gradual deterioration of slow systems, such as the soil. The problem of "large-scale environmental change" meant that all biological systems were affected.

However, Odén concluded that the consequences of acidification could not yet be fully foreseen. Technical and administrative measures must be based on reliable data, so "we should be concentrating strongly on eliminating the main cause, which in this case is the excessive sulfur dioxide concentration in the atmosphere."

In the longer 1968 article, Odén gave more details on the transboundary nature of air pollution, arguing that we "find the lowest pH values when the air that reaches Sweden, and from which precipitation is formed, has passed over the sources of acidification in central and Eastern Europe" (quoted in Lundgren 1998, 205). The Scandinavian countries were receiving air pollution emitted in central Europe. The situation was "similar to that of slow but steady chemical warfare" (quoted in Lundgren 1998, 206) and had become worse in the 1960s, when the United Kingdom, continental Europe, and Sweden "solved" the sulfur problem by building tall smokestacks. This "solution" simply transformed local problems into regional ones. To create real solutions, international cooperation was needed.

What Odén presented was not certain knowledge, but a new picture, a broad overview of a complex emergent problem. He explicitly discussed uncertainties, limitations, and lack of evidence. For example, he admitted that he could not quantify Sweden's own contribution to the air-polluted precipitation in comparison with what arrived from abroad (Lundgren 1998, 205–206). What he presented was a rough assessment. In relation to the important issue of effects, Odén explicitly claimed a lack of knowledge. He did, however, believe that acid precipitation and direct fallout would eventually result in the acidification of soil and water; the time frame would depend on the ecosystems' buffering capacity. It was also clear, he stated, that this acidification would in some way or another affect botanic conditions (Lundgren 1998, 206–207). Moreover, he noted that scientists were unable to explain the relationship between fish death and water pH or the relationship between increasing leaching of mercury and soil acidification. Odén knew that his picture was incomplete and that more details had to be filled in.

Swedish scientists quickly accepted the new picture presented by Odén. He was criticized on only one point—namely, forest damage, an issue concerning which he explicitly admitted lacking knowledge. Odén talked about the acidification of forest soil as being due to increased levels of sulfur in the atmosphere. Swedish scientists argued at this point that this outcome was highly improbable and that, on the contrary, sulfur was an important nutrient for forests. One professor in forest botany concluded: "I feel, therefore, that we need to be very careful in drawing conclusions about an adverse effect on forest growth resulting from the so-called acidification of the environment. Intensive research within this area of ecology, without too many prior hypotheses, is surely one of the most important tasks to enable us properly to understand our environment" (quoted in Lundgren 1998, 85–86).

The Odén findings became widely accepted outside the research community and the small group of government experts. *Dagens Nyheter*, which published Odén's article, presented follow-up articles almost every day in the autumn of 1967. Two weeks after the publication, Swedish prime minister Tage Erlander stated at a meeting, as quoted by the newspapers the next day, "It's certainly strange that no-one has previously noticed that our world is being poisoned by quantities of sulfuric acid that should have been easy to detect" (quoted in Lundgren 1998, 95). The next week the minister of finance made the issue of transboundary air pollution his own, stating in Parliament that "we've now also been told . . . that even if we overcome this problem in Sweden, this is no guarantee that we will escape its effects" (quoted in Lundgren 1998, 96). In May 1968, the issue was discussed in the Parliament; the minister of the environment also agreed with the transboundary thesis but was keen to highlight the existing expert conflicts over possible effects on forest growth and even concluded that factors other than air pollution were also important (Lundgren 1998, 120–121).

To sum up, politicians quickly learned the content of the Odén thesis and were well served by experts and the mass media. The first presentation of the findings was in fact made in a daily paper and the second in a government report. Moreover, uncertainties and lack of knowledge were explicitly discussed from the outset. Politicians and the public could not have been served better: firsthand information was conveyed in the largest daily by the scientist himself. No strong boundary between private and public discussions existed in this case.

However, when Swedish scientists presented the results internationally, massive criticism was raised. The day after the Odén article was published, the managing director of the Royal Swedish Academy of Engineering Sciences, a professor in chemistry, speaking at the academy's formal annual meeting, proclaimed Odén's discovery of acidification of precipitation to be one of the most important scientific contributions of the year (Lundgren 1998, 78). When this professor later presented the results to the Organization for Economic Cooperation and Development Scientific Cooperation Committee, however, "he was told to go to blazes" (Lundgren 1998, 103). The members of this forum were steadfast in their belief that tall smokestacks and dilution were a solution to the sulfur emission problem.

At the United Nations Conference on the Human Environment, held in Stockholm in 1972, the Swedish delegation presented the report *Air Pollution across National Boundaries*. The transboundary nature of air pollution was still not generally accepted, but one important result of the Stockholm conference was the adoption of a Declaration of Principles, one of which stated that nation-states had the responsibility to ensure that emissions in their own territory did not cause damage in other countries (Letell 2002, 35–36). The International Conference on the Effects of Acid Precipitation, held in Telemark, Norway, in 1976, almost 10 years after Odén's first publication, is often considered the time and place when transboundary air pollution was established as a scientific fact (Letell 2002, 41). Then in 1979 CLRTAP was signed by 33 parties.

Waldsterben as a Possible Effect of Transboundary Air Pollution

The Swedish government report on natural resources and environmental research that was published just one week after Odén's newspaper article in 1967 described in great detail forest damage due to sulfur dioxide air pollution (Lundgren 1998, 92). Examples of forest dieback in the former Czechoslovakia were presented, and their costs estimated. As noted, the effects on forest soil and forest growth were one of the few topics that aroused controversy in Sweden when Odén's discovery was presented. On the international scene, the issue of forest dieback was raised at the 1982 Stockholm Conference on the Acidification of the Environment, when evidence of the connection between sulfur dioxide (leading to altered soil chemistry) and forest damage was presented. The updated Swedish report at this conference cited this connection as a strong justification for international control strategies for sulfur dioxide emissions and viewed the international situation as supportive because the understanding of acidification had spread and been accepted throughout

Europe (Letell 2002, 48). Especially in the former Federal Republic of Germany, the fear of widespread forest dieback was mounting, which motivated national politicians to change the country's status from being a laggard state regarding international regulation of sulfur dioxide emissions in the 1970s to being a major proponent of such regulation in the 1980s. The German word for forest dieback, *Waldsterben*, was often used and became an indication of Germany's changing official opinion (Pleijel and Grennfelt 2007, 38; cf. Boehmer-Christiansen and Skea 1991).

In the early 1980s, scientists still did not agree on the causes of forest dieback. However, there was consensus that transboundary air pollution could cause the acidification of lakes and rivers thousands of kilometers away from its source, thereby killing salmon and trout at a distance (Roll-Hansen 2002, 15). But did forest dieback fit into this picture?

Scientists agreed that high concentrations of sulfur dioxides were causing severe forest damage in Czechoslovakia, the German Democratic Republic, and Poland. These countries in eastern and central Europe were heavily dependent on the use of brown coal with high sulfur content for energy, and it was evident that forest damage in these areas might be connected to local emissions (Pleijel and Grennfelt 2007, 38–39). What was disputed was whether concentrations of sulfur dioxide lessened by long-distance transport could cause widespread forest damage by changing soil conditions (Roll-Hansen 1994, 339). If transboundary air pollution could kill not only fish but also forests at a distance, this would mean that emissions from continental Europe could cause forest damage in the Nordic countries as well. The forest industry in the Nordic countries is crucial to their economies, so, unlike the freshwater fish issue, the forest issue was of great economic concern to the governments in these countries (cf. Roll-Hansen 2002, 13).

The discussion of serious and widespread forest damage, a new kind of environmental effect that had not previously been experienced, was presented in the German mass media. In 1981, the cover of an issue of *Der Spiegel* proclaimed, "The forest is dying" (quoted in Roll-Hansen 2002, 16–17). However, German forest scientists could not agree that long-distance transport of sulfur was an important cause of the problem. Some scientists were more convinced of a strong connection between observed effects on forests and transboundary air pollution, whereas others considered causes such as high concentrations of sulfur dioxide from local emissions, sea salt, drought, untimely frost, age, fungi, and insects as also important (Roll-Hansen 1994, 328–329). "The forest research establishment, traditionally concerned with the productivity and health of trees, was generally sceptical. Botanists and other scientists belonging to neighbouring disciplines were more willing to believe in a pending catastrophe due to acid rain" (Roll-Hansen 2002, 22).

The latter group became most attractive for journalists, and one of its best-known representatives was the German soil chemist Bernard Ulrich, who "developed a theory of soil acidification leading to aluminum poisoning" (Roll-Hansen 2002, 18). What happened was that one side in this controversy was given public attention in the mass media. This kind of publicly presented science—exposing one side but hiding the other and thereby transforming uncertainty into certainty—is what Collins calls "pathological science" (1988, 743).

In the Scandinavian countries, environmentalists and some leading politicians, researchers, and bureaucrats embraced the reports from the German public discussion. The book by the Swedish science journalist and environmentalist Bo Landin, *Om träd kunde gråta* (If Trees Could Cry), published in 1986, is an interesting example of publicly presented science. The book strongly criticized researchers who did not accept acid rain as the main cause of forest damage. The Norwegian Ministry of Environment financed a translation of the book, and the minister wrote the preface of the Norwegian edition (Roll-Hansen 1994, 327).

The discussion of forest dieback and not least the changed opinion among politicians and the public in the Federal Republic of Germany became the main reasons for the development of the First Sulfur Protocol, signed in 1985 and of the Protocol Concerning the Control of Nitrogen Oxides or Their Transboundary Fluxes (Nitrogen Oxides Protocol), signed in 1988 (Roll-Hansen 2002, 5). The experts connected to the recently (1979) ratified CLRTAP had proposed a first protocol on sulfur dioxide emission cuts, which would have been difficult to agree on without the strong public fear of widespread forest death.

But a more nuanced message based on uncertainties and expert disagreements was blocked by a certain journalist-produced message. This was most visible in Sweden, Norway, and the Federal Republic of Germany, the three leading countries sponsoring the First Sulfur Protocol. In the Scandinavian countries, "the acid rain issue" had long dominated the public environmental agenda, and in the Federal Republic of Germany the agenda was born out of the *Waldsterben* discussion.

This one-sided media message meant that the public was not well served. According to STS scholars, when uncertainties finally leak to the public, what should be expected to happen is a backlash leading to flip-flop thinking: certainty transformed into distrust and the rejection of science. This never happened with respect to the sulfur dioxide pollution issue, however, for two reasons.

First, there were several good reasons for cuts in sulfur dioxide emissions other than forest dieback caused by transboundary air pollution, such as acidification of rivers and lakes, which was causing widespread harm to fish (Roll-Hansen 2002, 5–6). This was also what Odén said in 1967, based on the uncertainties in assessing the effects.

Second, it was journalists and not researchers who presented a one-sided picture in which uncertainties and disagreements were hidden. Nevertheless, scientific experts giving policy advice took advantage of this situation because they were protected by a shield of strong and supportive opinion based on "pathological science." Most of the scientists, however, were not happy with the situation and tried to reach and convey a more nuanced view, but journalists had no interest in this view (Roll-Hansen 1994, 327). Uncertainties in connection to forest damage were, according to science advisors, not a reason for postponing cuts in sulfur dioxide emissions. The First Sulfur Protocol, based on the same flat-rate emission cuts (30 percent) for all, was not connected to forest damage or other specific effects; in fact, it was not based on science at all. Policymakers, advised by scientific experts, could base their assessments on the whole picture when they agreed on the overall direction to take—that is, cuts to sulfur dioxide emissions. It was not necessary to include detailed knowledge about forest damage.

Uncertainties and expert disagreement concerning effects on forest growth have dogged the field of transboundary air pollution from the start, and a firm connection has still not been established between causes and effects (Pleijel and Grennfelt 2007, 39–40).

Two Science-Based Policy Tools: The Critical-Loads Concept and the RAINS Model

At the end of the 1980s, when public interest in air pollution and acid rain was at its peak, the concept of critical loads was proposed as a tool for devising further control measures. The aim was to shift regulation from emissions to effects and, in connection, to use integrated assessment models to simulate costs and effects in order to achieve cost-effectiveness.

The concept of critical loads was initially developed in the research-and-development unit at the Swedish Environmental Protection Agency, and the Nordic Council of Ministers organized a workshop to assess the concept in April 1986 (Nilsson 1986). The conclusion of the workshop was that the concept should be further evaluated and scientifically supported. In Sweden in 1988, CLRTAP held a workshop to elaborate the concept; a critical load was defined as "a quantitative estimate of an exposure to one or more pollutants below which significant harmful effects on specified sensitive elements of the environment do not occur according to present knowledge" (Nilsson and Grennfelt 1988).

Despite the concept being considered "in its infancy" and the underlying scientific knowledge "weak," the attendees agreed on its "usefulness" (Nilsson and Grennfelt 1988, 4). Researchers first complained about oversimplification and the difficulties of scientifically defending appropriate critical-loads values, but they soon accepted the results obtained using this tool as giving at least a rough picture of how nature actually works.

Environmental groups supported the critical-loads concept. Their interest in acid rain was great, and in the wake of the 1982 Stockholm Conference on the Acidification of the Environment, on the tenth anniversary of the 1972 Stockholm Conference, they launched their Stop Acid Rain campaign. The critical-loads concept focused on sensitive ecological systems and "nil damage" and was therefore considered a favorable policy target for environmental groups arguing for precautionary-based policy measures (Sundqvist, Letell, and Lidskog 2002, 150; Asdal 2008, 129).

In 1988, the critical-loads concept was accepted as a basis for future regulations. The earlier conferences and workshops on the concept were important in this respect. The CLRTAP Task Force on Mapping's adoption of a standardized method for preparing critical-loads maps was also important. In 1994, the Second Sulfur Protocol was signed in Oslo. It is the first protocol based on the concept of critical loads, which means that effects and not emissions become the key when negotiating national commitments.

With the use of integrated assessment models such as RAINS, the critical-loads concept was put into practice; options for emission reductions could be identified in which an efficient reduction in critical-loads exceedance could take place (Tuinstra, Hordijk, and Kroeze 2006, 354). With the help of the model and its various components (called "modules"), acid compounds could be traced from point of emission to final destination in grid cells measuring 150 square kilometers (Tuinstra, Hordijk, and Amann 1999, 36). The International Institute for Applied Systems Analysis developed the RAINS model in collaboration with researchers from a range of disciplinary and national affiliations, most of them connected to CLRTAP. Three models were initially developed, but it was soon decided that RAINS should perform most of the analysis, and the other two should be used for comparison. This splitting of functions made it possible to reveal uncertainty explicitly, thereby "avoiding giving the false impression that there was one indisputable solution" (Tuinstra, Hordijk, and Amann 1999, 39).

During the development of the RAINS model, workshops were organized to provide input for improving the model and using it. At one such workshop, policymakers (i.e., negotiators from various European countries) recommended increasing the model's flexibility, arguing that "politicians need a menu to choose from: not a recipe" (Tuinstra, Hordijk, and Kroeze 2006, 357). Decision makers did not want to be told what to decide, but they saw the model as a tool for facilitating negotiations; if RAINS could provide alternatives to choose from, so much the better. Their recommendation resulted in a division of labor between experts and politicians, the latter giving input and making the final decisions and the former doing the calculations. A close collaborative working relationship was established in which experts, bureaucrats, and politicians focused on the same objectives and gathered around the same tools to achieve these objectives. "Typically the working group would request a particular set of simulations, study the results, and then request different or more refined simulations" (Tuinstra, Hordijk, and Amann 1999, 37). Scientific experts were close to the negotiators, and their policy tools facilitated and gave important direction to the negotiations.

The critical-loads concept and the RAINS model became tools of communication between scientific expertise, on one hand, and negotiators and political decision makers, on the other. A buffer body was thus established between science and politics (Wettestad 1999, 33), which enabled coordination of work that gave stability to the process (Tuinstra, Hordijk, and Kroeze 2006, 35).

Policy-oriented researchers and bureaucrats developed both the critical-loads concept and the RAINS model in a policy context, and the scientific and policy communities supported the work. "In one sense, it is surprising that the scientific community accepted these models so readily, as they are fairly crude compared with the analytical techniques of most scientific disciplines" (Tuinstra, Hordijk, and Amann 1999, 38).

As a vague concept and a rough model, critical loads and RAINS could be filled with content in every specific case (Sundqvist, Letell, and Lidskog 2002, 154). They were also assessed as appropriate and promising parts of an effect-based regulation focusing on ecological sensitivity and sustainability. Environmental groups liked the critical-loads concept because it set clear limits on what nature could withstand, and policymakers and negotiators liked the RAINS model because it presented clear alternatives from which to choose. These tools, from the time of their introduction to the present, have enhanced communication between science, policy, and environmental groups. Their general principles are easy to communicate: emit no more than nature can withstand, and do so cost-effectively. Some researchers were critical, however, pointing out oversimplification, arbitrariness, and uncertainties, yet they still accepted them as policy tools (Nordberg, Bull, Chrast, et al. 2004, 109). Most scientific experts and policymakers agreed on the existence of large uncertainties, but they also agreed on the models' usefulness. The most important conclusion to point out here, however, is that when the critical-loads concept and the RAINS model were developed and used, no strong boundaries between science and policy existed. What was lost later in the 1990s, compared with the situation when the first protocols based on flat-rate emission cuts were signed in the 1980s, was the close link to the public. Regulation was delegated to a small group of scientific experts and negotiators. In 1999, this process culminated in the Protocol to Abate Acidification, Eutrophication, and Ground-Level Ozone (Gothenburg Protocol), a multipollutant/multieffect protocol described as one of the most sophisticated protocols ever signed.

A New Boundary between Science and Policy: Particulate Matter and Public Health

After the success of the Gothenburg Protocol, some experts worried about the increased complexity of abatement strategies based on the critical-loads concept—regulating several compounds and effects in one single protocol—and argued that protocols must not be so "smart" that only a few experts can understand their mechanisms for distributing the signatories' commitments (Sundqvist 2003, 66). Such a configuration implies expert domination of the process, and a discrepancy between expert and lay understandings would lead to a vulnerable policy process.

Not much indicates that simplicity will replace complexity. On the contrary, over the past decade a shift has been noted in the driving forces of air pollution–abatement strategies, making regulation even more complex. Since the early years of CLRTAP, ecological effects such as acidified lakes, harm to fish populations, and forest damage have dominated scientist, policymaker, and public awareness, and regulation based on critical loads has been directed toward ecological endpoints. Today, however, the focus has shifted to health issues for two reasons. First, the successful implementation of the discussed environmental protocols has resulted in improved environmental quality. However, this success does not mean that the acidification of lakes and forest soils is a solved problem; it means that to maintain public interest in Europe it is no longer enough to point to less acid lakes located in rural areas in peripheral Scandinavian countries (Sundqvist 2003, 66). Second, new medical research demonstrates that air pollution causes public-health problems. Large-scale epidemiological studies indicate that air pollution contributes significantly to decreasing human health and increasing mortality in cities all over Europe (WHO 2007). Studies claim that two million people die prematurely every year due to air pollution (WHO 2006b, 5). Public health has quickly come to be seen as a main reason for further combating air pollution in Europe.

Among the air pollutants of interest—particulate matter (PM), ozone, nitrogen oxides, and sulfur dioxide—the greatest focus has been on PM. Local emissions are important sources of this pollutant, and the situation in "problem cities" such as Athens, Brussels, and London is explicitly mentioned. Vulnerable groups such as infants, children, elderly people, and those suffering from asthma, allergies, and other cardiovascular and respiratory diseases have been identified, and the WHO estimates that "worldwide, close to 6.4 million years of healthy life are lost due to long-term exposure to ambient particulate matter" (WHO 2003). However, for the Scandinavian countries, assessments indicate that about 66 percent of the premature deaths are related to long-distance transport of PM, which means 3,500 deaths per year in Sweden, equivalent to an average reduction of about seven months per individual life span (Forsberg, Hansson, Johansson, et al. 2005, 18).

The focus on health means that new kinds of expertise have entered the pollution regulation scene: medical scientists, epidemiologists, and public-health researchers. New organizations have to adapt to the existing regulatory framework; at the same time, the inclusion of these organizations means that the existing framework will inevitably change. The most important of these organizations is WHO. Is this new player on the environmental scene changing the relationship between science and policy within the CLRTAP framework, which until now has been characterized by close relationships and weak boundaries, where researchers and policymakers meet in the same arenas, exchange information, discuss problems, and develop shared concepts?

In 2006, WHO published a global update of the earlier guidelines on air quality presented in 1997 but officially published three years later (WHO 2006a). A great deal of research has been carried out over the past 10 years as part of a growing awareness among both researchers and policymakers of the health problems related to PM. PM has also been included in the regulatory framework of CLRTAP, though not yet as part of a protocol. This step has been considered necessary, although there has been much discussion of various uncertainties—for example, what size of PM to regulate, what PM characteristics are important, and by what mechanisms PM causes health problems (Grennfelt and Pleijel 2007, 25). In 1997, CLRTAP and WHO set up the Joint Task Force on the Health Aspects of Air Pollution to assess the health effects of long-distance air pollution and to guide future abatement strategies.

Most of the important assessment work is done by WHO expert groups and not by the Joint Task Force. In the first WHO study, experts were reluctant to give advice regarding levels and argued that no thresholds existed; they considered it a task for policymakers and not researchers to determine what risks and exposure levels were acceptable (Elvingson 2006, 4).

In contrast to earlier CLRTAP work, the interplay between CLRTAP and WHO is characterized by the cultivation of a clear boundary between science and policy. WHO strongly emphasizes scientific integrity and a determination to stand free from political and sectoral interests. The work on ecological effects carried out by scientific experts under CLRTAP is focused on providing policymakers with what they need and, not least, on presenting innovative ideas *concerning what policymakers should need*; then, in cooperation with policymakers, CLRTAP refines these ideas to make them integral parts of a science-based regulatory framework. WHO experts want to accomplish this task in a way that protects scientific integrity more strongly, the implication being that policymakers, not scientists, should make policy. What policy needs, according to these experts, is not close cooperation with science but good science delivered by detached scientists (Lidskog and Sundqvist 2007, 196). In other words, the scientific input should not be discussed by scientists and policymakers together. This opinion is close to IR scholars' view on the role of science in the policy process: science should provide neutral input to policy, and a strong boundary should be maintained between the two fields. The WHO experts' entrance on the scene thus brought something new to the CLRTAP process.

Health problems related to PM are front-page news. The quality of the air we breathe is of interest to everybody because bad air frightens most of us. At least in the Scandinavian countries, there is fairly widespread public understanding that PM comes from various sources and that the particles that enter our lungs at street level come both from the cars in front of us—from their exhaust pipes and wheels—and from long-distance transport. The medical effects are often discussed in terms of statistics, "statistical life expectancy," and vulnerable groups. The discussion of what size and what characteristics of particles cause what effects due to what mechanisms includes many uncertainties, but the picture seems easy to communicate: PM causes health problems. Since the *Waldsterben* debate in the 1980s, CLRTAP has had severe problems finding an object that can hold public attention; it has one now in PM.

Conclusions: Configuring the Science–Policy–Public Interplay

The history of transboundary air pollution regulation contains several different configurations of the interplay between science, policy, and the public. At the outset—the Odén case—science, policy, and the public were near each other, and strong boundaries did not exist. Scientific findings were presented in the daily press and were part of government investigations. Uncertainties and expert disagreements were not hidden in public presentations. This lack of detachment between science, on one hand, and policy and the public, on the other, supported the emergence of a shared view of a new problem. In the *Waldsterben* case, the uncertain and controversial discussion of the causes of forest dieback took two directions. The mass media conveyed a one-sided picture, including an emphasis on scientific certainty, which led to a strong and engaged public opinion supported by the Green movement's Stop Acid Rain campaign. Even though some scientific experts were unhappy with this one-sided public picture, they took advantage of the situation and supported the first flat-rate protocols and thereafter developed new tools for implementing science-based abatement strategies: the critical-loads concept and the RAINS model. These tools resulted in more complicated and expert-driven protocols (i.e., the Second Sulfur Protocol and the Gothenburg Protocol), but also in diminishing public engagement. Scientists were initially reluctant to contribute to policy and feared that their results would be oversimplified and misused in the regulatory process, but they soon accepted their contribution as a way of making policy. The public-health issue, not least in relation to PM, has raised new awareness among researchers and policymakers and is today widely discussed in the mass media. New groups of health experts have entered the field, and new patterns of interplay between science and policy have arisen—patterns that differ from those that developed in the 1980s and 1990s under CLRTAP. For the first time in the history of air pollution regulation, a boundary can be noted between scientists and policymakers regarding how to set limits. WHO scientists tell policymakers that science is of no use when it comes to deciding on specific limits. Human bodies have no pollutant thresholds; limits are bureaucratic constructs and as such have nothing to do with the scientific enterprise.

In short, the story of transboundary air pollution presents us with several distinct patterns of the relationship between science, policy, and the public. Interestingly, only a few of them are in accordance with the one that STS scholars consider dominant, which should lead to a crisis of expert credibility and misunderstanding in the public's perception of science. Strong boundaries between experts, on one hand, and policymakers and the public, on the other, have usually been avoided, as has a focus on certainty. The mass-media discussion of *Waldsterben*, the poor public understanding of regulation based on critical loads and RAINS, and the communication between medical experts and policymakers concerning health effects are the only clear examples of strong boundaries creating discrepancies between the understandings of different groups. Concerning certainty, the mass-media presentation of *Waldsterben* is the only example in which certainty has been in focus and uncertainties concealed.

As demonstrated in this chapter, the STS key findings do not give a correct picture of the development of transboundary air pollution regulation. Instead, results pertaining to the air policy case can be used to problematize the STS findings, which, counter to intentions, do not provide a general model of how experts act in public. STS needs to be more open to various configurations of the interplay between science, policy, and the public. A systematic overview of case studies should be done, investigating and comparing different kinds of interplay in different fields (a first step in this direction has been taken in Collins and Evans 2002, 2007). In this effort, it should be central to focus not only on cases of technocratic management leading to public distrust, in which experts speak with two voices—one used backstage and the other used center stage. Of course, failures are of great interest, and STS scholars have made important findings when investigating cases of failure, providing crucial insights into both the reasons for these failures and how to avoid them. But we also need to focus on cases characterized by cooperation, trustful relationships, and democratically framed issues.

The general model proposed by prominent STS scholars seems paradoxically to contradict the hallmark of STS—namely, the assumption that there is no given path for the development of expertise. Therefore, the analyst should be sensitive to various kinds of links between experts and laypeople. Concepts such as co-production and scientific citizenship have been developed to help us understand these important links (Elam and Bertilsson 2003; Jasanoff 2004; Lidskog 2008). There is thus no reason to base empirical studies on the assumption of strong boundaries between various groups or on experts focusing on certainty. If one does so, details, links, alternatives, and changes will remain hidden. It is reasonable to claim that STS scholars oversell their message. Even if they are right to criticize a strong focus on certainty—the false aura leading to discrepancy in the perceptions of different groups—they are wrong to claim that they have a general model of how experts act in public. Certainty and strong boundaries should be criticized wherever they are found, but it is not true that they are found everywhere.

Interestingly, air policy experts often characterize their work as science based and consider this basis the reason for their success. IR policy analysts understand this characterization as concerning scientific integrity, illustrating a demarcation between science and policy in which scientific results can "speak truth to power." As demonstrated in this chapter, this picture is one sided. In fact, success hinges on soft and flexible boundaries and on mutual understanding between experts and policymakers; the regulation is less scientific than policy analysts claim, and the experts themselves often highlight the soft boundaries and negotiable interaction between science and policy as an important part of science-based policy. As STS has heavily emphasized, this way of interacting—in which scientific expertise is not detached from the policy process—is a recipe for success. IR research has something to learn from this conclusion.

The important STS lessons to learn are that experts should avoid focusing on certainty and that there must be meeting places where actors from different fields can discuss and negotiate both their knowledge and their identities as actors. When such discussion and negotiation are done as part of expert work, science is enriched and becomes less isolated. A more realistic picture of science and its role in regulatory work—its power, limitations, and usability—can emerge. I hope that decision makers responsible for European air pollution strategies are aware of this picture and can continue to learn from STS findings about how science really works.

Note

1. See the Web site at http://www.unece.org/env/lrtap/30anniversary.html, accessed September 2010. The counterfactual question regarding what improvements would have taken place if no protocols calling for emission cuts had been agreed on is difficult to answer (see Wettestad, chap. 2 in this volume). Industrial changes and economic restructuring have obviously taken place independently of the convention, and reduced emissions of certain airborne pollutants can be explained in part by the expansion of nuclear power in Europe.

References

Andresen, Steinar, and Jon Birger Skjærseth. 2007. Science and Technology: From Agenda Setting to Implementation. In *The Oxford Handbook of International Environmental Law*, ed. Daniel Bodanski, Jutta Brunnée, and Ellen Hey, 182–202. Oxford, UK: Oxford University Press.

Asdal, Kristin. 2008. Enacting Things through Numbers: Taking Nature into Account. *Geoforum* 39 (1): 123–132.

Blowers, Andrew, and Göran Sundqvist. 2010. Radioactive Waste Management: Technocratic Dominance in an Age of Participation. *Journal of Integrative Environmental Sciences* 7 (3): 149–155.

Boehmer-Christiansen, Sonja, and Jim Skea. 1991. *Acid Politics: Environmental and Energy Policies in Britain and Germany*. London: Belhaven Press.

Callon, Michel, Pierre Lascoumes, and Yanick Barthe. 2009. *Acting in an Uncertain World: An Essay on Technical Democracy*. Cambridge, MA: MIT Press.

Collins, H. M. 1988. Public Experiments and Displays of Virtuosity: The Core-Set Revisited. *Social Studies of Science* 18 (4): 725–748.

Collins, H. M., and Robert Evans. 2002. The Third Wave of Science Studies: Studies of Expertise and Experience. *Social Studies of Science* 32 (2): 235–296.

Collins, Harry, and Robert Evans. 2007. *Rethinking Expertise*. Chicago: University of Chicago Press.

Collins, Harry, and Trevor Pinch. 1993. *The Golem: What You Should Know about Science*. Cambridge, UK: Cambridge University Press.

Collins, Harry, and Trevor Pinch. 1998. *The Golem at Large: What You Should Know about Technology*. Cambridge, UK: Cambridge University Press.

Collins, Harry, and Trevor Pinch. 2005. *Dr. Golem: How to Think about Medicine*. Chicago: University of Chicago Press.

Dovland, Harald, Richard Ballaman, and Jan Thompson. 2004. Introduction. In *Clearing the Air: 25 Years of the Convention on Long-Range Transboundary Air*

Pollution, ed. Johan Sliggers and Willem Kakebeeke, 1–30. New York: United Nations.

EC (Commission of the European Communities). 2001. *White Paper on Governance: Report of the Working Group "Democratising Expertise and Establishing Scientific Reference Systems."* Brussels: Commission of the European Communities, 2 July.

Elam, Mark, and Margareta Bertilsson. 2003. Consuming, Engaging, and Confronting Science: The Emerging Dimensions of Scientific Citizenship. *European Journal of Social Theory* 6 (2): 233–251.

Elvingson, Per. 2006. New WHO Air Quality Guidelines. *Acid News* 2: 1, 4–5.

Elvingson, Per, and Christer Ågren. 2007. Far from Target. *Acid News* 4: 10.

Forsberg, Bertil, Hans-Christen Hansson, Christer Johansson, Hans Areskoug, Karin Persson, and Bengt Järvholm. 2005. Comparative Health Impact Assessment of Local and Regional Particulate Air Pollutants in Scandinavia. *Ambio* 34 (1): 11–19.

Gehring, Thomas. 1994. *Dynamic International Regimes: Institutions for International Environmental Governance*. Berlin: Peter Lang.

Grennfelt, Peringe, and Håkan Pleijel. 2007. Air Pollution—a European Perspective. In *Transboundary Air Pollution: Scientific Understanding and Environmental Policy in Europe*, ed. Håkan Pleijel, 13–26. Lund, Sweden: Studentlitteratur.

Gross, Alan G. 1994. The Roles of Rhetoric in the Public Understanding of Science. *Public Understanding of Science* 3 (1): 3–23.

Hilgartner, Stephen. 2000. *Science on Stage: Expert Advice as Public Drama*. Stanford, CA: Stanford University Press.

Irwin, Alan. 1995. *Citizen Science: A Study of People, Expertise, and Sustainable Development*. London: Routledge.

Irwin, Alan. 2006. The Politics of Talk: Coming to Terms with the "New" Scientific Governance. *Social Studies of Science* 36 (2): 299–320.

Irwin, Alan, and Brian Wynne. 1996. Introduction. In *Misunderstanding Science: The Public Reconstruction of Science and Technology*, ed. Alan Irwin and Brian Wynne, 1–17. Cambridge, UK: Cambridge University Press.

Jasanoff, Sheila. 2003. Technologies of Humility: Citizen Participation in Governing Science. *Minerva* 41 (3): 223–244.

Jasanoff, Sheila, ed. 2004. *States of Knowledge: The Co-Production of Science and Social Order*. London: Routledge.

Latour, Bruno. 1987. *Science in Action: How to Follow Scientists and Engineers through Society*. Cambridge, MA: Harvard University Press.

Latour, Bruno. 1998. From the World of Science to the World of Research. *Science* 280 (5361): 208–209.

Leach, Melissa, Ian Scoones, and Brian Wynne, eds. 2005. *Science and Citizens: Globalization and the Challenge of Engagement*. London: Zed Books.

Letell, Martin. 2002. *Acid Narratives: National Security and the Configuration of SO_2*. STS Research Reports 5, Section for Science and Technology Studies. Gothenburg: University of Gothenburg.

Levy, Marc A. 1993. European Acid Rain: The Power of Tote-Board Diplomacy. In *Institutions for the Earth: Sources of Effective International Environmental Protection*, ed. Peter M. Haas, Robert O. Keohane, and Marc A. Levy, 75–132. Cambridge, MA: MIT Press.

Lidskog, Rolf. 2008. Scientised Citizens and Democratised Science: Re-Assessing the Expert–Lay Divide. *Journal of Risk Research* 11 (1–2): 69–86.

Lidskog, Rolf, and Göran Sundqvist. 2007. Regulating European Air: The Co-Production of Science and Policy. In *Transboundary Air Pollution: Scientific Understanding and Environmental Policy in Europe*, ed. Håkan Pleijel, 177–201. Lund, Sweden: Studentlitteratur.

Lundgren, Lars J. 1998. *Acid Rain on the Agenda: A Picture of a Chain of Events in Sweden, 1966–1968*. Lund, Sweden: Lund University Press.

Miller, Steve. 2001. Public Understanding of Science at the Crossroads. *Public Understanding of Science* 10 (1): 115–120.

Nilsson, Jan, ed. 1986. *Critical Loads for Nitrogen and Sulphur: Report from a Nordic Working Group*. Miljørapport 1986:11. Copenhagen: Nordic Council of Ministers.

Nilsson, Jan, and Peringe Grennfelt, eds. 1988. *Critical Loads for Sulphur and Nitrogen*. Report from a workshop held at Skokloster, Sweden, 19–24 March 1988. Copenhagen: Nordic Council of Ministers.

Nordberg, Lars, Keith Bull, Radovan Chrast, Oddmund Graham, Andrzej Jagusiewicz, Peter H. Sand, Arne Tollan, and Henning Wuester. 2004. The Role of the Secretariat: Building the Protocol Tree. In *Clearing the Air: 25 Years of the Convention on Long-Range Transboundary Air Pollution*, ed. Johan Sliggers and Willem Kakebeeke, 97–117. New York: United Nations.

Odén, Svante. 1967. Nederbördens försurning (The Acidification of Precipitation). *Dagens Nyheter*, 24 October.

Odén, Svante. 1968. *The Acidification of Air Precipitation and Its Consequences in the Natural Environment*. Ecology Committee Bulletin no. 1. Stockholm: Swedish National Research Council.

Pleijel, Håkan, and Peringe Grennfelt. 2007. European Air Pollution—the Historic Legacy. In *Transboundary Air Pollution: Scientific Understanding and Environmental Policy in Europe*, ed. Håkan Pleijel, 27–53. Lund, Sweden: Studentlitteratur.

Roll-Hansen, Nils. 1994. Science, Politics, and the Mass Media: On Biased Communication of Environmental Issues. *Science, Technology, & Human Values* 19 (3): 324–341.

Roll-Hansen, Nils. 2002. *Ideological Obstacles to Scientific Advice in Politics? The Case of "Forest Death" from "Acid Rain."* Rapportserien no. 48. Oslo: Makt-og Demokratiutredningen.

Shapin, Stephen. 1995. Cordelia's Love: Credibility and the Social Studies of Science. *Perspectives on Science* 3 (3): 255–275.

Sundqvist, Göran. 2002. *The Bedrock of Opinion: Science, Technology, and Society in the Siting of High-Level Nuclear Waste*. Dordrecht, Netherlands: Kluwer Academic.

Sundqvist, Göran. 2003. Recovery in the Acid Rain Story: Transparency and Credibility in Science-Based Environmental Regulation. *Journal of Environmental Policy and Planning* 5 (1): 57–79.

Sundqvist, Göran, Martin Letell, and Rolf Lidskog. 2002. Science and Policy in Air Pollution Abatement Strategies. *Environmental Science & Policy* 5 (2): 147–156.

Tuinstra, Willemijn, Leen Hordijk, and Markus Amann. 1999. Using Computer Models in International Negotiations: The Case of Acidification in Europe. *Environment* 41 (9): 33–42.

Tuinstra, Willemijn, Leen Hordijk, and Carolien Kroeze. 2006. Moving Boundaries in Transboundary Air Pollution: Co-Production of Science and Policy under the Convention on Long-Range Transboundary Air Pollution. *Global Environmental Change* 16 (4): 349–363.

Wettestad, Jørgen. 1999. *Designing Effective Environmental Regimes: The Key Conditions*. Cheltenham, UK: Edward Elgar.

WHO (World Health Organization). 2003. *Health Aspects of Air Pollution with Particulate Matter, Ozone, and Nitrogen Dioxide*. Report on a WHO Working Group. EUR/03/5042688. Bonn: WHO. Available at http://www.euro.who.int/document/e79097.pdf.

WHO (World Health Organization). 2006a. *Air Quality Guidelines Global Update 2005*. Copenhagen: WHO Regional Office for Europe.

WHO (World Health Organization). 2006b. *WHO Air Quality Guidelines for Particulate Matter, Ozone, Nitrogen Dioxide, and Sulfur Dioxide. Global Update 2005. Summary of Risk Assessment*. Geneva: WHO. Available at http://www.who.int/phe/health_topics/outdoorair_aqg/en.

WHO (World Health Organization). 2007. *Health Relevance of Particulate Matter from Various Sources*. Report on a WHO Workshop, Bonn, Germany, 26–27 March. Copenhagen: WHO Regional Office for Europe.

Wynne, Brian. 1993. Public Uptake of Science: A Case for Institutional Reflexivity. *Public Understanding of Science* 2 (4): 321–337.

Wynne, Brian. 2001. Creating Public Alienation: Expert Cultures of Risk and Ethics on GMOs. *Science as Culture* 10 (4): 445–481.

Wynne, Brian, and Ulrike Felt. 2007. *Taking European Knowledge Society Seriously. Report of the Expert Group on Science and Governance to the Science, Economy, and Society Directorate, Directorate-General for Research, European Commission*. Brussels: European Communities.

8

Co-producing Policy-Relevant Science and Science-Based Policy: The Case of Regulating Ground-Level Ozone

Rolf Lidskog and Håkan Pleijel

Few environmental issues on the international political agenda are unsupported by a body of scientific research. It is virtually impossible for an environmental condition to be successfully transformed into a political problem without scientific support in the form of data and analysis. At the same time, the relationship between science and policy is not simple: some environmental problems become politically prioritized despite considerable scientific uncertainty (e.g., forest decline in the 1980s, discussed in chapter 7), but scientific evidence supports other environmental problems that are still low priorities in practice (e.g., the detrimental effect of many air pollutants on materials, including cultural heritage).

Science and policy have often been described as two discrete but interrelated entities, scientific findings being delivered to policymakers, who then act on their basis. As emphasized in this volume, a closer look reveals a complex relationship between science and policy. Science and policymaking are not discrete activities related through a one-way process, extending from scientific discovery of a problem to politically formulated environmental goals and then to governmental strategies for implementing these goals. Instead, the relationship between science and policy is one of reciprocity, the two intermingling in a complex and dynamic process. Through interaction and negotiation between various actors, science and policy are co-produced (Jasanoff and Wynne 1998).

A prerequisite for any air pollution policy is that what constitutes "clean"—or "unclean"—air should be defined and operationalized. How the category "un/clean air" is framed and thereby made measureable and governable is decisive when analyzing science–policy relationships in the air pollution field. In this chapter, we focus on a specific aspect of un/clean air—namely, ground-level ozone.[1] We chose the issue of ground-level ozone primarily because it is seen as an urgent threat to human health and the environment and is supported by a long history of research and policy. Scientific understanding of the problem, however, has undergone dramatic change, with great implications for the formulated policy. Moreover, the co-production of science and policy with reference to this environmental problem has been poorly researched. In fact, social scientists have neglected the ground-level ozone issue, in sharp contrast to their treatment of a number of other environmental problems.

Our overall aim is to determine how science and policy have been co-produced in this particular case. Our guiding research questions are: How is the problem of ground-level ozone scientifically conceptualized? What fundamental changes can be traced in the evolution of the problem definition? What are the regulatory implications of the various problem definitions? What political considerations have influenced the process, leading to a changed understanding of the problem?

The empirical material we used consists of official documents and records of discussions from workshops organized by the United Nations Economic Commission for Europe (UNECE) Convention on Long-Range Transboundary Air Pollution (CLRTAP, adopted in 1979). This institutional setting has been decisive for scientific discussions of ozone exposure. Even though the European Union (EU) has gradually become an important actor in developing air pollution policy in Europe, CLRTAP—with its organizational setting, institutionalized knowledge, and research networks—remains central to scientific discussions of ground-level ozone (Siebenhüner, chap. 4 in this volume).[2] By analyzing scientific discussion and the results of workshops arranged by CLRTAP, we can explore expert opinion regarding relevant—that is, desirable and achievable—abatement strategies and investigate the political considerations and implications of this expert discussion. Investigating fundamental changes in scientific understanding in this context leads to exploring the importance of science in environmental regulation and of political expediency in the expert community.

The chapter is divided into five parts, this introduction being the first. The second part further develops the analytical framework and describes how the science–policy relationship has been configured within CLRTAP. The third part investigates the evolution of scientific understanding of the environmental effects of ground-level ozone, finding various phases in the scientific conceptualization of the problem, all having different implications for the regulation of ground-level ozone and its precursors. The fourth part analyzes certain aspects of this history to trace how science and policy have been co-produced in developing a scientific understanding of the environmental effects of ground-level ozone. In the fifth and concluding part, we draw some general conclusions concerning how science and policy have been co-produced in this specific case, resulting in science-based policy and policy-relevant science.

Science-Based Policy and Policy-Relevant Science

Nature becomes knowable via various social practices. In environmental policy, it becomes known primarily through science. Even though the lay understanding and various other kinds of knowledge—fueled by media coverage and public discussion—are important in putting an environmental problem on the political agenda, these understandings and claims also need a scientific underpinning. In most cases, it is through measurement instruments, monitoring stations, experiments, data collection, mapping, and modeling that environmental problems become objects for political intervention (Asdal 2007).

Science not only plays a crucial role in discovering environmental problems and analyzing their causes and effects; it also plays a central role in proposing solutions and political action. Solutions are proposed when science consciously and explicitly tries to influence policymaking. At other times, the scientific influence on policy is more implicit as science makes certain options relevant and others irrelevant in how it frames and conceptualizes environmental issues (Fischer 2003; Hajer and Wagenaar 2003; Lidskog, Soneryd, and Uggla 2009). How science represents an environmental problem opens doors to certain modes of action at the same time that it closes others.

The production and use of scientific knowledge are widely recognized as essential elements of environmental standard setting (Jasanoff and Wynne 1998; O'Riordan, Cooper, Jordan, et al. 1998). By defining the criteria by which environmental degradation and health risks are assumed to be avoided or mitigated, a standard determines what are considered "good" and "bad" environments—in this case, clean and unclean air. Environmental standards include such concepts as critical levels. Measuring the levels of various substances (e.g., ground-level ozone and particulate matter), to determine whether they exceed established air quality guidelines, makes it possible to state that air is clean or unclean. How an environmental problem is represented and measured—provided this representation is seen as scientifically sound and the political community deems it valid—has great implications for its regulation.

At the same time, measurement is developed in a social context, in which scientists themselves take account of political considerations about how to design policy-relevant knowledge and in which powers other than scientific ones also influence and shape how measurement develops. It is completely misleading to see scientific knowledge that supports policy processes as produced at a distance from political activities and only thereafter being communicated to political actors. On the surface, however, science and policy are often presented as separate activities in terms of both social practices and organizational sites (Latour 1998; Hilgartner 2000). Scrutiny of this view reveals a more dynamic interdependence in which science and policy are co-produced in the sense that policy influences the production and stabilization of knowledge, while knowledge simultaneously supports and justifies that policy (Jasanoff 2004a). As demonstrated later in this chapter, scientific activities are formally separated from political negotiations but at the same time are substantially integrated with them. Science and policy mutually create an action space populated by governable entities. Exactly how science and policy interact in various contexts is an empirical question. It is therefore of great interest to explore how scientifically based measurements are produced and integrated into air pollution policies—in this case, the regulation of ground-level ozone. It is also of great interest not to limit the investigation only to how science–policy relationships are configured, but to include how they are publicly presented, as either separated or integrated activities.

Regulating Air Pollution

Compared with many other environmental regimes, CLRTAP (together with the Montreal Protocol on Substances That Deplete the Ozone Layer) is considered a success (Underdal and Hanf 2000; Wettestad 2002; Dovland, Balaman, and Thompson 2004; Dimitrov 2006; Lidskog and Sundqvist, chaps. 1 and 12 in this volume). This success is often explained by referring to the close interaction between scientists and policymakers in the CLRTAP institutional setting.

It is simultaneously emphasized, however, that the reason for the success is that science and policy were separated: scientists in working groups, international coordination programs, and task forces were allowed to develop sound science without needing to consider national interests. An example of this view can be found in a book published on the twenty-fifth anniversary of the convention, which repeatedly emphasizes that CLRTAP succeeded in integrating science into negotiations while keeping it separate from policy (Nordberg, Bull, Chrast, et al. 2004, 109).

The convention aims to reduce emissions of airborne transboundary pollutants in the UNECE region, supported by the development of scientific effect criteria, such as the critical-load concept. Scientific effect criteria are today the basic norms within CLRTAP and offer a mechanism for creating a science-based environmental policy (Wettestad 2000, 116). CLRTAP is one of the most strongly science-based regimes in existence, its formulation and implementation being unthinkable without strong support from scientific assessments. As asserted earlier and demonstrated later, although this description seems superficially correct, our point is that science-based policy and policy-relevant science are two sides of the same coin, both of which result from interaction and negotiation between the various actors on each side. The scientific definition of un/clean air cannot be separated from the production of air pollution policy. Behind what may seem to be the formally separated activities of science and politics stands close interaction and integration.

CLRTAP's Institutional Setting: Political Aspects of Scientific Organization

CLRTAP comprises both international agreements (protocols) and a structure of bodies for environmental monitoring and modeling, scientific assessment, and policy development. A dense network of scientists and policymakers has long been involved in CLRTAP, giving rise to a kind of collective memory—a shared history and understanding of the problems at stake—and to mutual trust.

CLRTAP's Executive Body is a treaty body independent of the UNECE. At the same time, the UNECE offers a platform for intergovernmental cooperation. During the Cold War, it was the key organization working on complicated East–West issues. In fact, until the Gorbachev era and ultimately the dissolution of the Soviet Union in 1991, the UNECE was one of only a few shared fora for countries on both sides of the Iron Curtain.

CLRTAP has at the same time been heavily dependent on economic support from individual nation-states. Not only in its initial years but throughout its existence, CLRTAP has been dependent largely on individual nation-states' willingness to contribute to its work. Parties to the convention have voluntarily provided secretariat staff for periods up to one year; lead countries or hosting countries have generally (and voluntarily) borne the costs of coordinating the activities of the international program centers; and single nations or groups of nations have sponsored workshops and scientific meetings (Hettelingh, Bull, Chrast, et al. 2004, 73).

The working groups and task forces are led by the parties to CLRTAP and are normally chaired by an expert nominated by a lead country; nation-states select experts who—together with other, non-nominated researchers—participate in the discussion. A central mechanism for policy–science interaction is the close and confidential relationship between the secretariat and the chair, who is selected based on both scientific merit and political expediency. The secretariat carefully selects chairs and vice chairs; during the Cold War, selection was made from both sides of the Iron Curtain, there being a tacit agreement that a departing staff member from one side should be replaced by another from the same side (Nordberg, Bull, Chrast, et al. 2004, 98).

This structure—the nation-state as an important actor providing economic support, coordinating scientific activities, and selecting experts to be involved in the scientific assessment work—did not politicize the convention work in the sense that it resulted in divergent national policies. Instead, the convention's design was seen as a way to separate science from policy, with the aim of effectively influencing international negotiations. As the book on CLRTAP's twenty-fifth anniversary puts it, "While the experts in such bodies were nominated by governments (and by accredited NGOs), they were there to contribute their expert knowledge and not to represent the political position of their countries" (Nordberg, Bull, Chrast, et al. 2004, 109). Despite the fact that the experts were in part selected by national governments, they were not seen as representing those nations, but instead as operating on a scientific and universalistic basis. Because a number of political considerations impinged on the organization of CLRTAP's structure and work, it was possible to see the convention as devoted to the issue of transboundary air pollution without siding with any particular national interest. The convention mobilized scientific and technical arguments to facilitate negotiations and propose workable solutions, a strategy seen as a main determinant of its success.

CLRTAP at first sight seems a somewhat paradoxical convention because it was consciously designed to maximize its political influence by separating policy and science. By this design, national interests were subsumed to the benefits of scientific findings, and in negotiations CLRTAP was seen as delivering trustworthy advice about un/clean air. However, two matters merit consideration. First, what is visible on the surface may be the result of less visible processes; that is, the separation of science and policy is often the result of cooperation between science and policy. Second, if such a separation is made, channels must be constructed to transmit and translate understandings from one area to another; otherwise, there can be no exchange or influence between science and policy.

Mechanisms for Making Science Matter

The European Monitoring and Evaluation Programme (EMEP) is decisive for CLRTAP's functioning and may even be seen as its backbone (Schneider and Schneider 2004, 31). The program was established in 1977 and now encompasses some 100 monitoring stations in Europe—measuring, for example, sulfur dioxide, nitrogen dioxides, and ground-level ozone. EMEP provides a mechanism for producing standardized emission data, key material with which to evaluate the effects of various abatement strategies (Munton, Sooros, Nikitina, et al. 1999; Wettestad 2002). It has also fostered the development of uniform abatement strategies and shared understanding and modeling of transboundary air pollution flows.

EMEP currently has four centers and three task forces, all of which report annually to the EMEP Steering Body, which in turn reports annually to the CLRTAP Executive Body. The meetings between the Steering Body and the coordination centers are kept at a distance from diplomatic and political venues; for example, their meetings were initially often held outside of United Nations buildings, sometimes even in hotels frequented by Soviet EMEP members (Schneider and Schneider 2004, 37).

Scientists invented the critical-loads and critical-levels concepts as ways to evaluate the environmental effects of airborne pollutants. Although the earlier understanding measured not only emissions, but also to some extent their effects, it did so in an unqualified way, without differentiating between environmental targets of higher or lower sensitivity. Tracing critical levels or loads for various ecosystems could help make abatement strategies more effective in terms of the environmental gain per invested abatement effort. However, assessing effects—including estimating the cost-effectiveness of various measures—is very complicated. It requires not only knowledge of airborne pollutants, but also detailed knowledge of the sensitivities of various ecosystems (e.g., soils, lakes, crops, and forests). In that sense, the critical-load/level approach meant that air pollution policy by necessity became science based (Lidskog and Sundqvist 2002; Sundqvist, Letell, and Lidskog 2002).

However, simply producing quantitative estimates of critical levels for various pollutants and ecosystems is of no help to policymakers when they are negotiating relevant, cost-effective abatement strategies. They need help interpreting these data so they can develop ways to abate air pollution. CLRTAP's mapping manual ("Manual on Methodologies and Criteria for Modeling and Mapping Critical Loads and Levels; and Air Pollution Effects, Risks and Trends") was developed as a tool for modeling and mapping critical levels and loads and their exceedances and later for dynamically modeling acidification. It standardized the methods for deriving data with which to assess effects and risk. CLRTAP's International Cooperative Program Modeling and Mapping Task Force (lead country: Germany) developed the manual, with support from the Coordination Centre for Effects (located in the Netherlands). The *Mapping Manual* serves as an important means for making the effects of air pollution understandable and meaningful to politicians and citizens.

The organization of thematic and accredited workshops within CLRTAP is a central method for discussing, negotiating, and distributing knowledge. As demonstrated later, the definition of un/clean air cannot be separated from the production of air pollution policy. To be policy relevant, science must be sensitive to political considerations and discussions, and in order to develop science-based policy decision makers must mobilize expert resources to measure and abate air pollution.

It has been crucial that changes in the understanding of the effects of ground-level ozone were not restricted to the scientific community but were communicated within CLRTAP and integrated into its institutional mechanisms. Scientific representations must be made intelligible and communicated to diverse communities of practice; to do that, they must adapt to a range of practices (cf. Jasanoff 2004b). Policy-relevant science needs science-based policy, and CLRTAP provided an institutional framework for this requirement. Thematic workshops organized within the CLRTAP framework were an important means for producing policy-relevant science and for making science-based policies possible. By means of monitoring activities and critical levels, un/clean air was identified and measured, and by means of integrated assessments and mapping technologies the geographical consequences of current emissions were presented, and maps were produced for future scenarios. This process presented governable entities to nation-states, which had to make decisions concerning abatement strategies. In the following two sections, we look in greater detail at the scientific work concerning ground-level ozone within CLRTAP to trace how political considerations not only were included in the organizational design of the scientific work but became part of the scientific deliberations as well.

Representing Ground-Level Ozone

Early observations of the effects of ozone or other strongly oxidizing agents on plants (Knight and Priestly 1914) and animals (Hill and Flack 1911) were based on artificial experimental exposures. In the United States in the late 1940s, the large-scale air quality problem of ground-level ozone and other photochemical pollutants was discovered first in California (Middleton, Kendrick, and Schwalm 1950) and soon afterward in other states (Heggestad 1991). These findings, however, had little impact in Europe. In the 1970s, ozone monitoring started in some countries, and a few experimental investigations were conducted, in part using bioindicators developed in the United States, where the effects of ozone on plants were being noted (Ashmore, Bell, and Reily 1978). It was not until the early 1980s that the problem of ground-level ozone first attracted broader scientific and political attention in Europe (Grennfelt 1984).

One reason for this delay was that in Europe the problem of freshwater acidification dominated discussion of transboundary air pollution at this time. Chemical changes in freshwater and their effects on fish as well as the transboundary nature of the pollutant fluxes were established scientifically in the 1970s, which was critical for the establishment of CLRTAP in 1979 and its first protocol on sulfur emission reductions in 1984.

Ground-level ozone gradually came to be included in CLRTAP, not least because it was seen as a potential explanation of the widely discussed forest dieback in West Germany and neighboring countries (Ashmore, Bell, and Rutter 1985). It was first formally included in the convention in the 1988 Sofia Protocol Concerning the Control of Emissions of Nitrogen Oxides or Their Transboundary Fluxes, which focused largely on how to reduce acidification and eutrophication; the contribution of nitrogen oxides to photochemical pollution was stated as one of the environmental drivers of that protocol.

Three years later the 1991 Geneva Protocol Concerning the Control of Emissions of Volatile Organic Compounds or Their Transboundary Fluxes went a step further, and under its terms some countries (e.g., Canada and Norway) defined so-called Tropospheric Ozone Management Areas in which to implement emission reductions, thus underpinning the importance of ground-level ozone.

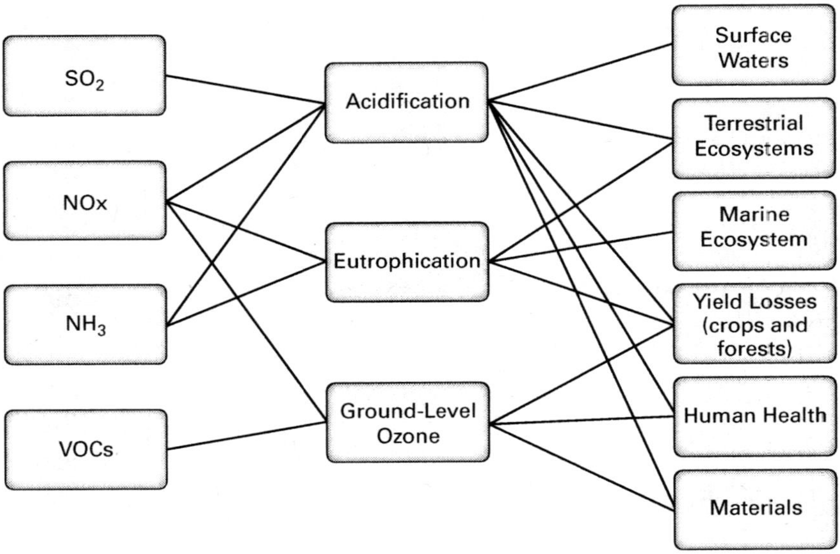

Figure 8.1
Air pollutants and environmental problems covered by the Gothenburg Protocol (SO_2 = sulfur dioxide; NO_x = nitrogen oxides; NH_3 = ammonia; VOCs = volatile organic compounds). *Source*: UNECE 1999, 2.

The Protocol to Abate Acidification, Eutrophication, and Ground-Level Ozone of 1999 (Gothenburg Protocol), establishes that ground-level ozone emissions are one of three environmental targets addressed (see figure 8.1). Both groups of ozone-forming substances—nitrogen oxides and volatile organic compounds (VOCs)—are regulated in the protocol. Ground-level ozone was a key focus in the preparation of the Gothenburg Protocol, and its effects on both health (i.e., life years lost and premature deaths) and vegetation were explicitly considered in the negotiations leading up to it.

Developing a Scientific Understanding of Ozone

Ground-level ozone became an object of regulation in the late 1980s, and demands were made for scientific ways to measure ozone exposure based on critical levels of exposure. Because ground-level ozone was included in CLRTAP, several workshops were organized to conceptualize, define, and put the problem in a form suited for political consideration and action. In addition to critical loads for sulfur and nitrogen deposition—which relate mostly to acidification and eutrophication—

CLTRAP began to develop definitions of "critical levels" for gaseous pollutants (UNECE 1988).

Critical levels were defined as levels of pollutants above which adverse effects on vegetation were likely to occur, according to present knowledge. Determining these levels has made it possible to define what should be seen as un/clean air with regard to specific substances. Workshops on interpreting critical levels have been held seven times under the auspices of CLRTAP (table 8.1). Discussion at these workshops indicates a gradual shift in priorities between different issues and altered understandings of the environmental effects of ozone.

It was only at the two first workshops that pollutants other than ozone were considered in any substantial sense; after these workshops, sulfur dioxide was not prioritized because very substantial emissions reductions were under way and expected to continue. Although nitrogen oxides and ammonia were still believed to have significant environmental effects (not

Table 8.1
Workshops within CLRTAP Related to Critical Levels of Gaseous Air Pollutants

Site and Year	Pollutants Considered	Plant Ozone Exposure	Proceedings
Bad Harzburg, Germany, 1988	SO_2, O_3, NO/ NO_2, and NH_3	Concentration	UNECE 1988
Egham, United Kingdom, 1992	SO_2, NO/NO_2, NH_3, and O_3; direct effects of acid rain and clouds	Concentration	Ashmore and Wilson 1993
Bern, Switzerland, 1993	O_3	Concentration	Fuhrer and Achermann 1994
Kuopio, Finland, 1996	O_3	Concentration	Kärenlampi and Skärby 1996
Gerzensee, Switzerland, 1999	O_3	Concentration/flux	Fuhrer and Achermann 1999
Hindås, Sweden, 2002	O_3	Flux/concentration	Karlsson, Selldén, and Pleijel 2003
Obergurgl, Austria, 2005	O_3	Flux/concentration	Wieser and Tausz 2006

Note: SO_2 = sulfur dioxide; NO/NO_2 = nitrogen oxide/dioxide; NH_3 = ammonia; and O_3 = ozone. "Concentration" refers to the amount of ozone in the air, "flux" to the actual uptake of ozone.

least in terms of eutrophication and acidification), they were considered less important than ozone in terms of direct effects on vegetation on a European scale. Therefore, critical levels of sulfur dioxide, nitrogen oxides, and ammonia have not been emphasized in negotiations on air pollution emissions reductions in Europe, and ground-level ozone is currently considered the most important toxic gaseous pollutant.

Three Generations of Critical Levels for Ozone Effects

Abatement strategies must measure the effects of various airborne pollutants; they do so by establishing critical levels and in the case of ground-level ozone by using an exposure index. Three generations of critical levels can be discerned in the case of ground-level ozone: a first generation of average ozone exposure values, a second of ozone concentration integrated over time, and a third of ozone uptake by plants. These indices are the result of an evolving scientific understanding of how ozone affects vegetation, which also has significant political implications for ozone regulation.

The first-generation approach was used in formulating CLRTAP's first set of critical levels for gaseous pollutants, developed in 1988. The results of several experiments were pooled, and the minimum pollutant concentrations required to obtain effects were identified on different temporal scales (e.g., daily or monthly). A short exposure requires a higher concentration to have a significant effect, whereas long-term exposure has chronic effects at lower average concentrations. Critical levels were based on the simple toxicological concepts of time of exposure and average concentration values. The introduction of critical levels for ozone served to justify reductions; CLRTAP stated that ground-level ozone exposures were too high in Europe and that to avoid damage to plants measures had to be taken so that critical levels would not continue to be exceeded. However, because CLRTAP provided no detailed information about the effects of exposure, it could not serve as an instrument for effect-based abatement strategies.

In the early 1990s, researchers suggested a new kind of critical level for ozone based on accumulated exposure, representing a move to the second generation of critical levels for ozone. A critical-level workshop devoted to discussing ozone effects on plants and elaborating on the concept of accumulated exposure was held in 1993. An ozone exposure index was established at the workshop: the capture of accumulated exposure over a threshold concentration of 40 parts per billion (ppb) of ozone during daylight hours for a specified period (e.g., three months for crops and six months for forest trees), called "Accumulated Exposure over Threshold 40 ppb" (AOT40). It was demonstrated, based on experiments in several countries that yield effects on wheat conformed to a common relationship with AOT40.

In 1996, after further scientific discussion at a workshop in Kuopio, Finland, AOT40-based critical levels were included in the CLRTAP *Mapping Manual*. The second generation of critical levels had become institutionalized, guiding the political regulation of ozone.

The AOT40 exposure index was also used in the deliberations preceding the negotiation of the 1999 Gothenburg Protocol. From a scientific point of view, the AOT40 exposure index represented a substantial improvement over earlier concepts. The use of a concentration threshold was reasonable considering the evidence that plants have a certain ozone detoxification capacity. Furthermore, scientists saw the specification of an integration time for ozone, which is realistic in relation to a plant's life cycle, as another improvement.

However, most scientists believe that a physiologically relevant representation of ozone damage to plants should quantify the ozone uptake (Grünhage and Jäger 1994; Pleijel, Danielsson, Pihl Karlsson, et al. 2000); the AOT40 concept largely ignores this factor and is sensitive only to ozone concentration in the air near plants.[3] It was clearly indicated in the 1996 version of the CLRTAP *Mapping Manual* that AOT40 should be used only for general risk assessment and that a more sophisticated analysis would be required to estimate actual effects.

The policymakers involved in CLRTAP asked for successively more sophisticated analyses of the relationships between ozone exposure and effects because further reducing the emissions of ozone precursors implied costlier abatement strategies. If science could estimate the economic cost of ozone exposure, the estimates could be used in arguing for further reduction of nitrogen oxide and VOC emissions. However, because the AOT40 concept was not designed to assess actual yield loss, for political reasons there was a need to develop a more complex scientific understanding that enabled the calculation of effects.

In response to this need, development began on a third generation of critical levels for ozone based on plant ozone uptake, making use of ozone flux-based assessment methods. The last three workshops on critical levels (1999, 2002, 2005) were devoted largely to flux-based ozone risk assessment. Focusing on ozone uptake made it easier to discuss the cost-effectiveness of abatement measures because this topic included the effects of ozone exposures.

Early reports clearly indicated that irrigated plants suffer greater ozone damage during ozone episodes than nonirrigated ones. This difference was explained by the fact that moist soil promotes stomatal opening and thus ozone uptake. Current discussion of climate change has also enhanced the development of flux-based methods. The new methods explicitly consider the effects of temperature, humidity, and other environmental factors on plants' ozone uptake. Flux-based methods for ozone risk assessment are now incorporated into the CLRTAP *Mapping Manual* (UNECE 2004). Like AOT40, the flux-based methods include a threshold to represent plants' detoxification capacity.

The preceding description of the evolution of the critical-level concept may be seen as a rational success story in which scientific progress rendered older understandings obsolete by establishing new, more complex ones. The main task seems to be finding a proper balance between scientific complexity and political prescription because too much complexity makes it difficult for science to serve as a knowledge base for political action. At the same time, though, policymakers require more detailed knowledge to be able to develop cost-effective abatement strategies. As demonstrated in the next section, however, the development of scientific understanding in this case has not been smooth, and the replacement of the established critical-level index has not occurred without contestation and considerations of political expediency.

Scientific Developments and Political Considerations

The CLRTAP has doubtless succeeded in mobilizing researchers, creating networks, and fostering science–policy interactions with the aim of reducing emissions of certain airborne compounds. Nevertheless, scientists both inside and outside the research networks organized by CLRTAP have asked whether the chosen strategies are scientifically defensible. Some have even claimed that scientists have adapted too much to political expediency and that they have gone too far in reducing scientific complexity when defining what should be considered clean or unclear air in order to facilitate political action (Bull 2000, 61; cf. Lidskog and Sundqvist 2004). As described earlier, the case of ground-level ozone demonstrates the opposite: growing scientific complexity made the issue harder to understand for politicians and negotiators. This does not mean that the science was developed in isolation from political considerations, but that even in this case scientific development and political expediency were deeply involved in defining what should be considered unclean air and how to abate air pollution. In this section, we first describe the scientific considerations used to choose an exposure index for measuring ground-level ozone uptake. Second, we briefly discuss the institutional reasons why the scientific community has not agreed on how to develop a more advanced exposure index. Third, we turn to the broader issue of what should be protected: What effects should be considered when estimating the effects of ground-level ozone?

Political Considerations in Scientific Assessments

Within CLRTAP, experts shared the understanding that an index based on uptake was better than one based on concentration. However, how uptake should be scientifically represented was heatedly discussed. At the 1993 Bern workshop, there was discussion of whether AOT30 or AOT40 was more suitable for measuring ozone exposure. In this discussion, the expert community was geographically divided, Sweden and Finland promoting AOT30, but Switzerland, Spain, and Greece arguing that AOT40 would be a better critical-level index. This split occurred because adoption of AOT40 would exclude substantial parts of northern Europe's ozone exposure, which was considered to have adverse effects, whereas southern Europe felt that AOT30 would result in overly large exceedances of the critical level, the cut-off concentration used (30 ppb) being too close to the natural background of ground-level ozone. This conflict was later solved by introducing the ozone flux concept.

AOT40 was adopted three years later at the 1996 workshop in Kuopio, and later that year it was included in the *Mapping Manual*. However, the controversy continued, but now concerning how to establish the critical level for ground-level ozone effects on crops: Should it be 3,000 or 5,000 ppb hours?[4] Finally, in part based on statistical considerations, 3,000 ppb hours was included in the *Mapping Manual* as the critical level that should not be exceeded to avoid risk of negative ozone effects on crops.

At the 1999 workshop in Gerzensee, Switzerland, the flux (i.e., uptake) concept was introduced as a new way to measure the environmental effects of ground-level ozone. However, doubts were raised as to whether the benefits of using a flux index would outweigh the advantages of using AOT40. The flux approach represents a more complex calculation for modelers and other actors, but it has a stronger foundation in toxicology. Furthermore, there were also warnings that available data were inadequate to support the development of a flux concept. The scientific community could not reach any conclusion as to whether to use a flux-based critical level, and the ensuing two workshops were devoted to discussing the possibilities and problems associated with a flux-based concept.

At the 2002 workshop in Hindås, Sweden, delegates from Germany and other countries proposed that the maximum permissible ozone concentration (MPOC) should be used as the critical level and included in the *Mapping Manual.* Other delegates opposed this proposal because they considered it a step backward—in certain respects a return to the type of critical levels defined in 1998. The 2004 revision of the *Mapping Manual* included flux-based critical levels for crops, and the MPOC was marginalized and given only a minor position.

The 2005 workshop in Obergurgl, Austria, discussed how the flux concept could be refined and applied in practice and finally reached broad agreement on the relevance and use of a flux-based concept. Attention was at the same time devoted to how to apply the concept, not least concerning what time perspective should be used and how to represent the differences in climatic conditions and plant types in various parts of Europe. The use of generalized methods for measuring critical levels of ozone implied that local variations would not be taken into consideration. It was claimed that for forests and to some extent crops the use of more locally sensitive flux concept variants was needed.

Despite CLRTAP's view of itself as fostering scientific work and excluding considerations of political expediency concerning specific nation-states, the experts occasionally seemed to include national and regional considerations when evaluating and advocating specific ozone uptake indexes. In addition, competition between research groups also served as a driving force in discussions of how to describe ozone exposure.

Geographical Distribution and Institutional Inertia

A central implication of replacing an index based on concentration (AOT40) with one based on ozone uptake (flux) is that it revises the geographical distribution of risk of ozone damage to vegetation. In Europe, the highest ozone exposures are found in certain parts of the Mediterranean region and south-central Europe, regardless of whether exposure is measured in terms of concentration or flux (Simpson, Ashmore, Emberson, et al. 2007). The gradient of estimated risk over Europe, from the Mediterranean region to southern Scandinavia, is much larger based on the AOT40 approach than on the flux-based approach. The scientific explanation of this difference is that the influence of dry soil and air in the Mediterranean restricts plant gas exchange, whereas the relatively favorable conditions for ozone uptake in northern and western Europe mean that the risk of effects on vegetation have a different geographical distribution from that suggested by AOT40.

The relative importance of emission reductions of ozone precursors thus varies geographically depending on which approach is taken. As mentioned, a flux-based approach seems most relevant from a narrowly scientific viewpoint, but it has the disadvantage of being more complex and needing more input information (e.g., meteorological variables), and its application depends more on modeling. There are also various opinions on exactly how to express and apply the flux-based approach on different geographical scales. Furthermore, the current way of monitoring air pollution via the regional network of monitoring stations within EMEP, measuring the concentration of ground-level ozone, can easily be used to estimate the AOT40 concentration-based exposure index. The flux approach is based on a biological dose–response model that needs data on both ozone concentrations and meteorological conditions and is appropriate for estimating ozone uptake. An important challenge facing EMEP modelers is how to represent the biologically relevant ozone uptake of plants, an effort that is in progress (Simpson, Ashmore, Emberson, et al. 2007).

Not all researchers welcome a flux-based concept of critical levels. First, the AOT40 concept has already been widely adopted, and some scientists and policymakers are reluctant to revise a measure that is well established and still has certain merits. Second, the new concept requires more input data (e.g., temperature, solar radiation, and air and soil humidity) than does the AOT40 concept. Mapping using the new concept will thus be more time-consuming and complex and will call for large-scale integrated assessment models. Third, there are competing approaches to assessing ozone flux for the estimation of plant-level ozone concentration and, based on that, the risk of ozone damage (e.g., Grünhage, Krause, Bender, et al. 2001). Fourth, and most important, the uptake/flux concept implies revising the geographical distribution of risk and thereby the distribution of incentives for political action among European countries. The choice of ozone exposure index may well alter the estimated cost-effectiveness of various abatement strategies. The new flux-based concept creates a new map of the distribution of effects and a need to revise current abatement strategies.

Hence, revised scientific understanding does not easily mesh with existing political achievements, and it can be a long process to redraw a map that the political community has already embraced, such as the AOT40-based risk map.

What to Protect—and Why?

The priorities of air pollution–abatement efforts in Europe have changed considerably, shifting from acidifying compounds, such as sulfur dioxide, to particulate matter and ground-level ozone. The convention originally focused strongly on the need to reduce sulfur emissions. However, because of economic restructuring (e.g., Britain's electricity and coal mining industries), technical development (the introduction of low-sulfur fossil fuels and the development of desulfurization equipment for coal-burning plants), and policy development, sulfur emissions have gradually decreased (Boehmer-Christiansen 2000, 280). The problem of elevated sulfur dioxide concentrations seems to be fairly close to a solution, apart from the considerable and increasing sulfur emissions from shipping. At the same time, evidence of the importance of ozone among the remaining pollutants has accumulated.

There has also been a shift concerning what ecosystems should be considered when evaluating the negative effects of ground-level ozone. In the 1990s, policymakers and integrated assessment modelers paid more attention to ozone effects on crops than to effects on forest trees and seminatural vegetation, mainly because agro-ecosystems are in many ways simpler and therefore easier to study (they are dominated by one species, which is typically harvested once every year) and because scientific understanding of ozone effects was more developed for this ecosystem than for other, more complex vegetation types. However, the huge subsidies that support European agriculture today have led to less political interest in crop loss, and the potentially considerable effects of ozone on forest production have achieved greater prominence, a shift also supported by the accumulation of new scientific evidence (Pleijel and Karlsson 2007).[5]

Furthermore, the effects of pollutants on biodiversity have led to increased attention being paid to how ozone affects seminatural vegetation. The available scientific data with which to judge the extent of this problem are still limited, although substantial information on this problem has accumulated over the past 10 years. Besides these reasons, the development of the United Nations Framework Convention on Climate Change (Climate Convention) also implies a considerable shift in focus and emphasis. Despite its being an important greenhouse gas, ozone has largely been excluded from discussions and negotiations under the auspices of the Climate Convention, and the Kyoto Protocol attributes no global warming potential value to tropospheric ozone. However, ozone is currently being reintroduced into discussion of greenhouse gases (e.g., Dentener, Stevenson, Ellingsen, et al. 2006; IPCC 2007), perhaps because many of the important difficulties in quantifying the contribution of ozone to greenhouse warming (related to the fact that ozone concentration varies much more in time and space than do the concentrations of most other greenhouse gases) have been overcome by progress in atmospheric modeling.

Parallel to the shift from crops to forests, there has also been a shift from ecosystems to human health (Pleijel and Grennfelt 2007; Sliggers 2004). Ground-level ozone is seen now as an urgent threat, not only to the environment but to human life as well. According to current research, ozone exposure can result in inflammation and irritation of the respiratory tract, increased lung susceptibility to toxins and microorganisms, and premature death (EEA 2007, 7).[6] The background material for the Gothenburg Protocol (1999) states that successful implementation of the protocol means cutting in half the number of days with excessive ozone levels. The estimated consequences of this reduction would be approximately 2.3 million fewer life-years lost due to the chronic effects of ozone exposure in 2010 than in 1990, 47,500 fewer premature deaths resulting from ozone and particulate matter in the air, and 44 percent lower exposure of vegetation to excessive ozone levels compared with 1990 levels.

Current policies seem to lead to much smaller reductions in health effects than assumed in the Gothenburg Protocol (WHO 2009, 86). In addition, any improvements achieved may be counteracted in part by an increase in hemispherical background ozone concentrations over this century (Prather, Gauss, Berntsen, et al. 2003) caused by rising global methane concentrations and global emissions of other tropospheric ozone precursors. This increase does not mean, however, that the political community will prioritize ground-level ozone.

Yet the EU's growing importance in European cooperation on air pollution regulation implies increasing attention paid to health effects. In particular, urban air quality has become a serious problem and has received increasing EU attention. Twenty to thirty percent of the European urban population currently lives in cities where EU air quality limit values for ozone and nitrogen dioxide have been exceeded at urban background monitoring stations. Large-scale epidemiological studies indicate that air pollution contributes significantly to worsening human health and increasing mortality in cities all over Europe (Gryparis, Forsberg, Katsouyanni, et al. 2004). Exposure to ground-level ozone causes some 20,000 premature deaths and 14,000 respiratory hospital admissions annually in Europe (WHO 2009, 62).

However, the current emphasis on health arose not only because of the seriousness of the problem of urban air quality, but also because of a belief that people in general care about the health aspects of air pollution. Focusing on human health is thus also part of a strategy aimed to create public support for further abatement strategies. The EU Air Quality Framework Directive explicitly focuses on the effects of air quality on human health (EC 2008; Nikolaou 2003). Reviewers have found that this focus has successfully increased public concern in a way that has led to stronger demands for further pollution-control measures (Sundqvist and Letell 2005; Lidskog and Sundqvist 2007). Public health is accordingly considered the main driving force of future abatement work in Europe (Grennfelt, Moldan, Alveteg, et al. 2001, 10; Sliggers 2004).

Besides adverse health effects, there is at least one more important reason why human health has received growing attention: cost-effectiveness. The development of science-based abatement strategies allows for the formulation of cost-effective political policies. For the very first time in CLRTAP, monetary cost–benefit analysis played a role in negotiating an international environmental agreement, the Gothenburg Protocol (Sliggers 2004). The cost–benefit analysis demonstrated that almost all countries benefited from implementing the abatement strategies required by the protocol. In calculations regarding the economics of meeting the emission ceiling, the economic benefits are estimated to outweigh the costs: the expected costs are approximately 70 billion euros per year for Europe as whole, whereas the benefits of meeting the emission ceiling are estimated to be approximately 200 billion euros a year. Most of these benefits result from the significant reduction in impairment of human health caused by ozone and particulate matter. Including health costs in the cost–benefit analysis has created economic incentives for political action (at least from a long-term perspective) in almost all European countries (Sliggers 2004).

Political considerations are not the only nonscientific factors influencing scientific work and discussions in the CLRTAP setting. External factors—such as nation-states' evaluation of the costs of economic damage, the prioritization of other environmental goals, and perceptions of what confers public credibility or even results in public mobilization—

have strongly influenced the scientific work prioritized by the experts involved in the convention work.

Conclusion: Making Ground-Level Ozone Governable

In this chapter, we have discussed the establishment of a new object for environmental regulation: ground-level ozone. As emphasized here, this problem is not defined in a sociopolitical vacuum. Instead, there is co-production of the governable space of air pollution, a space in which scientific indicators and political considerations influence each other, resulting in policy-relevant science and science-based policy.

In the development of a policy for abating ground-level ozone, scientific representations—ways to measure un/clean air in terms of critical levels of ground-level ozone—were elaborated that also served to mobilize arguments for political action. The ozone issue, by its attachment to the broader issue of long-range transboundary air pollution, had access to the institutional mechanisms already developed within CLRTAP. The workshops on this issue considered how to represent un/clean air in a way that would make it intelligible to stakeholders and policymakers. Defining exposure indices for ozone uptake, producing empirical data at monitoring stations, and developing technologies for mapping its detrimental effects made ground-level ozone visible, measurable, understandable, and governable.

Emphasizing the threat that ground-level ozone posed to society created political incentives for action. As described in this chapter, there has been a shift in what threats are considered most serious—from threats to ecosystems (in the form of crops, forests, and seminatural vegetation) to threats to human health, as currently emphasized. The economic aspects of these threats—for example, loss in crop yields and forestry as well as ill health—are also emphasized today.

An environmental problem is not presented solely via instruments such as monitoring technologies and mapping activities; rather, these instruments are also important in shaping the problem. Ground-level ozone—like all other environmental problems—calls for social practices that make it visible, institutional mechanisms that disseminate knowledge about it, and strategies that mobilize support for the need to regulate it. To begin with, there was no well-defined environmental problem, ready to be acted on. Instead, a decisive part of the regulatory process was first to define the object at stake in a way that made it governable. In this process, various actors interacted, negotiated, and struggled to determine the most appropriate way to measure the problem and the most efficient strategies to abate it.

CLRTAP efforts have succeeded in making science matter by separating scientific work from political expediency. Political actors designed a setting for science–policy interaction in which science was both politically trusted by the nation-state (whose role was to select scientific representatives) and publically trusted because it had no explicit national mandate or outspoken national interests. The result was policy-relevant science that could simultaneously be presented as nonpolitical, at least in the sense of being separated from specific national interests. The convention's secretariat has been seen as an apolitical body, and with its ability to wield scientific and technical arguments it has facilitated negotiations and the possibility of achieving workable solutions. By organizing expert groups and international scientific collaborations, the secretariat has reviewed available scientific knowledge and supported scientific assessments.

The secretariat facilitates effective communication between experts and decision makers. It formally separates scientific activities from political negotiations, but it has also substantially succeeded in integrating them. Cost-effective approaches, critical loads and levels, EMEP, and integrated-assessments—all are based on scientific achievements, and all have been adopted in the negotiations owing to interaction and communication between scientists and policymakers. CLRTAP task forces have been open to national experts as well as to policymakers and nongovernmental organization representatives. The task forces have also organized training workshops for policymakers to brief them on the state of the art as well as on the costs and effectiveness of possible policy strategies (Maas, Amann, ApSimon, et al. 2004, 94). The result can be seen in protocols that not only are based on scientific knowledge but are constantly dependent on scientific inputs to find cost-effective ways to abate long-range air pollution. Thus, a type of regulation has been constructed that science and policy have been pivotal in creating and in which science and policy remain important for further development and implementation.

Not only is policymaking open to scientific inputs and influence, but work in the expert groups also involves political considerations. As demonstrated here, the experts did take different standpoints that in some, but not all, cases were explainable with reference to their different national allegiances. In addition, communication with policymakers and nongovernmental organization representatives means that the groups have also been somewhat open to political concerns and preferences.

To sum up, how the environment is scientifically measured—providing this measurement is seen as scientifically valid and that the political community deems it trustworthy—has strong implications for how the environment is regulated. At the same time, measurement develops in a social context in which scientists themselves make political calculations about how to design policy-relevant knowledge, and, in response, powers other than purely scientific ones influence and shape how environmental problems are represented.

Notes

1. The problem of ground-level ozone (i.e., 0–10 kilometers elevation) should not be confused with that of stratospheric ozone, a different environmental problem regulated by the Montreal Protocol (1987), which concerns the phasing out of chlorofluorocarbons and other ozone-depleting substances.

2. Whereas CLRTAP organizes its own workshops and has established its own networks of scientists, the EU's air pollution policy is characterized by collaboration with scientific bodies outside its institutional setting, such as the World Health Organization. For discussion of the difference between the EU and CLRTAP uses of science, see Letell 2006; Lidskog and Sundqvist 2007; and Selin and VanDeveer, chapter 3 in this volume.

3. A plant's gas exchange takes place mainly through the stomata, microscopic pores on the leaves that plants can open and close. The stomata's role is to allow carbon dioxide uptake for photosynthesis and to prevent excessive water vapor loss. Thus, the stomata are closed at night when there is no photosynthesis in the absence of light. The AOT40 index took account of this process by excluding nighttime exposure. However, there are also other aspects to consider when discussing ozone uptake. Stomatal opening is restricted when the air is dry because the driving force of water loss from leaves is high in such situations. Stomata are also closed when the soil is dry in order to prevent unreplenishable water loss from the plant. Hence, conditions characterized by long daylight hours and humid conditions in soil and air are favorable for stomatal opening, whereas darkness and water deficit restrict stomatal opening. The AOT40 index included only one quantitative element when considering ozone uptake by the plant—namely, night exposure—ignoring other aspects of stomatal opening.

4. The term *ppb hours* refers to the number of hours the threshold is exceeded (in this case 40-ppb ozone) multiplied by the amount of the exceedance in ppb; for example, 3 hours at 50 ppb is equivalent to 30 ppb hours (10 ppb exceedance × 3 hours).

5. This situation can obviously change rapidly; on a global level, increased demands are being placed on crop production because of the growing middle class and changed eating habits in China and India. In addition, larger areas may be used for ethanol production, which means that ozone effects on crops may be reintroduced as an important aspect in formulating abatement strategies.

Finally, climate change may imply a loss of areas suitable for agriculture, which may result in renewed interest in European agricultural production.

6. Short-term studies demonstrate that ozone has adverse health effects in terms of impaired pulmonary function, lung inflammation, lung permeability, respiratory symptoms, increased medication usage, morbidity, and mortality; its long-term effects are less clear (EEA 2007, 29).

References

Asdal, Kristin. 2007. Enacting Things through Numbers: Taking Nature into Account/ing. *Geoforum* 39 (1): 123–132.

Ashmore, Mike, Nigel Bell, and C. L. Reily. 1978. A Survey of Ozone Levels in the British Isles Using Indicator Plants. *Nature* 276: 813–815.

Ashmore, Mike, Nigel Bell, and Jack Rutter. 1985. The Role of Ozone in Forest Damage in West Germany. *Ambio* 14 (2): 81–87.

Ashmore, Mike, and Robert B. Wilson. 1993. *Critical Levels of Air Pollutants for Europe.* Background papers prepared for the United Nations Economic Commission for Europe Workshop on Critical Levels, Egham, UK, 23–26 March 1992. London: UK Department of the Environment, Air Quality Division.

Boehmer-Christiansen, Sonja. 2000. The British Case: Overcompliance by Luck or Policy? In *International Environmental Agreements and Domestic Politics: The Case of Acid Rain*, ed. Arild Underdal and Kenneth Hanf, 279–312. Aldershot, UK: Ashgate.

Bull, Keith. 2000. What Are the Needs for the Revisions of the Protocols and Strategies on Transboundary Air Pollution? In *Workshop on Future Needs for Regional Air Pollution Strategies: Saltsjöbaden 10–12 April 2000,* ed. Peringe Grennfelt, Lars Lindau, and Catarina Sternhufvud, 59– 64. TemaNord 2000, no. 57. Copenhagen: Nordic Council of Ministers.

Dentener Frank, D. Stevenson, K. Ellingsen, T. van Noije, M. Schultz, M. Amann, C. Atherton, et al. 2006. The Global Atmospheric Environment for the Next Generation. *Environmental Science & Technology* 40 (11): 3586–3594.

Dimitrov, Radoslav S. 2006. *Science and International Environmental Policy: Regimes and Nonregimes in Global Governance.* Lanham, MD: Rowman and Littlefield.

Dovland, Harald, Richard Balaman, and Jan Thompson. 2004. Introduction. In *Clearing the Air: 25 Years of the Convention on Long-Range Transboundary Air Pollution*, ed. Johan Sliggers and Willem Kakebeeke, 1–30. New York: United Nations.

EC (Commission of the European Communities). 2008. *Directive 2008/50/EC of the European Parliament and of the Council of 21 May 2008 on Ambient Air Quality and Cleaner Air for Europe.* 11.6.2008 EN Official Journal of the European Union L 152/1. Brussels: Commission of the European Communities.

EEA (European Environment Agency). 2007. *Air Pollution in Europe 1990–2004*. Copenhagen: EEA.

Fischer, Frank. 2003. *Reframing Public Policy: Discursive Politics and Deliberative Practices*. Oxford, UK: Oxford University Press.

Fuhrer, Jurg, and Beat Achermann. 1994. *Critical Levels for Ozone*. A United Nations Economic Commission for Europe Workshop Report. Schriftenreihe der FAC Liebefeld no. 16. Brussels: UNECE.

Fuhrer, Jurg, and Beat Achermann. 1999. *Critical Levels for Ozone: Level II*. Swiss Agency for the Environment, Forests, and Landscape (SAEFL), Environmental Documentation no. 115. Bern, Switzerland: SAEFL.

Grennfelt, Peringe, ed. 1984. *The Evaluation and Assessment of the Effects of Photochemical Oxidants on Human Health, Agricultural Crops, Forestry, Materials, and Visibility: Proceedings of an International Workshop, February 29–March 2, 1984*. Gothenburg: Swedish Environmental Research Institute.

Grennfelt, Peringe, Filip Moldan, Mattias Alveteg, Per Warfvinge, and Harald Sverdrup. 2001. Critical Loads: Is There a Need for a New Concept? *Water Air and Soil Pollution Focus* 1 (1–2): 21–27.

Grünhage, Ludger, and Hans-Jürgen Jäger. 1994. Influence of the Atmospheric Conductivity on the Ozone Exposure of Plants under Ambient Conditions: Considerations for Establishing Ozone Standards to Protect Vegetation. *Environmental Pollution* 85 (2): 125–129.

Grünhage, Ludger, Gerhardt H. Krause, Barbara Köllner, Jürgen Bender, Hans-Joachim Weigel, Hans-Jürgen Jäger, and Robert Guderian. 2001. A New Flux-Orientated Concept to Derive Critical Levels for Ozone to Protect Vegetation. *Environmental Pollution* 111 (3): 355–362.

Gryparis, Alexandros, Bertil Forsberg, Klea Katsouyanni, Antonis Analitis, Giota Touloumi, Joel Schwarz, Evangelia Samoli, et al. 2004. Acute Effects of Ozone on Mortality from the "Air Pollution and Health: A European Approach" Project. *American Journal of Respiratory and Critical Care Medicine* 170: 1080–1087.

Hajer, Maarten, and Hendrik Wagenaar, eds. 2003. *Deliberative Policy Analysis: Understanding Governance in the Network Society*. Cambridge, UK: Cambridge University Press.

Heggestad, Howard E. 1991. Origin of Bel-W3 and Bel-B Tobacco Varieties and Their Use as Indicators of Ozone. *Environmental Pollution* 74 (4): 264–291.

Hettelingh, Jean-Paul, Keith Bull, Radovan Chrast, Heinz-Detlef Gregor, Peringe Grennfelt, and Wojciech Mill. 2004. Air Pollution Effects Drive Abatement Strategies. In *Clearing the Air: 25 Years of the Convention on Long-Range Transboundary Air Pollution*, ed. Johan Sliggers and Willem Kakebeeke, 59–84. New York: United Nations.

Hill, Leonard, and Martin Flack. 1911. The Physiological Influence of Ozone. *Proceedings of the Royal Society of London. Series B, Containing Papers of a Biological Character* 84: 404–415.

Hilgartner, Stephen. 2000. *Science on Stage: Expert Advice as Public Drama*. Stanford, CA: Stanford University Press.

IPCC (Intergovernmental Panel on Climate Change). 2007. *Intergovernmental Panel on Climate Change: Fourth Assessment Report*. Geneva: IPCC. Available at http://www.ipcc.ch.

Jasanoff, Sheila. 2004a. The Idiom of Co-Production. In *States of Knowledge: The Co-Production of Science and Social Order*, ed. Sheila Jasanoff, 1–12. London: Routledge.

Jasanoff, Sheila. 2004b. Ordering Knowledge, Ordering Society. In *States of Knowledge: The Co-Production of Science and Social Order*, ed. Sheila Jasanoff, 13–45. London: Routledge.

Jasanoff, Sheila, and Brian Wynne. 1998. Science and Decision-Making. In *Human Choice and Climate Change*, vol. 1: *The Societal Framework*, ed. Steve Rayner and Elisabeth L. Malone, 1–87.Columbus, OH: Battelle Press.

Kärenlampi, Lauri, and Lena Skärby. 1996. *Critical Levels for Ozone in Europe: Testing and Finalizing the Concepts*. United Nations Economic Commission for Europe Workshop Report. Kuopio, Finland: University of Kuopio, Department of Ecology and Environmental Science.

Karlsson, Per-Erik, Gun Selldén, and Håkan Pleijel, eds. 2003. *Establishing Ozone Critical Levels II: UNECE Workshop Report. Göteborg, Sweden 19–22 November, 2002*. IVL Report B 1523. Gothenburg: Swedish Environmental Research Institute.

Knight, R. C., and J. H. Priestly. 1914. The Respiration of Plants under Various Electrical Conditions. *Annals of Botany* 28: 135–161.

Latour, Bruno. 1998. From the World of Science to the World of Research. *Science* 280 (5361): 208–209.

Letell, Martin. 2006. *Governable Air: Studies on the Science and Politics of Air Pollution in Europe*. Science and Technology Studies Research Report no. 12. Gothenburg: Gothenburg University.

Lidskog, Rolf, Linda Soneryd, and Ylva Uggla. 2009. *Transboundary Risk Governance*. London: Earthscan.

Lidskog, Rolf, and Göran Sundqvist. 2002. Science within Environmental Regimes: The Case of LRTAP. *European Journal of International Relations* 8 (1): 77–101.

Lidskog, Rolf, and Göran Sundqvist. 2004. From Consensus to Credibility: New Challenges for Policy-Relevant Science. *Innovation: European Journal of Social Science Research* 7 (3): 205–226.

Lidskog, Rolf, and Göran Sundqvist. 2007. Regulating European Air: The Co-Production of Science and Policy. In *Transboundary Air Pollution: Scientific Understanding and Environmental Policy in Europe*, ed. Håkan Pleijel, 177–202. Lund, Sweden: Studentlitteratur.

Maas, Rob, Markus Amann, Helen ApSimon, and Willemijn Tuinstra. 2004. Integrated Assessment Modelling: The Tool. In *Clearing the Air: 25 Years of the*

Convention on Long-range Transboundary Air Pollution, ed. Johan Sliggers and Willem Kakebeeke, 85–96. New York: United Nations.

Middleton, John T., J. B. Kendrick, and H. W. Schwalm. 1950. Injury to Herbaceous Plants by Smog or Air Pollution. *Plant Disease Reporter* 34: 245–252.

Munton, Don, Marvin Sooros, Elena Nikitina, and Marc A. Levy. 1999. Acid Rain in Europe and North America. In *The Effectiveness of International Environmental Regimes: Causal Connection and Behavioural Mechanisms*, ed. Oran R. Young, 155–248. Cambridge, MA: MIT Press.

Nikolaou, K. 2003. Air Quality in European Urban Areas and the New EC Directives. *Journal of Environmental Protection and Ecology* 4 (2): 477–482.

Nordberg, Lars, Keith Bull, Radovan Chrast, Oddmund Graham, Andrzej Jagusiewicz, Peter H. Sand, Arne Tollan, and Henning Wuester. 2004. The Role of the Secretariat: Building the Protocol Tree. In *Clearing the Air: 25 Years of the Convention on Long-Range Transboundary Air Pollution*, ed. Johan Sliggers and Willem Kakebeeke, 97–118. New York: United Nations.

O'Riordan, Timothy, Chester L. Cooper, Andrew Jordan, et al. 1998. Institutional Frameworks for Political Action. In *Human Choice and Climate Change*, vol. 1: *The Societal Framework*, ed. Steve Rayner and Elisabeth L. Malone, 345–439. Columbus, OH: Battelle Press.

Pleijel, Håkan, Helena Danielsson, Gunilla Pihl Karlsson, Johanna Gelang, Per Erik Karlsson, and Gun Selldén. 2000. An Ozone Flux-Response Relationship for Wheat. *Environmental Pollution* 109 (3): 453–462.

Pleijel, Håkan, and Peringe Grennfelt. 2007. European Air Pollution: The Historic Legacy. In *Transboundary Air Pollution: Scientific Understanding and Environmental Policy in Europe*, ed. Håkan Pleijel, 27–53. Lund, Sweden: Studentlitteratur.

Pleijel, Håkan, and Per-Erik Karlsson. 2007. Ozone—Compound with Many Faces. In *Transboundary Air Pollution: Scientific Understanding and Environmental Policy in Europe*, ed. Håkan Pleijel, 123–150. Lund, Sweden: Studentlitteratur.

Prather, Michael, Michael Gauss, Terje Berntsen, Ivar Isaksen, Jostein Sundet, Isabelle Bey, Guy Brasseur, et al. 2003. Fresh Air in the 21st Century? *Geophysical Research Letters* 30 (2): 1–4.

Schneider, Toni, and Jurgen Schneider. 2004. EMEP: Backbone of the Convention. In *Clearing the Air: 25 Years of the Convention on Long-Range Transboundary Air Pollution*, ed. Johan Sliggers and Willem Kakebeeke, 31–44. New York: United Nations.

Simpson, David, Mike R. Ashmore, Lisa Emberson, and Jari-Petteri. Tuovinen. 2007. A Comparison of Two Different Approaches for Mapping Potential Ozone Damage to Vegetation: A Model Study. *Environmental Pollution* 146 (3): 715–725.

Sliggers, Johan. 2004. Blue Skies for Ever. In *Clearing the Air: 25 Years of the Convention on Long-Range Transboundary Air Pollution*, ed. Johan Sliggers and Willem Kakebeeke, 149–167. New York: United Nations.

Sundqvist, Göran, and Martin Letell. 2005. Configuring the Urban Traveller: Local Adaptation of European Air Policy. *Acta Sociologica* 48 (3): 205–219.

Sundqvist, Göran, Martin Letell, and Rolf Lidskog. 2002. Science and Policy in Air Pollution Abatement Strategies. *Environmental Science & Policy* 5 (2): 147–156.

Underdal, Arild, and Kenneth Hanf, eds. 2000. *International Environmental Agreements and Domestic Politics: The Case of Acid Rain*. Aldershot, UK: Ashgate.

UNECE (United Nations Economic Commission for Europe). 1988. *ECE Critical Levels Workshop*. Final Draft Report. Bad Harzburg, Germany: UNECE, 14–18 March. http://www.unece.org/env/wge/1988BadHarzburg.zip.

UNECE (United Nations Economic Commission for Europe). 1999. *Protocol to Abate Acidification, Eutrophication, and Ground-Level Ozone: Heralding 20 Years of International Cooperation to Reduce Air Pollution*. Geneva: UNECE. Available at http://www.iiasa.ac.at/~rains/dutch/pollueng.pdf.

UNECE (United Nations Economic Commission for Europe). 2004. *Revised Manual on Methodologies and Criteria for Mapping Critical Levels/Loads and Geographical Areas Where They Are Exceeded*. Berlin, Ökodata GmbH. http://www.oekodata.com/icpmapping/index.html.

Wettestad, Jørgen. 2000. The ECE Convention on Long-Range Transboundary Air Pollution: From Common Cuts to Critical Loads. In *Science in International Environmental Regimes: Between Integrity and Involvement*, ed. Steinar Andresen, Tora Skodvin, Arild Underdal, and Jørgen Wettestad, 95–121. Manchester, UK: Manchester University Press.

Wettestad, Jørgen. 2002. *Clearing the Air: European Advances in Tackling Acid Rain and Atmospheric Pollution*. Aldershot, UK: Ashgate.

WHO (World Health Organization). 2009. *Health Risks of Ozone from Long-Range Transboundary Air Pollution*. Copenhagen: WHO.

Wieser, Gerhard, and Michael Tausz. 2006. *Proceedings of the Workshop Critical Levels of Ozone: Further Applying and Developing the Flux-Based Concept*. 15–19 November 2005, Obergurgl, Tyrol, Austria. Vienna, Austria: Federal Research and Training Centre for Forest, Natural Hazards and Landscape (BFW).

III

Citizens and Involvement

9

Citizen Engagement with the Politics of Air Quality: Lessons for Social Theory, Science Studies, and Environmental Sociology

Steven Yearley

Air pollution is in many ways an exemplar of environmental problems. Pollution of the air from factory emissions, household heating, and vehicles is a very clear example of an externality. It is a burden imposed on others—often the general urban population—through the acts of individuals and companies who are not charged (or not charged much) for the impacts they cause. At first sight, it makes the logic of environmental impacts and pollution control manifestly apparent.

However, air contamination is an exemplary problem also in that a single label is used to give a name to something with many dimensions and variations. Some pollutants, such as sulfur dioxide, are reasonably obvious and can be directly harmful to humans, wildlife, and buildings. They can be tasted or sensed. Others are much more indirect. Low-level ozone is generally produced through oxidization reactions involving other pollutants in the presence of sunlight. It is not directly emitted by any specific polluting company or individual. And though it often arises in cities, it may also spread over the countryside. Canonical forms of air pollution consist of gases, but other pollutants of the air are solids so small that they are airborne. Some polluting substances are released in enormous quantities—the exhaust of millions of vehicles in Los Angeles or Beijing, for example—whereas others are emitted from small point sources only. Many pollutants decay relatively quickly in the atmosphere, often because they are so reactive: nitrogen dioxide, for example, is typically scavenged within hours, even if each molecule is typically replaced by a newly emitted one. But others can stay in the atmosphere for a long time, such as the aptly named persistent organic pollutants (POPs), which may endure for years, though they do not necessarily stay in the atmosphere all that time. Air pollution—though it often bears a single name—is not a single problem or phenomenon.

Furthermore, the exact consequences of air pollution are also difficult to specify. For example, there is a widespread assumption in the West that air quality is associated with the apparent growth in the incidence of childhood asthma, though the links between presumed cause and effect have proven hard to pin down. Similar anxieties about childhood respiratory problems are now familiar in India and other rapidly developing countries also (see, for illustration, Baviskar, Sinha, and Philip 2006, 193).

Air quality problems have also become international. In part, this international quality is a consequence of trade and economic development: in short, economic globalization. Air quality in China has declined just as China has taken on the manufacturing opportunities that have drifted away from Europe and North America. Extremely rapid industrialization, relatively weak regulation, and unmodernized energy-production methods have given rise to unprecedented urban smog in Chinese cities. It appears that the improvements in air quality in London and New York have been almost directly matched by declining air standards in Beijing and Shanghai. But air pollution problems have also been internationalized through air movements themselves. The whole rationale for the United Nations' work on the transboundary movement of air pollutants (which is the prime stimulus for this chapter) was that even sovereign nations cannot govern their own air quality because they are unable to prevent other countries' pollution from entering their own air space. As successive stages in the Convention on Long-Range Transboundary Air Pollution (CLRTAP) agreements indicate, it has been found that more and more forms of airborne pollutants are able to cross national boundaries and influence ambient air quality (see VanDeveer 2005, 27ff.). The POPs are just among the latest additions to the list of internationally active forms of pollution. Low-level ozone, too, has been found to cross national boundaries with impunity, circulating with weather systems in unexpected ways and thwarting attempts at national control of air quality.

The international air movements are indicative of the complex geographies of air quality: US pollution may end up in Canada, British pollution in Norway. But there are also local geographies of distribution and impact. The national politics of air pollution until relatively recently turned largely on two principal issues; the national (or regional) government set air quality standards, and people then questioned whether the air quality targets were stringent enough and whether there was adequate monitoring to ensure that the authorities and relevant watchdogs knew if the targets were being met. Into the 1990s, arguments in Britain and much of the European Union, for example, centered on the issue of whether "safe" levels were in fact safe or not (particularly in light of the possible synergistic effects of different pollutants) and on the question of where monitoring stations were located. There were disputes over the height at which monitoring equipment should be placed because small children might be closer to vehicle exhausts than adults and adult-height detectors. People were also divided over how near to (or far from) the roadside sampling should be conducted. Some urban monitoring equipment was found to be located in pedestrianized areas where vehicle pollution would likely be underrepresented. However, this approach to assessing air quality was short lived because, aided by increased computing power and the devolution of political responsibility, air-quality governance was moved to a more local level. As part of this new regulatory framework, city authorities needed to have a better sense of the distribution of air quality across urban areas than previously. All that was needed formerly was that air quality should generally meet national targets at a small number of particular locations where air quality was monitored. Now the requirement was that councils should assume responsibility for the quality of air throughout the urban area. Without enormously increasing the burden of monitoring, the obvious way to achieve this objective was by investing in computer programs that could model the state of the air across urban areas (Yearley, Cinderby, Forrester, et al. 2003). Air quality was interpreted as a spatially variable issue, and previous notions of a national threshold appeared outdated and unrealistic. New media were developed for communicating air quality information to drivers, cyclists, and urban populations.

In the United States, similar conclusions were being arrived at from an additional source. Arguments about environmental injustices became prominent, and by the 1990s activists and academic analysts had become interested in the geographical distribution of environmental "bads." Air pollution was one such issue, and the focus switched to the air quality systematically enjoyed by groups from different ethnic and class backgrounds. The mobility of air pollution led to complications in such assessments (unlike in assessments of clearly situated waste-disposal sites, which were the "classical" environmental injustice issue), but the idea that disadvantaged communities were often left to cope with contaminated air and other potential harms close to polluting sites became widespread. Local communities even developed techniques for sampling local air quality in the absence of official measures: they used standardized "buckets" (sealed containers with a valve) that could be evacuated and then filled with local air and sent off for testing. The South African group Groundwork now espouses this approach and provides the following history:

> The community based "bucket" air sampling system was developed in California in 1994, when local residents living near oil refineries and chemical plants decided to do something to put an end to the pollution in their neighbourhood. These residents sought expert assistance to develop a system of independent monitoring of air pollution. Together they designed a simple air monitoring device, called a "bucket," to take grab air samples. Subsequently, a US NGO, Communities for a Better Environment (CBE), organized a regional effort to provide "buckets" to five industrial communities in the San Francisco Bay area.[1]

Groundwork and numerous other community-based groups have followed this approach and are drawing attention to likely inequalities in people's exposure to polluted air.

A related question over whose space and whose air the city is deemed to encompass was raised starkly in the case of Delhi in the late 1990s. With the support of many in the middle class, a pioneering lawyer managed to persuade the Supreme Court of India to issue directives that led to the closing of numerous industrial outfits in Delhi in the name of the public good for the city. But the action stirred heated controversy because the closures impacted many urban lower-class industrial workers. In their review of this case, Amita Baviskar, Subir Sinha, and Kavita Philip raise a series of questions about the framing of the issue:

> The pursuit of the "public interest" deprived a large section of Delhi's working class of their means of subsistence. Environmental benefits—clean air and water—were obtained at the cost of losing *working environments*, resources that sustained some of the most vulnerable citizens of Delhi. How did the middle class succeed in presenting health and hazard, beauty and order as environmental concerns that superseded the welfare of Delhi's working class? Why were the environmental priorities of workers—jobs, food, and shelter—overlooked? Why were workers not represented in the decision-making process that led to the factory closings? Why was action routed through the judiciary when the Indian government has an extensive administrative setup for monitoring and regulating pollution? Was the closure of factories the most effective way of improving air quality? (2006, 193, original emphasis)

The argument used by campaigners against the closures was that the middle-class groups were pursuing their own objectives under the guise of the public interest. The campaigners felt that the latter groups' position was blinkered, as indicated by the fact that the middle class was uninterested in curbing traffic—a source of air pollution that it was very happy to tolerate. The middle class seemed to want to achieve a cleaner city by targeting only forms of air pollution that did not impact on its own lifestyles and direct economic interests. The case was further complicated because it seemed that many of the legal directives were rather blunt in their consequences, leading to the closure not only of grossly polluting plants but also of installations that fell within the same legal classification as polluting ones but that were in practice comparatively innocuous. In this way, numerous relatively nonpolluting working-class jobs were lost while wealthier people's lives were left unaffected.

According to these social dimensions and interpretations of air contamination issues, I reflect here on the implications of air quality analyses for core issues in social theory, science studies, and environmental sociology. By turns, my argument focuses on insights that debates over air quality can offer into the politics of "citizen engagement"; the understanding of the nature of links between technological innovation and social change; and the way that environmental problems are often treated as though they were fundamentally about "risks."

Transboundary Air Quality and the Politics of Public Engagement

In the past 15 years, there has been a fast-growing debate—in the science and technology studies (STS) literature, in the policy literature, and even in policy practice—about the scope for and role of public engagement in scientific and technical knowledge making and decision taking (for recent overviews, see Evans and Collins 2008; Stirling 2008). Much of this debate has centered on environmental issues: nuclear energy policy, genetically modified organisms, water quality. In some cases, as with the European Union's Water Framework Directive, public participation has been mandated for public bodies. As is well known, the debate has revolved largely around two issues: (1) What role should the engaged public have? and (2) What methods should be used to capture the public's contribution?

The first of these issues is readily understood. Put another way, the question asks: Once one agrees that publics should have a role, does that role extend to evaluating technical knowledge, or does it amount to something else, such as a consideration of ethical and societal dimensions of the issue only? This dilemma was graphically illustrated in the United Kingdom in the early 2000s by a body the government established to advise on methods for the disposal of radioactive waste from the nuclear industry: the Committee on Radioactive Waste Management. This committee had an innovatively mixed membership in an attempt to embody the idea that environmental decisions should result from a broad representation of views. But this body began to divide exactly around this question of what role the public's input should play. Some committee members took the view that only scientists could advise on the fundamental nuclear issues, but others adopted the idea that a variety of members with a range of backgrounds could take a holistic view that would include technical and scientific matters. The group found that these different interpretations were irreconcilable and that the divisions could be overcome only once the membership was changed: one member resigned, and another was asked to leave. The two former members then attacked the committee for having wasted time on what they saw as pointless public consultations.[2] This case is certainly dramatic, but it emphasizes the point that, with little unanimity in policy circles about this first question regarding public engagement, there has inevitably remained doubt also about the second question regarding the methods to be used to capture citizen insights.

In a series of papers, colleagues and I have investigated these issues in relation to local air quality models, arguing that in these cases (at least) residents and other local-interest representatives have a genuine ability to participate in knowledge making (see the overview of these studies in Yearley 2006a). We propose this argument both because these actors possess substantive knowledge about ambient air quality at a finer scale than official monitoring systems commonly manage and because they have insights into the social practices that underlie some assumed inputs into official systems. For example, modeling systems typically use averaged traffic survey data as the basis for emissions modeling, but local cyclists know which routes especially dirty-engined buses travel and where buses are left idling (of course, transport managers and bus operators will likely know, too, but they may be less motivated to share their insights). Modeling systems also make assumptions about emissions from point sources (factories and so on), and they typically assume that factories emit at the maximum permitted level—official modelers may not be institutionally free to model the idea that factories routinely flout the law. Factory employees and factories' neighbors, though, have a different view regarding what factory managers do, founded on their experience-based knowledge about the likelihood that emissions are actually checked at night or during the weekend and in official holiday periods.

Briefly expressed, the conclusion that my colleagues and I advanced was that lay participants can readily operate as a kind of peer reviewer of the local air quality models. The idea is not to privilege local people's accounts over professional modelers' determinations, but to devise a forum in which local stakeholders can be offered and claim a voice.

This overall debate about public participation in technical decision making has been slowly progressing, with commentators refining their stances and coming to take more nuanced positions. However, nearly all of the debate has focused on factors internal to knowledge making: whether laypeople have expert-like knowledge and how generalizable their insights are; whether they should be seen as experts or as peer reviewers or as special advisers only. The case of the CLRTAP process allows us to observe something different: that factors quite extrinsic to the politics of citizen engagement in knowledge-making practices may impact the scope and credibility of public participation.

Let me explain what I mean by this statement in the context of recent discussions about ways to advance the well-known and generally acclaimed CLRTAP process. In particular, my interest focuses on examining whether there is potential to extend CLRTAP through the consideration of initiatives for public engagement (see Sundqvist, Letell, and Lidskog 2002). With regard to the question of whether citizen perspectives can be brought to bear on this policy process, developing the knowledges that ordinary people possess and devising ways to pool their various forms of knowledge, it appears that CLRTAP—for quite understandable historical reasons—developed in such a way as to minimize the obvious scope for participation of the sort promoted in the case of local-authority modeling in Britain or the United States. In order to emphasize the legitimacy of the treaty's international aspects and to minimize the extent to which it stumbled into the realms governed by national sovereignty, it was agreed that CLRTAP was to focus only on long-range, transboundary pollutants (see Levy 1993). Through the emission of such pollutants one might do damage to one's neighbors without at all meaning them any harm, and steps to reduce these kinds of harm would simultaneously mend fences while (potentially) improving the state of the environment (I say "potentially" here because it appears that it was only long-range, *transboundary* pollution that some signatories tried to stop, with the USSR, for example, reported early on to have moved some of its long-range pollution-emitting facilities deeper within the country so that emissions were no longer transboundary, even if they were still intense and long range). This meant that the treaty's business was confined to being about pollutants that came from a considerable distance and typically from outside a national territory.

The United Nations' publicity material on the successive generations of the treaty emphasized the ways in which people's lives can be affected by pollution that comes stealthily from far away to acidify their lakes or rot their public buildings or impair their children's health. One widely distributed leaflet marking 20 years of the CLRTAP (United Nations 1999) rather beguilingly used personalized vignettes. The stories ran roughly as follows: Sven used to fish in his local lake with his father, but when he came to teach his daughter how to use a lure, fish numbers had steeply declined. Matilde had been coming to the state park all her life. She remembered how tall the plants grew and how prodigiously the flowers bloomed when she was a young woman. These days plant growth is very restricted, the trees lack vigor, and the park is much less appealing. In both cases, attractive local features of the environment have been compromised by pollution that has come from far away—through acidification of lakes caused by low-pH precipitation, for example.

To sell its message, the CLRTAP document understandably stressed that these problems were not primarily of local origin but resulted from pollutants transported in from far away. For present purposes, however, the interesting thing is that—at the same time—the problems these stories told of were exactly the kind of environmental difficulty where, on the face of it, citizen input would be least valuable.

There are at least three factors at work here. First, though urban pollution may often be the problem (and thus in some way comparable to the British case mentioned earlier), the sufferers are not the originators of the problem, so the solution cannot be directly in their own hands. It does not stem from their city's buses or primarily from local factories. Local people cannot easily be empowered to resolve their own air pollution problems. Second, the CLRTAP promotional literature stressed the extent to which the environmental pollutants operate almost by stealth, imperceptibly causing problems that become apparent late on, after years of invisible damage. Laypeople accordingly cannot expect to be good detectors of these problems. Their lay understandings in the matter of barely perceptible gases that build up and cause problems at very slow rates are likely much less than in the matter of conspicuous urban pollution from bus engines or industrial plant. A pollutant that is virtually imperceptible and that accumulates over periods of years seems almost designed to evade informal public practices of monitoring (Rodrigues 2006). Even if such environmental changes are detected and monitored, it is inevitably hard to attribute the changes conclusively to imperceptible causes. Finally, because the sources of pollution are far away and are located in other cultures with different legislative contexts, patterns of industry, agriculture and economic life, and even customs, the understandings of people in the polluted area may not relate at all well to the exigencies of everyday life in the places that are the sources of the problem. People in one locality may not even have much insight into the everyday concerns and motivations of the people who are causing the long-range pollution from which they are suffering. Local people may, with some justification, suppose they know more about factory management practices in their area than do local-government factory inspectors or university air pollution modelers. But it is hard to sustain the claim that these same people also know about the practices of a foreign city hundreds or thousands of kilometers away. For all these reasons, then, CLRTAP looks like an experts' charter.

The key point here is that CLRTAP was not designed to limit or minimize public participation, but the political dynamics that shaped the treaty happened also to have an impact on its permeability to active citizen engagement. In such a case, even if officials and the parties are interested in participatory initiatives, it is unclear that the rationale for those initiatives could be the same as in the earlier examples of local urban air management. Of course, there may be a role for public engagement in discussions about the value for money obtained, or how worthwhile these forms of international pollution abatement are compared with local initiatives. And public consultations might lead to more interest in the distributional consequences of existing air quality arrangements under CLRTAP and other agreements. Nonetheless, it seems that here is a case of an air quality policy in which public engagement is not easily converted into a form of extended peer review. This suggests that although public engagement in air quality management has been shown to be practical and potentially beneficial, the exact nature of public engagement in relation to the CLRTAP process will require further investigation and quite possibly the development of novel methods. At a theoretical level, the conclusion is that though most of the debate has been about what sorts of insight or expertise local and laypeople can have, in this case the scope for public participation is shaped extrinsically—by the nature of the agreements that international political forces can foster—rather than by considerations internal to lay knowledge makers' epistemological and other qualifications.

The Ecological Modernization of Air Quality

Let us turn now from STS perspectives to those from environmental sociology. The theory of ecological modernization offers itself as a comprehensive theoretical perspective for environmental sociology. Of most significance to present concerns, ecological modernization stands out from the constructionism commonly associated with STS approaches because it argues and assumes that there can be a progressive rationalization of environmental policy in the direction of modernization. As Arthur Mol helpfully expresses this idea, "Ecological modernization indicates the possibility of overcoming the environmental crisis while making use of the institutions of modernity and without leaving the path of modernization. The project aims at 'modernizing modernity' by repairing a structural design fault of modernity: the institutionalized destruction of nature" (1995, 37). Such modernization produces economic gains and ecological rewards simultaneously. Ecological modernization appears to offer a rationalized track along which environmental policy travels, a vision quite at odds with the contingency of constructionist views.

Though a modernization perspective can be adopted as an interpretation of how environmental policy should be, for authors such as Mol (1997, 140) and Gert Spaargaren (1997) it stands as an account of how social practices in fact are. Along with Mol and Spaargaren, Albert Weale (1992) and Joseph Huber (2004) assert that, in specific ways, commercial development and rising environmental standards and performance go hand in hand. In particular, such authors argue that in specific cases the theory of ecological modernization actually accounts for the way in which reforms have been put in place: "[The] theory has proved valuable in elucidating *how* the 'environment' moves into the process of chemical production and consumption and transforms it" (Mol 1995, 391, original emphasis). Cases such as those of low-odor (or otherwise less dependent on organic solvents) paints or less and less environmentally damaging agropesticides become paradigmatic for this analysis. There is, so to speak, a ratchet effect that drives environmental reform in a single direction and makes it cumulative. Moreover, once one country establishes demonstrable environmental improvements through the pursuit of a policy and shows that the policy can be adopted without causing economically disastrous consequences, there is enormous pressure on others to follow. If limitations on types of light bulb result in reduced carbon emissions without undermining the market for electrical goods and without leaving the poor with no domestic lighting, or if one country reduces demands on landfill by obligatory recycling measures without destroying the construction and packaging sectors, others will (in some strong sense) have to follow. The "betterness" of the policy is indubitable. Of course, this is not to imply that ecological modernization guarantees that enough reform can be achieved to avert major ecological problems. The claim is a weaker one than that, but it still amounts to the strong assertion that technological development, industrial policy, and environmental improvement can pull in the same direction without undermining ordinary citizens' standard of living.

Recent work by William Freudenburg (2005) fruitfully explores this empirical claim using statistical data from US toxic release inventories to examine firms' performance across the United States, both across and within sectors. Freudenburg found that firms' footprints were surprisingly diverse even within the same sector. Using as a form of null hypothesis the idea that all firms within a sector would be driven by legislative and cost pressures to have more or less the same environmental profile, he found that the actual distribution of performance was enormously skewed. Measuring this skew in terms of the Gini coefficient, he determined that the data displayed "a higher level of disproportionality in emissions than in the incomes of the most unequal nations ever studied" (2005, 96). This result was highly unexpected. It admits of two interpretations in relation to ecological modernization; both views are interesting, but they tend to be in conflict. On the one hand, and as Freudenburg himself notes, this work indicates that there is much scope for environmental improvement through modernizationist remedies. Given how vastly discrepant plant performances are within the same industry, a great deal of environmental improvement is clearly attainable just by having the worst installations raise their game to the level of, say, the average ones. With such a skewed distribution of emissions, nearly all of the worst problems are caused by a relative handful of firms. On the other hand, the existence of such variability in environmental performance *within* sectors in one of the world's freest and most competitive economies suggests that not much pressure is exerted by the forces of modernization themselves, at least in the United States. The ratchet effect has not driven all firms to converge toward (relatively) high standards of ecological conduct.

Our question accordingly becomes: What does the evidence from the CLRTAP process suggest about ecological modernization as an explanatory theory? At first sight, the history of air quality policies seems to fit well with the modernizationist position. In the United Nations Economic

Commission for Europe region, for example, there have been successive moves to modernize different forms of air contamination; as advances have been made in the handling of one pollutant, other pollutants have been targeted. Follower countries have adopted technological innovations in the light of the leading states' performance. And there has been a great deal of agreement about what the environmental improvement targets should be, even if it seems that air quality improvements in Western countries have not been continuing as strongly as had been anticipated (see Environmental Audit Committee 2010, 5). Countries have by and large pursued similar goals with regard to similar pollutants, and cases of countries' striking off in rival regulatory directions are rare. But the point here is that this process—though it resembles the modernizationist "script"—has been driven not by a market logic or by regulatory rationalization, but by the logic of international treaties and agreements, by considerations of diplomacy and international face saving. The fact that countries comply with the treaty by moving the sources of pollution away from borders to distant inland areas—thus cutting transboundary pollution but not pollution per se—is only the most dramatic illustration of this point.

The ironic conclusion is that the appearance of monodirectional ecological modernization has been produced not so much by the forces of modernization themselves (where "wins" in profitability are matched by "wins" in ecological performance), but by successive steps to harmonize air-quality performance across an increasing range of pollutants. These steps have clearly been intended to produce environmental "wins," but the issue of economic "wins" has been largely beside the point; it has not been the driving force. Of course, scientific research is deployed to give direction to the CLRTAP process, and there is administrative pressure to tidy up any anomalies that appear, but neither of these elements is enough to demonstrate that the process is driven by time's modernizationist arrow (see Jasanoff 1996; Yearley 1996, 117–121).

Elsewhere in the literature, ecological modernization is presented less as a theory of contemporary social change than as a label for an emerging policy discourse or ideology. Though Weale appears to embrace the concept as an explanatory one, he also describes it as an ideology (1992, 75–79). He notes that it does not amount to a "coherent, well-formulated doctrine"; rather, there are many differences of interpretation (78). On this view, ecological modernization is attractive precisely because it "forms a category of discourse that is a flexible and powerful instrument for criticising the assumptions built into the first-wave of environmental protection. [And it thus] became appealing to many members of the policy elite in European countries and international organizations during the 1980s" (78).

Maarten Hajer (1995) further advances this general interpretation of ecological modernization as a flexible interpretative construct. He observes how it became a portmanteau term describing a collection of policies and outlooks. A variety of actors in various countries and supranational bodies could assent to the term without necessarily agreeing in detail about the policy proposals to which their shared commitments should allegedly lead. In his detailed tracking of the way that policies about acidification were developed in light of avowed commitments to ecological modernization, he indicates how the two elements were mutually constructed. In so doing, Hajer also demonstrates how a constructionist approach to ecological modernization may be more empirically rewarding than an approach that treats ecological modernization at face value as a theory of social development. Such a verdict coincides well with this section's main conclusion—namely, that although the CLRTAP appears to exhibit the logic of ecological modernization, the key impulses to its development have not been the factors identified by ecological modernization authors. A seemingly modernizationist policy trajectory need not have a modernizationist explanation.

Air Quality Problems and the Conceptualization of Environmental Issues as Risks

Strikingly absent in the account of air pollution with which I opened this chapter is the language of risk. In recent years and particularly in the English-speaking world after 1992, Ulrich Beck's work characterizing contemporary society as a "risk society" (Beck 1992) has been pervasive in the way social theorists and many sociologists have sought to analyze the environment. Of course, several high-profile and environmentally significant topics have been readily conceptualized in these terms, notably genetically modified crops in agriculture and foodstuffs as well as the renewed impetus lately given to nuclear power. Both of these issues do seem to exhibit the logic of risk. In these two examples, society is taking on a technology where both the likelihood of its malfunctioning and the consequences of any such malfunction are not known with precision. If we think, as conventional risk analysis does, of risk as "magnitude of the harm multiplied by the probability of the harm occurring," then we can see that in such significant cases neither the extent of the harm nor the probability of the consequences is well known. Indeed, both variables may be highly disputed (on this risk approach and its inevitable shortcomings, see Yearley 2005, 130–134). Thus, the notion that one can attach an objective, numerical "risk" figure to such novel technologies seems highly unrealistic, if not downright misleading.

Followers of Beck experience a complex relationship to the notion of risk. It is important to their analysis to show that modern technologies are indeed risky and that (most) environmental concerns are a form of risk concern. At the same time, they also harbor strong reservations about the way that specific risks are customarily calculated. Writers from within the STS tradition, by contrast, tend to focus more exclusively on the way estimates of risks are worked out, but they do so without subscribing to the view that contemporary society or culture is pervasively characterized by risk concerns. State agencies that have to deal with risks and their administration ironically find themselves in a situation not that far removed from Beck's "risk society." Though they are repeatedly made aware of shortcomings with particular risk calculations, they have not been able to break away from the language of risk—as many scholars such as Sheila Jasanoff have demonstrated (see, e.g., Jasanoff 1986). Officials need to find an apparently universalistic basis for dealing with regulatory problems, particularly in the United States, where executive agencies such as the Environmental Protection Agency can be taken to court by the companies they regulate if the basis for their judgments is not universalistic and not backed by scientific rationales. Agencies are thus repeatedly drawn to the idea that a risk-based methodology is the correct and "rational" approach to such policy matters: risk offers a common currency for dealing with myriad forms of harms and dangers and converts them all into aspects of the "same" thing. Agency staff know that such an approach has weaknesses, but they don't know what other tools they can turn to without forsaking the idea of an objective and universalistic regulatory approach. Thus, even the issue of climate change is increasingly presented within this framework, particularly in policy circles and by environmental economists. By tolerating the continued emissions of enormous amounts of carbon dioxide into the atmosphere, we are taking—so the analysis proposes—a risk with the future of the climate. On this view, we should take climate change seriously precisely because the statistical risks to which it gives rise are so large. So pervasive is this approach that many commentators now appear to treat environmental issues and risk issues as synonymous.

In this respect, the conspicuous thing about air quality issues—particularly as they have been addressed in the CLRTAP process—is the relative absence of the discourse of risks. This is not to claim that one cannot recouch air quality problems as worries about risks: for example, the connection between atmospheric pollutants and ill health is, in part at least, a statistical one. As recent European Union approaches have done, one can talk about the way that poor air quality is associated with increased risks of respiratory infection. Thus, the logical unsuitability of risk language does not account for its rarity. The empirically telling point is that the risk idiom has not been the favored one for discussing the politics of air pollution, neither in local air quality controversies nor in international fora such as those surrounding the CLRTAP.

This point indicates that the common equation of environmental worries with risk concerns—which sometimes passes as the orthodoxy in social theory—is not empirically justified (on the phenomenological [in]adequacy of this orthodoxy, see also Yearley 2006b). Indeed, as the Delhi example illustrates, disputes may frequently be expressed in terms of harms—with injuries or offenses caused to people, their employment chances, and their environments—than in terms of risks. And as the citizen "bucket" air-sampling practice suggests, the focus may be on justice and fairness, not on risks or on risk–benefit ratios. The evidence presented here is of course limited and capable only of showing that *not all* environmental anxieties are presented and handled in terms of risks. However, given that environmental activism, treaty making, and legislation around air quality issues have made up a very large component of recent environmental policy and practice, it is tempting to suggest that if one drew up the list of arenas in which risk is not the preferred idiom for talking about environmental harms, that list would not be significantly shorter than the list of cases where the risk idiom readily fits. Of course, this nonrisk list can only be a "thought experiment"; one cannot draw up a definitive list in an agreed or objective way. But the point remains: the conspicuous absence of the language of risk in this important environmental policy arena suggests that social theorists who treat all environmental worries as risk concerns are not characterizing the phenomenon in an empirically justified way.

Conclusions

This chapter has taken the opportunity offered by studies of the international politics of air quality, in particular studies of CLRTAP, to reflect on the way that environmental topics are treated in leading currents within the social sciences. In this sense, the chapter has tended to operate differently from most others in this book, which apply social science to air pollution issues. I have tried to use the politics of the air to evaluate social science. A key issue that has emerged is the importance of the international context in determining the applicability of various social scientific approaches. Thus, with regard both to ecological modernization theory and STS work on public engagement, I found that the cross-national aspect of the CLRTAP process had a decisive influence. In the latter case, the international dimension of the treaty process impacted in an unintended way on the scope for public participation. The negotiations that developed during the treaty process led to agreements focusing exclusively on aspects of air quality regulation that were relatively inaccessible to public participation. Public engagement was not resisted or purposely omitted, but the form that agreements took tended to marginalize precisely those factors that are usually used to justify public participation in scientific processes and science-based policymaking. This connection between treaty design and public engagement has not been extensively investigated in the literature, but it may turn out to be very consequential for the future development of citizen engagement in environmental knowledge making, particularly in relation to global environmental issues, including some aspects of climate change.

For the case of ecological modernization, it similarly turned out that the demands of international treaty making had a strong impact on the supposed "ratchet effect" within the modernization of environmental performance. As Stacy VanDeveer has dryly noted, "Air pollution becomes transboundary only when it crosses political borders, which are constructed largely irrespective of ecosystem boundaries and meteorological phenomena. . . . [T]he political scale of air pollution issues is not determined by natural phenomena, but by social institutions" (2005, 27). A direct corollary of this point is that air pollution policy is unlikely to be made in the kind of national context that many ecological modernizationist arguments presuppose. In this case, the power of a diplomatic ratchet—the desire not to offend neighbors needlessly and not to be seen on the international stage as an environmentally weak country—has been apparently every bit as strong as the economically rational ratchet to which ecological modernization authors typically point. It becomes clear that the international diplomatic dimension has a strong influence, this time on the "logic" of modernization. In some cases, centers of pollution are mere tens of kilometers from borders; in others, the distance may be thousands of kilometers. The supposed "ratchet" effect will operate differently in these different situations. The interplay between diplomatic and rationalistic ratchets merits further attention, as does the possibility that in international fields (such as transboundary air pollution) the diplomatic ratchet may generally be the stronger kind.

Finally, in this review I noted that the discourse of risk seems notably absent in claims made around international air quality problems. This absence implicitly throws doubt on the assumption—now almost the orthodox view within sociology and in many policy circles—that environmental worries are all effectively risk worries. It indicates that environmental concerns are not simply reducible to risk concerns. In all these ways, it turns out that the case of air quality politics offers fertile material for examining the value of dominant social scientific approaches to environmental issues and for refining our understanding of current trends in social theory, science studies, and environmental sociology.

Notes

1. From Groundwork's Web site at http://www.groundwork.org.za/Pamphlets/bucketff.asp.

2. See the report at http://www.guardian.co.uk/science/2005/apr/24/energy.greenpolitics from the UK newspaper *The Observer*, 24 April 2005.

References

Baviskar, Amita, Subir Sinha, and Kavita Philip. 2006. Rethinking Indian Environmentalism: Industrial Pollution in Delhi and Fisheries in Kerala. In *Forging Environmentalism: Justice, Livelihood, and Contested Environments*, ed. Joanne Bauer, 189–256. Armonk, NY: M. E. Sharp.

Beck, Ulrich. 1992. *Risk Society*. London: Sage.

Environmental Audit Committee of the UK House of Commons. 2010. *Air Quality: Fifth Report of Session 2009–10*. London: Her Majesty's Stationery Office.

Evans, Robert, and Harry Collins. 2008. Expertise: From Attribute to Attribution and Back Again? In *The Handbook of Science and Technology Studies*, ed. Edward J. Hackett, Olga Amsterdamska, Michael Lynch, and Judy Wajcman, 609–630. Cambridge, MA: MIT Press.

Freudenburg, William R. 2005. Privileged Access, Privileged Accounts: Toward a Socially Structured Theory of Resources and Discourses. *Social Forces* 84 (1): 89–114.

Hajer, Maarten. 1995. *The Politics of Environmental Discourse: Ecological Modernization and the Policy Process*. Oxford, UK: Oxford University Press.

Huber, Joseph. 2004. *New Technologies and Environmental Innovation*. Cheltenham, UK: Edward Elgar.

Jasanoff, Sheila. 1986. *Risk Management and Political Culture*. New York: Russell Sage Foundation.

Jasanoff, Sheila. 1996. Science and Norms in Global Environmental Regimes. In *Earthly Goods: Environmental Change and Social Justice*, ed. Fen O. Hampson and Judith Reppy, 173–197. Ithaca, NY: Cornell University Press.

Levy, Marc A. 1993. European Acid Rain: The Power of Tote-Board Diplomacy. In *Institutions for the Earth: Sources of Effective International Environmental Protection*, ed. Peter M. Haas, Robert O. Keohane, and Marc A. Levy, 75–132. Cambridge, MA: MIT Press.

Mol, Arthur. 1995. *The Refinement of Production: Ecological Modernization Theory and the Chemical Industry*. Utrecht, Netherlands: International Books.

Mol, Arthur. 1997. Ecological Modernization: Industrial Transformations and Environmental Reform. In *The International Handbook of Environmental Sociology*, ed. Michael Redclift and Graham Woodgate, 138–149. Cheltenham, UK: Edward Elgar.

Rodrigues, Maria Eugénia. 2006. From the Mundane to the "Out of Place": Exploring the Dynamics of Public Participation in Environmental Monitoring." Paper presented to Research Committee 24 of the International Sociological Association, World Congress of Sociology, Durban, South Africa, July.

Spaargaren, Gert. 1997. *The Ecological Modernization of Production and Consumption: Essays in Environmental Sociology*. Wageningen, Netherlands: University of Wageningen.

Stirling, Andy. 2008. "Opening Up" and "Closing Down": Power, Participation, and Pluralism in the Social Appraisal of Technology. *Science, Technology & Human Values* 33 (2): 262–294.

Sundqvist, Göran, Martin Letell, and Rolf Lidskog. 2002. Science and Policy in Air Pollution Abatement Strategies. *Environmental Science & Policy* 5 (2): 147–156.

United Nations. 1999. *The UN in Your Daily Life: A Breath of Fresh Air: UN/ECE Convention on Long-Range Transboundary Air Pollution*. Geneva: United Nations.

VanDeveer, Stacy D. 2005. European Politics with a Scientific Face: Framing, Asymmetrical Participation, and Capacity in LRTAP. In *Assessments of Regional and Global Environmental Risks: Designing Processes for the Effective Use of Science in Decisionmaking*, ed. Alexander E. Farrell and Jill Jäger, 25–63. Washington, DC: Resources for the Future Press.

Weale, Albert. 1992. *The New Politics of Pollution*. Manchester, UK: Manchester University Press.

Yearley, Steven. 1996. *Sociology, Environmentalism, Globalization*. London: Sage.

Yearley, Steven. 2005. *Making Sense of Science*. London: Sage.

Yearley, Steven. 2006a. Bridging the Science–Policy Divide in Urban Air-Quality Management: Evaluating Ways to Make Models More Robust through Public Engagement. *Environment and Planning C* 24 (5): 701–714.

Yearley, Steven. 2006b. How Many "Ends" of Nature: Making Sociological and Phenomenological Sense of the End of Nature. *Nature and Culture* 1 (1): 10–21.

Yearley, Steven, Steve Cinderby, John Forrester, Peter Bailey, and Paul Rosen. 2003. Participatory Modelling and the Local Governance of the Politics of UK Air Pollution: A Three-City Case Study. *Environmental Values* 12 (2): 247–262.

10

Framing Air Pollution and Health Problems: How to Include Stakeholder Perspectives?

Arthur Petersen, Leendert van Bree, and Willemijn Tuinstra

The interface between science, policy, and society has been the topic of study of a whole body of political science and social studies of science literature (see also Lidskog and Sundqvist, chaps. 1 and 12 in this volume, for a broader discussion).

Social scientists do not arrive at an unequivocal view on the science–policy–society interface, neither on how it works nor on how it should work. Nevertheless, important questions are addressed in these analyses that provide the context of our chapter.

We analyze a case study on science–policy–society interactions in the field of air quality and health using a perspective on the interface that was developed for the PBL Netherlands Environmental Assessment Agency (see the next section). Past and recent improvements in environmental quality, including air quality, are the result of weighing ecological, economic, and social interests in decision making. Despite this success, air pollution is still one of the major environmental factors affecting human health (Brunekreef and Holgate 2002), and concern over the substantial health impact remains high. In addition, large investments have to be put into further emission reductions to decrease the remaining risks, which may trigger debate on "value for money." The science–policy–society interplay in the area of air pollution becomes, therefore, even more important when governments want to develop health-effective and cost-efficient control strategies and measures that are transparent and sound and that carry public support. As we show in this chapter, the way in which complex air pollution issues are framed is a crucial factor determining the quality of communication between science, policy, and other stakeholders (see also Lidskog and Pleijel, chap. 8 in this volume). We argue that scientific advisers should use specific tools, such as those employed at the PBL Netherlands Environmental Assessment Agency, to address this issue.

Bringing scientists, policymakers, and the public together is a difficult and contentious process, as analysts have observed, for instance, in the field of air quality and health. Those who hold the ideal that these parties should be allies in efforts to promote more sustainable approaches in environment and health to bring the exposure and health risk down to healthy or acceptable levels often conclude that the interface between the players is underdeveloped or functions poorly (Samet and Lee 2001; Ginsburg and Cowling 2003). However, the different parties involved typically behave in a more antagonistic than cooperative manner. On the one hand, scientists may not view end users as legitimate clients for their research results. And on the other hand, end users may not perceive the research community as the producer of relevant information for decision-making processes. In fact, end users often desire a level of certainty that science cannot offer (Matanoski 2001). Analyses of the science–policy–society interface should pay attention to these tensions.

The case discussed in this chapter is the AIRNET Thematic Network project, a European multistakeholder project that has aimed to facilitate a network between policy, stakeholders, public, and scientific communities and to link air pollution health risk research with the various end users' needs. One of the starting points for AIRNET was that players consider important not only the content of air pollution and health issues, but also the process and mechanisms by which the interface operates. The envisioned interplay and cooperation between scientists, policymakers, and other stakeholders is based on the premise that they all play a legitimate and necessary role in the generation, dissemination, and evaluation of policy options. The use of models and techniques to assess the various impacts and the application of aggregated impact indexes to support scenario analyses and outlook types of assessments need broad support from scientists, policymakers, and stakeholders in order to be effective in policymaking. On the one hand, knowledge is needed of the various stakeholders' different information requirements in order to make the outcomes as informative and targeted as possible. On the other hand, agreement is needed on problem definition and framing, various risk characterization aspects, treatment of uncertainties, decision rules, preferred methodology and assessment models, and necessary input data in order to increase confidence and support for the outcomes and their implications for decision making. What surprised the initiators of the AIRNET project was that the framing of the air pollution and health problem in particular turned out to be an issue of contention. With the benefit of hindsight, we should add here that the stated premises of the

AIRNET project were not completely followed from the beginning (see the analysis of the AIRNET project in the case study section). We discuss the experiences of the AIRNET project in the context of the important role of framing environmental policy issues, and we present elements of practical guidance to deal with uncertainty and framing of environmental policy problems in cooperation with various stakeholders.

First, we delineate the theoretical and normative commitments of two operational tools for policy analysis introduced at the PBL Netherlands Environmental Assessment Agency. These tools embody a particular view on the science–policy–society interface. We highlight the importance of framing according to this view. We then introduce the AIRNET project and show how it involved stakeholders by organizing multistakeholder "AIR-NETwork days." We evaluate this operationalization by applying one of the mentioned tools and in the last section draw some wider conclusions, including a reflection on the tools themselves.

Two Tools to Deal with the Science–Policy–Society Interface

The PBL Netherlands Environmental Assessment Agency, an independent government agency that acts as a boundary organization between science and policymaking at Dutch, European, and global levels, has adopted a view on the science–policy–society interface that stimulates some level of democratization of expertise by offering its analysts some tools to incorporate stakeholder perspectives on uncertainties and risks (see van Asselt 2000; Petersen 2006; van der Sluijs, Petersen, Janssen, et al. 2008; Hage, Leroy, and Petersen 2010).

According to this view, two critical issues shape the science–policy–society interface. First, there is a need for better communication of scientific information, including uncertainty information, to those who need it, ask for it, or have the right to know about it. Second, there is a need for scientists and policymakers to better understand stakeholder interests and perspectives (their "problem framings"). Both issues point to a need for an explicit "management" of science–policy–stakeholder interplay, while recognizing that the interplay should be managed as an open process.

Guidance for Dealing with Uncertainty and Stakeholder Participation

In order to help environmental assessors to deal better with the science–policy–society interface, in particular with uncertainty and the framing of policy problems, the PBL Netherlands Environmental Assessment

Agency, together with Utrecht University, developed the *Guidance for Uncertainty Assessment and Communication* series (Janssen, Petersen, van der Sluijs, et al. 2003; Petersen, Janssen, van der Sluijs, et al. 2003; van der Sluijs, Risbey, Kloprogge, et al. 2003; van der Sluijs, Janssen, Petersen, et al. 2004; see also Petersen 2006 and van der Sluijs, Petersen, Janssen, et al. 2008). In this series, six parts of environmental assessments are distinguished: (1) problem framing; (2) stakeholders' involvement; (3) selection of indicators; (4) appraisal of the knowledge base; (5) mapping and assessment of relevant uncertainties; and (6) reporting of the uncertainty information.

In the view on the science–policy–society interface that underlies the uncertainty guidance, the shaping of this interface is strongly related to the framing of the policy problem. In the first step of the guidance, assessors outline the problem and its context and history by identifying major issues, past work, the level of contention, and the (expected) role of the assessment in the policy or decision-making process. They are explicitly asked to consider various views/perspectives on the problem and to pay attention to the problem's interconnectedness with other problems. They should be specific on what knowledge is needed with regard to the problem and into which research questions this knowledge is translated. Relevant aspects that are not dealt with in these research questions may need to be indicated. Moreover, it should be outlined what role the study is expected to play in the policy process and what the relation the study has with previous studies on the subject (policy context and problem history).

The uncertainty guidance sections dealing with stakeholder involvement are also directly related to shaping the science–policy–society interface because these sections help delineating the people involved in structuring and dealing with the policy problem. Here, the main parties (stakeholders/actors) and their views and roles with respect to the problem are identified as well as the aspects of the problem about which they disagree. On the basis of all this information, it has to be decided in assessments *if*, *how* (e.g., in formulating research questions, contributing information/data, evaluating findings/results), and *when* (at the beginning, during, after) one should involve *which* stakeholders in the study. The uncertainty guidance offers some hints and actions based on problem characteristics, as is illustrated in table 10.1.

After the *Guidance for Uncertainty Assessment and Communication* series was published in 2003, the PBL Netherlands Environmental Assessment Agency decided to publish the more detailed *Stakeholder*

Table 10.1
Example of Problem Characteristics and Associated Suggestions for Stakeholder Involvement

Problem Characteristics	Suggested Stakeholder Involvement
• There is no consensus about policy goals with respect to the problem and/or about the direction in which solutions need to be found.	Involve the stakeholders or their views in defining and framing the problem and selecting the indicators; be explicit about the limited scope of the study and its results.
• Decision stakes are high.	• Be transparent and open: let stakeholders "take a look behind the scenes" (in all stages of the study); aim for a broad composition of the advisory panel; involve stakeholders in the review of the study.
• There is no consensus about the (type of) knowledge required to solve the problem.	• Discuss and—if possible—use knowledge produced or put forward by stakeholders (including other research institutes); motivate the chosen approach (especially the choices about involving certain scientific disciplines) and state the potential limitations; signal and discuss the controversies with respect to the knowledge base and account for deviating theories and approaches to the problem; provide for external review.
• Major uncertainties exist regarding the behavior of the (natural and social) system(s) under study.	• If feasible, use knowledge and information produced or put forward by stakeholders, including knowledge and information derived from nonscientific sources, in order to be able to come up with a study of the required quality. When communicating intermediary and final results, be specific on the lack of knowledge and clearly state its consequences for the quality and the scope of the conclusions. Provide for external review or even counterexpertise. Deliver a clear mapping of the uncertainties.

Sources: Janssen, Petersen, van der Sluijs, et al. 2003; Petersen, Janssen, van der Sluijs, et al. 2003; van der Sluijs, Risbey, Kloprogge, et al. 2003.

Participation Guidance (Hage and Leroy 2008; see also Hage, Leroy, and Petersen 2010) to its toolkit for environmental assessors. This guidance is organized around a number of questions:

1. Why do you want participation?
2. What should the participation be about?
3. Who do you want to involve?
4. How much participation do you want?
5. What form are you choosing?

One of the most important things when organizing participation is to formulate clear aims. The PBL Netherlands Environmental Assessment Agency does not consider stakeholder participation in science-for-policy as an aim in itself for each project. But for cases in which stakeholder participation is found important, the *Stakeholder Participation Guidance* offers advice on how to organize it.

All these guidance documents distinguish between different types of policy problems (see the next section) and indicate that uncertainties and stakeholder participation in each should be dealt with differently. Important in this respect is a characterization of the problem at hand in terms of a number of features: level of consensus on policy goals regarding the problem, type of knowledge needed, decision stakes, and uncertainties involved. This characterization can lead to useful suggestions for stakeholder involvement and can serve as an initial step in deciding on an appropriate level and form of participation for the assessment.

Framing and Types of Policy Problems

The issue of framing plays a central role in the guidance documents from the PBL Netherlands Environmental Assessment Agency. People define or frame particular problems in different ways based on their knowledge and underlying views of the world (Rein and Schön 1996; see also de Vries and Petersen 2009). Framing is the process of limiting the scope of an (environmental) problem under consideration by choosing what sorts of knowledge and actions are relevant. Framing determines what should be the appropriate data, what analytical methods and indicators are needed, and the direction of solutions involved (Keating and Farrell 1998). The framing of the problem also determines who are the people involved in, for example, scientific assessment processes, ideas about the "stakes" involved, and determinations of the appropriate policy process to be followed (Farrell, VanDeveer, and Jäger 2001). Steven Yearley (chap. 9 in this volume) notes, for instance, that the transboundary character of the Convention on Long-Range Transboundary Air Pollution (CLRTAP) and its scope and organization impact on the scope and credibility of (local) public participation.

"Framing" thus describes the notion that different audiences perceive or "frame" air pollution in different ways. It is important to acknowledge differences in perception or framing of air pollution issues in order to understand different audiences' priorities and to integrate scales and foci—for example, whether the focus is on ecosystems or health problems, urban air pollution or transboundary air pollution, international negotiation or local decision making (see, e.g., Yearley, chap. 9 in this volume). The framing of the issue determines which indicators are used to express or measure risk, effects, costs, policy progress, and policy action (see also Lidskog and Pleijel, chap. 8 in this volume). The indicators themselves require information, and the framework determines what kind of scientific and institutional infrastructure is needed to provide the information. In turn, the scientific and institutional infrastructure available may guide the way in which future aspects of the problem are dealt with. Thus, framing is an important aspect in the "co-production of knowledge and social order" (cf. Jasanoff and Wynne 1998 and Jasanoff 2004; see also Lidskog and Sundqvist, chaps. 1 and 12 in this volume).

When differences between frames are not articulated explicitly, the parties might have difficulty understanding each other. Also misunderstandings and disagreements can arise about what information or which actor is relevant to the problem. Indicators or solutions from one frame may not be recognized as being relevant in another frame.

Communication between the different groups in the science–policy interface can be improved if the existence of different problem frames is acknowledged. For some environmental policy problems, this difference is more evident than for others. Matthijs Hisschemöller and Rob Hoppe (1996) have developed a typology to discern between so-called structured and unstructured problems (see figure 10.1). In each of the different problem types, the role of scientists also varies. When agreement is assumed on the societal norms and values pertaining to a policy problem and on the kind of knowledge used to reach a solution, the policy problem is considered "structured." The preferred mode of providing scientific advice to policy in this case is technocratic—that is, scientists can claim full authority in deciding on the solution to the problem. At the other extreme, when there is acknowledgment of a clash of values related to the problem and disagreement on what scientific knowledge

Consensus about relevant knowledge	Agreement on relevant norms and values: No	Agreement on relevant norms and values: Yes
No	**Unstructured problem** science as problem recognizer	**Moderately structured problem (ends)** science as advocate
Yes	**Moderately structured problem (means)** science as mediator	**Structured problem** science as problem solver

Figure 10.1
Four types of policy problems with associated roles for scientific advice. *Sources:* Hisschemöller and Hoppe 1996; Hisschemöller et al. 2001. See also Petersen 2006.

should be used to solve the problem, the problem is taken to be "unstructured," and scientists can at most serve as problem recognizers. In the latter case, scientific advisors should not aspire to more than helping the other actors in learning to deal with the value plurality and uncertainty; the technocratic mode of giving advice is out of the question in this case.

When dealing with unstructured problems, scientific advisors, in doing their boundary work, must claim a different kind of authority based on their ability to assess and communicate uncertainty and analyze the different perspectives on the problem—that is, the different ways the problem can be "framed." Environmental policymaking is an area in which policy problems of all different types can be observed. Climate change, biodiversity loss, and especially the broader policy problem of "sustainability" reside in the class of unstructured problems. Some of the more locally confined problems, such as ionizing radiation, belong to the class of structured problems. Air pollution and health problems can often be considered to lie somewhere in between. Following Hisschemöller and Hoppe's (1996) typology, such "in-between" problems can be called "moderately structured," with either consent on the values and norms ("ends") that pertain to the problem or consent on the possible solutions ("means").

Not all problems in the field of air pollution and health can be considered moderately structured; some problems can be taken to be unstructured. An example of the latter is the issue of ground-level ozone, as Rolf Lidskog and Håkan Pleijel discuss in chapter 8. Strong interests are involved, and different actors hold widely varying views on the goals that should be set for policies aimed at reducing ground-level ozone concentrations. Scientists currently cannot do more than alert people about the potential problem.

The way uncertainties are framed also differs between the problem types (e.g., van der Sluijs, Risbey, Kloprogge, et al. 2003). If the problem is structured, statistical analysis and reporting of uncertainty ranges are typically a suitable strategy. If the problem is moderately structured (means), the value-ladenness of assumptions and scenario uncertainty especially need to be addressed. If the problem is moderately structured (ends), weaknesses in the knowledge base and recognized ignorance become very relevant. Finally, if the problem is unstructured, recognized ignorance, weaknesses in the knowledge base, and value-ladenness of assumptions are at play. In the remainder of this chapter, we present and evaluate a case study that highlights the importance of framing by applying the *Stakeholder Participation Guidance.*

Case Study: The European AIRNET Thematic Network Project

The European AIRNET Thematic Network[1] (2002–2005) was a multistakeholder project in the field of air pollution and health effects (van Bree, Fudge, Tuomisto, et al. 2007). The main objective was to create a widely supported basis for public-health policy related to improving air quality in Europe and regulatory needs to achieve that goal. The specific objectives were to (1) develop an interpretation framework for the (final and intermediate) result of research supported by the fourth and fifth European Union Framework Programs (FP4, FP5) as well as by nationally funded studies; (2) collect, discuss, and interpret the (final and intermediate) results of research supported by FP4 and FP5 as well as by nationally funded studies; and (3) draw policy-relevant recommendations from the activities mentioned.

AIRNET and the project's Science–Policy Work Group in particular functioned as a structure and forum to get the science–policy–society interplay working for air pollution and health ("AIRNET" 2003). Within AIRNET, the scientific input required to better meet the end users' needs was discussed to support authoritative institutions to start building a more targeted and sustainable policy (Sanderson, Fudge, Totlandsdal, et al. 2006). The objective of the AIRNET Science–Policy Work Group was to facilitate an interactive communication and review forum on air pollution and health. The forum's aim was to discuss and interpret the outcomes of air pollution health risk research and impact assessments and to link these outcomes with the various end users' needs, policy issues, and abatement strategies and control measures. Thus, it formed a bridge-building network between policymakers, publics, scientific communities, and other stakeholders on end-user relevant research outcomes and their implications (Totlandsdal, Fudge, Sanderson, et al. 2007). One of the items covered by AIRNET was framing the complex issues of air pollution's effects on health. In the remainder of this section, we describe and evaluate the way the AIRNET project dealt with different problem frames and with stakeholders.

AIR-NETwork Days: Structure and Major Outcomes

AIRNET made headway in increasing science–policy–stakeholder interactions by generating multistakeholder AIR-NETwork days in Europe. These meetings, each lasting one day, were organized to actively involve local stakeholders in four countries: the Netherlands, Sweden, Hungary, and Spain. Each meeting was organized to get a snapshot of the air pollution and health issues in countries representing western, northern, central/eastern, and southern Europe, respectively.

Because European regions are different in several respects, at these network days all communication activities were tailored to the national/local target groups' specific needs. At the same time, though, the network days in the four different countries were organized in a similar way. Stakeholders participating in the meetings ranged from patient organizations and doctor organizations to representatives of the oil industry and transport organizations to national and local policymakers and scientists. The network days were attended on average by 60 to 100 participants. The meetings were typically organized by a small team consisting of one or two members of the AIRNET management team, who were local air pollution and health experts. This team was assisted by people from a local consultancy firm, who advised on the communication and facilitation aspects of the meeting or actually facilitated the meeting itself or both.

The first step for each organization team, including the national AIRNET coordinator and the local communications agency, was to identify representative stakeholders and target groups in a specific region, such as researchers, patient organizations and health professionals, non-governmental organizations, and local and national policymakers. The second step was to prepare and conduct focus-group discussions or interviews with those representative stakeholders and target groups to establish frames of reference for communication, areas of interest, and participants' likely expectations of and possible contributions to the AIR-NETwork day. At the AIR-NETwork day itself, work formats included roundtable discussions, a speaker's corner, contact boards, "silent wall discussions," and posters.

The goal of these AIR-NETwork activities was to bridge the gap between scientists, policymakers, and other stakeholders, resulting in a more profound exchange of each other's interests and information on issues relating to air pollution and health. Key problems and messages arising from these activities focused on several region-specific aspects of (1) air quality management and policy; (2) air quality modeling, monitoring, and assessment; and (3) air quality research. (The AIR-NETwork day model, the key messages from it, as well as the region-specific air pollution and health problems are described in more detail in Van Bree, Fudge, Tuomisto, et al. 2007.)

Topics of Major Discussion

The four network days were organized along the same basic principles, and the content and available work formats varied little from country to country. Overall, at all network days the principal focus of discussion was traffic-related air pollution, urban transport, and future air quality standards. Although the topics of interest varied relatively little between countries, the key messages that emerged from each workday were different in scope. For instance, issues specific to certain European regions were evident. In Sweden, the effects of wood burning and spring dust were important topics. In Hungary, many of the discussions focused on the health effects of ragweed exposure. Participants at the Netherlands workday placed a greater focus on actions for the government and policymakers.

Discussions related to policy, decision-making priorities, and the value of regional meetings reinforced AIRNET's activities. Overall, for all regions of Europe, improved communication between scientists, decision makers, and stakeholders was seen as highly desirable to increase the effectiveness of decision-making processes for environmental health improvement.

Participant Feedback

In general, participant feedback from the network days was positive. The majority of the participants at each network day felt that the day's objectives, to exchange knowledge and strengthen networks, were achieved. The majority also felt it was valuable to hold such events in the future.

Evaluation of Stakeholder Involvement in the AIRNET Project

By making use of the PBL Netherlands Environment Assessment Agency's new *Stakeholder Participation Guidance*, we can study stakeholders' involvement in the AIRNET project more closely and evaluate it following its five guiding questions.

Aims for Participation

Project initiators framed the original aim of involving stakeholders in the classical public understanding of science paradigm: if scientists were able to provide information to decision makers and stakeholders in a format that they could understand, the decision-making process would automatically improve. The intention was to maintain a strict boundary between scientists and other societal actors and to have one-way communication of scientific results across the interface. Nevertheless, the stakeholders were envisaged as playing a role in identifying the policy-relevant scientific questions that should be answered by the scientists. The project's principal aim was to create a network.

The AIRNET team quickly learned from the interaction with stakeholders that, besides the creation of a network itself, another explicit project aim should be to explore different perspectives and knowledge needs on air pollution and health problems, such as the specific health effects of various sorts of air pollution (road dust, traffic emissions, local heating, desert dust, and primary versus secondary particles), and the communication of the air pollution problems to responsible local government officials to help initiate targeted air pollution–control policy. Therefore, a questionnaire was designed to get a feeling for the involved actors' different perspectives and knowledge needs (Sanderson, Fudge, Totlandsdal, et al. 2006), resulting also in an understanding of the needs for better policy instruments and research programs.

Content of Participation

In terms of content, the project was ultimately focused on drawing clear and informative policy-relevant conclusions from the air pollution and health science. That is also why both the Commission of the European Communities (European Commission) was very interested in the project—the commission framed it as one of the key projects providing input for its Thematic Strategy on Air Pollution and associated Air Quality Framework Directive (van Bree, Fudge, Tuomisto, et al. 2007)—and many important stakeholders found it important to participate, thinking that through their participation they could have an influence on the commission's proposals for new policy, get scientific feedback on their specific questions, and fill their information needs.

Because the air pollution/health issue is a moderately structured or sometimes even unstructured policy problem, it was to be expected that the stakeholders would bring in a wide variety of perspectives on both values and knowledge. Some industry parties who were afraid of stringent policy measures zoomed in on the uncertainties that still exist surrounding the causal links between air pollution and health. This raising of uncertainties led to a more balanced presentation of the scientific basis to a wider audience; that is, it forced the scientists involved to make very clear both what was known and what was still uncertain.

Stakeholders Involved

A wide array of stakeholders was involved in the AIRNET project. First of all, many scientists from the public and private sector participated. A second main group was policymakers: senior officials of the European Commission (those drafting new policies and legislation) as well as national and local policymakers. Politicians were interested in the project. However, the participation of members of the European Parliament was very limited (for instance, only 3 out of 105 members returned their questionnaire; see Sanderson, Fudge, Totlandsdal, et al. 2006). A third group included sector-specific stakeholders covering the chemical industry, the oil industry, and the transport sector and represented by European-level organizations. The automobile industry, a crucial player, did not participate despite several invitations. A fourth group contained environmental, doctor, and patient organizations. A fifth group, individual citizens, was involved only in low numbers at the AIR-NETwork days. And finally, journalists were involved as well.

From this list of stakeholder categories, we can conclude that the AIRNET project succeeded in bringing together many parties involved in the science–policy–society interface. Whether policymakers from the European Commission have been able to make much use of the specific AIRNET results (which were targeted more to further research in the scientific community) remains unclear. The main outcome was that many of the parties were brought together with the scientists for the first time, which helped to generate a certain level of support for the policy measures proposed by the European Commission at a later stage.

Level of Participation

Because the stakeholders were assigned responsibilities for the production of report chapters, a high level of participation (codecision) was reached in the project. Although the initial project design envisaged a one-way communication of summarized research results to a wider nonspecialist audience (a noninteractive level of participation), the project management adapted the project to become much more interactive, which ultimately resulted in jointly written products that reflected the participants' various framings of the problem.

Participation Methods

AIRNET developed the concept of national network days, which stimulated active involvement and dialog among the different stakeholders. Within the AIR-NETwork day model, equal responsibility was given for participation: every person had the chance to contribute something. The interactive format (open roundtable, speaker's corner, silent wall discussion) helped to facilitate this participation process. Although the air pollution and health issues are big and the meeting time short, the AIRNET project can be seen as a beginning or continuation of the dialog (Totlandsdal, Fudge, Sanderson, et al. 2007). In addition, the AIRNET experience can also function as a working model for other projects aiming for the same sort of science–policy–society interplay.

Overall, during the AIR-NETwork days both proven knowledge and "good thoughts" were freely discussed without the weight of the printed word. From a transparency and clarity point of view, the AIRNET project managers thought that these types of meetings are important (if not necessary) because many different stakeholders hold different or competing interests and because complex air pollution and health issues can be framed in different ways. Such meetings give those who may not as yet be plugged into a network an opportunity to learn from others experience and offer their own experiences in the field.

The lesson learned during the project was that stakeholder mechanisms do not appear by themselves: a conscious design and professional facilitation are needed to develop a stakeholder process that leads to a real dialog and mutual learning among participants.

Lessons from the AIRNET Project about the Science–Policy–Society Interface

Had the *Stakeholder Participation Guidance* documents been available and used during the set up and performance of the AIRNET project, the issue of different framings among stakeholders could have come to the fore earlier and could have led to a more reflective way of thinking about stakeholder involvement from the beginning. However, the positive conclusion about the AIRNET project and stakeholder involvement is that this lesson was learned during the project and the setup was adapted accordingly.

The AIRNET project started mainly from a "technocratic" view on the science–policy–society interface and paid less attention to the socially constructed nature of this interface ("boundary work"). During the project, however, a more "democratic" view was developed in which different stakeholders' framings of air pollution and health problems could be given a much more important role. Although a joint production of knowledge was foreseen in the project's overall organizational setup—with particular stakeholders already put in charge of the production of report chapters—the methods to do the co-production were developed after the project had already started. Note that the "co-production" of reports by involved participants (scientists, policymakers, and other stakeholders) in the AIRNET project was no "co-production of science and social order" as understood elsewhere (Jasanoff and Wynne 1998; Jasanoff 2004; see also Lidskog and Sundqvist, chaps. 1 and 12 in this volume). As described earlier, the project originally aimed at contributing to the evaluation of policy options for European policymaking. However, the final results were rather targeted toward further research in the scientific community and were rather distant from the final policy proposals tabled by the European Commission for the Thematic Strategy on Air Pollution. Note also that the new insights and views on problem frames in the project emerged through the interaction of "new" groups of decision makers and stakeholders (e.g., the European Federation of Asthma Patients) rather than through interaction between, for example, the project management and policymakers from the European Commission. Thus, the science–policy–society interface in the AIRNET project in the end was "located" at a different level than where the project management and the European Commission might have envisaged it originally. The engagement of new groups and enhancement of new networks through AIRNET, however, might have influence on future policy developments.

To conclude, several important lessons can be drawn from the AIRNET project with respect to the shaping of the science–policy–society interface in a multistakeholder network. The wish to create an overarching science–policy network in Europe was historically generated by scientists themselves. The participation of the few stakeholders at the beginning and the identification of their information gaps rapidly resulted in a need to (1) increase their number, (2) incorporate different levels of stakeholders, and (3) try to identify their information needs more thoroughly. AIRNET also learned that the science–stakeholder interface is interpreted differently by the different players, is not naturally occurring (that is, is actively constructed by boundary work), does not work by itself, and requires a great deal of planning, structure, and energy (e.g., the AIR-NETwork days). Furthermore, with respect to dissemination of research findings, scientists found it hard to write in a concise and nonspecialist way; therefore, AIRNET provided guidelines and contracted science journalists to help. Because effective communication between stakeholders also appeared crucial, AIRNET contracted a communication agency to help to increase its effectiveness in that direction. Finally, organization of a network requires considerable planning; there was no "recipe" available on how to run a network optimally. Knowing what one wishes to achieve is not sufficient; one should also know how one wishes to achieve it. In fact, many participants also recognized that the science–policy–society interface must be considered a continuous (learning) process rather than a static consultation type of interaction mechanism. It also appeared difficult to get all "players" actively involved; therefore, defining a clear task and responsibility for participants seems of crucial value. AIRNET has been greatly appreciated for its efforts in building such a network in Europe.

Other lessons were learned. First, when a project aims to "improve" the science–policy–society interface, significant attention should be paid beforehand to what the "improvement" specifically consists of and where this interface is located. The AIRNET project initially paid less attention to the why and how of stakeholder participation. Second, with tools such as the PBL Netherlands Environmental Assessment Agency's guidance documents, which emphasize the importance of framing, the stakeholder perspectives on uncertainties and risks can be more readily brought into the picture.

Conclusions

In efforts to deal with the science–policy–society interface and to frame complex issues in the field of air pollution and health, several characteristics of policy problems should be taken into account: the level of consensus on relevant norms and values pertaining to the policy problem; the level of certainty about relevant knowledge; and especially the way uncertainties (ranging from technical and natural-scientific uncertainties to societal perspectives) are dealt with in the decision-making arena, which varies for different types of policy problems.

From a transparency and clarity point of view, knowledge uncertainty is a dominant feature in present-day science for policy, as becomes especially obvious in environmental assessment. The presence of this uncertainty constitutes the background to all activities shaping science–policy–society interfaces and acknowledges the need for continuous negotiations and updates. Many social scientists who have studied the relationship between environmental science and decision making conclude that these two sets of activities cannot be neatly separated. In order to stimulate practitioners who work at the science–policy–society interface to become more self-reflective, the PBL Netherlands Environmental Assessment Agency developed the *Guidance on Uncertainty Assessment and Communication* series and *Stakeholder Participation Guidance*. The general thrusts of these tools are to stimulate reflection among policy advisors on the science–policy–society interface, in particular the way problems are framed, and to help them create effective structures and mechanisms to shape the work at the boundary between the different domains.

This chapter presented only a few air pollution examples and demonstrated the experience of stakeholder networking in the European AIRNET Thematic Network project. With the guidance tool for stakeholder participation, we found that the issue of framing is important in the context of meaningfully engaging broader groups—broader than the usual circles of policymakers and industry stakeholders—in discussions about knowledge and action needs in relation to air pollution and human health. As Yearley notes (in chap. 10), CLRTAP's "framing" and organization indicate that it is not necessarily equipped to facilitate public participation. Processes like those experimented with in the AIRNET project can offer ways to deal with stakeholders' differentiated involvement in knowledge production. These differentiated interfaces between science, policy, and society will enhance the effectiveness of environmental regimes.

Note

1. See http://airnet.iras.uu.nl. AIRNET was coordinated by the Institute for Risk Assessment Sciences (B. Brunekreef), Utrecht University, Utrecht, Netherlands, and the PBL Netherlands Environmental Assessment Agency (L. van Bree), Bilthoven, Netherlands.

References

AIRNET. 2003. *AIRNET News* 2 (October). Available at http://airnet.iras.uu.nl.

Brunekreef, Bert, and Stephan T. Holgate. 2002. Air Pollution and Health. *Lancet* 360:1233–1242.

de Vries, Bert J. M., and Arthur C. Petersen. 2009. Conceptualizing Sustainable Development: An Assessment Methodology Connecting Values, Knowledge, Worldviews, and Scenarios. *Ecological Economics* 68: 1006–1019.

Farrell, Alexander E., Stacey D. VanDeveer, and Jill Jäger. 2001. Environmental Assessments: Four Under-Appreciated Elements of Design. *Global Environmental Change* 11: 311–333.

Ginsburg, Eric O., and Ellis B. Cowling. 2003. Future Directions in Air Quality, Science, Policy, and Education. *Environment International* 29: 125–135.

Hage, Maria, and Pieter Leroy. 2008. *Stakeholder Participation Guidance for the Netherlands Environmental Assessment Agency*. 3 vols. Bilthoven: Netherlands Environmental Assessment Agency and Radboud University Nijmegen.

Hage, Maria, Pieter Leroy, and Arthur C. Petersen. 2010. Stakeholder Participation in Environmental Assessments: The Case of the Netherlands Environmental Assessment Agency. *Futures* 42: 254–264.

Hisschemöller, Matthijs, and Rob Hoppe. 1996. Coping with Intractable Controversies: The Case for Problem Structuring in Policy Design and Analysis. *Knowledge and Policy* 8: 40–60.

Hisschemöller, Matthijs, Rob Hoppe, Peter Groenewegen, and Cornelis J. H. Midden. 2001. Knowledge Use and Political Choice in Dutch Environmental Policy: A Problem-Structuring Perspective on Real Life Experiments in Extended Peer Review. In *Knowledge, Power, and Participation in Environmental Policy Analysis*, ed. Matthijs Hisschemöller, Rob Hoppe, William N. Dunn, and Jerome R. Ravetz, 437–470. New Brunswick, NJ: Transaction.

Janssen, Peter H. M., Arthur C. Petersen, Jeroen P. van der Sluijs, James S. Risbey, and Jerome R. Ravetz. 2003. *RIVM/MNP Guidance for Uncertainty Assessment and Communication: Quickscan Hints & Actions List*. Bilthoven: Netherlands Environmental Assessment Agency.

Jasanoff, Sheila. 2004. Ordering Knowledge, Ordering Society. In *States of Knowledge: The Co-Production of Science and Social Order*, ed. Sheila Jasanoff, 13–45. London: Routledge.

Jasanoff, Sheila, and Bryan Wynne. 1998. Science and Decisionmaking. In *Human Choice and Climate Change*, vol. 1: *The Societal Framework*, ed. Steve Rayner and Elisabeth L. Malone, 1–87. Columbus, OH: Batelle Press.

Keating, Terry J., and Alexander E. Farrell. 1998. *Problem Framing and Model Formulation: Local Knowledge in Air Pollution Research and Policy*. Environment and Natural Resources Program Discussion Paper E-98-11. Cambridge, MA: Kennedy School of Government, Harvard University.

Matanoski, Genevieve M. 2001. Conflicts between Two Cultures: Implications for Epidemiological Researchers in Communicating with Policy Makers. *American Journal of Epidemiology* 154: S36–S42.

Petersen, Arthur C. 2006. *Simulating Nature: A Philosophical Study of Computer-Simulation Uncertainties and Their Role in Climate Science and Policy Advice*. Antwerp: Het Spinhuis. Available at http://hdl.handle.net/1871/11385.

Petersen, Arthur C., Peter H. M. Janssen, Jeroen P. van der Sluijs, James S. Risbey, and Jerome R. Ravetz. 2003. *RIVM/MNP Guidance for Uncertainty Assessment and Communication: Mini-Checklist & Quickscan Questionnaire*. Bilthoven: Netherlands Environmental Assessment Agency.

Rein, Martin, and Donald A. Schön. 1996. Frame-Critical Policy Analysis and Frame-Reflective Policy Practice. *Knowledge and Policy* 9 (1): 85–104.

Samet, Jonathan M., and Nora L. Lee. 2001. Bridging the Gap: Perspectives on Translating Epidemiological Evidence into Policy. *American Journal of Epidemiology* 154: S1–S3.

Sanderson, Eric G., Nina Fudge, Annike I. Totlandsdal, Bert Brunekreef, and Leendert van Bree. 2006. Stakeholder Needs for Air Pollution and Health Information. *Journal of Toxicology and Environmental Health. Part A* 69: 1819–1825.

Totlandsdal, Annike I., Nina Fudge, Eric G. Sanderson, Leendert van Bree, and Bert Brunekreef. 2007. Strengthening the Science–Policy Interface: Experiences from a European Thematic Network on Air Pollution and Health (AIRNET). *Environmental Science & Policy* 10: 260–266.

van Asselt, Marjolein B. A. 2000. *Perspectives on Uncertainty and Risk: The PRIMA Approach to Decision Support*. Dordrecht: Kluwer Academic.

van Bree, Leendert, Nina Fudge, Jouni Tuomisto, and Bert Brunekreef. 2007. Closing the Gap between Science and Policy on Air Pollution and Health. *Journal of Toxicology and Environmental Health. Part A* 70: 377–381.

van der Sluijs, Jeroen P., Peter H. M. Janssen, Arthur C. Petersen, Penny Kloprogge, James S. Risbey, Willemijn Tuinstra, and Jerome R. Ravetz. 2004. *RIVM/MNP Guidance for Uncertainty Assessment and Communication: Tool*

Catalogue for Uncertainty Assessment. Utrecht, Netherlands: Utrecht University.

van der Sluijs, Jeroen P., Arthur C. Petersen, Peter H. M. Janssen, James S. Risbey, and Jerome R. Ravetz. 2008. Exploring the Quality of Evidence for Complex and Contested Policy Decisions. Environmental Research Letters. Available at http://iopscience.iop.org/1748-9326/3/2/024008/pdf/1748-9326_3_2_024008.pdf.

van der Sluijs, Jeroen P., James S. Risbey, Penny Kloprogge, Jerome R. Ravetz, Silvio O. Funtowicz, Coral Quintana, Angela G. Pereira, et al. 2003. *RIVM/MNP Guidance for Uncertainty Assessment and Communication: Detailed Guidance*. Utrecht, Netherlands: Utrecht University.

11

Governance of Air Quality and Stakeholder Engagement: Lessons and Experience from International Cases

John Forrester, Kevin Hicks, Johan Kuylenstierna, Julie Simon, Carolyn Snell, Michael J. Chadwick, Dieter Schwela, and Lisa Emberson

This chapter is about stakeholder engagement in the governance and management of air pollution control at several levels of governance: the realm of strategic policy formulation, translation of that strategic policy into policy measures, implementation of these policy measures, and policy evaluation—in other words, all the stages of what is often referred to as the "policy cycle." It looks at four examples of institutional structures set up to provide frameworks within which different stakeholders have been engaged to differing degrees in addressing air pollution problems. These examples come from a variety of projects that we have been involved in and that led to what might be referred to as the good governance of air quality.

The chapter makes the point that within these practitioner contexts—operating at different levels of governance—definitions of the term *stakeholder* are often confused and confusing: thus, cross-case comparison of the term is problematic. Rather than try and subsume this problem in a generalized theory, we emphasize this divergence and propose a new rationale for multistakeholder engagement related to a holistic understanding of the complexity of the process of governance. Despite the fact that air pollution is considered an exemplary scientific problem, air pollution governance is not. Air pollution itself (e.g., nitrogen oxides, sulfur oxides, particulate matter, and so on) may be identified and measured within technical remits, but air quality action plans and policies are made within a complex system of political decision making. Thus, the governance of air pollution becomes a problem in governance rather than a problem in science policy and requires multistakeholder engagement because "complex problems, i.e., problems with many solutions that are quite different in execution and rankable in quality of outcome, may be solved better by a diverse team of competent individuals than by a [less diverse] team composed of the best individual problem solvers" (Carpenter, Folke, Scheffer, et al. 2009).

Therefore, the term *stakeholder* rightly has several meanings across the cases. A stakeholder can be a policy stakeholder, or the end user of scientific evidence (as is commonly seen across the cases given here). Industry (or some other polluter, such as car drivers) can also be a stakeholder—the ones who have eventually to implement many of the regulations or those who incur the costs of the economic instruments instituted to effect the policy goals. The term can also simply refer to "the public"—that is, the assumed end beneficiaries of policies. Indeed, in some examples given here, the term *stakeholder* is used to describe scientific practitioners and experts as well as other actors. Thus, a stakeholder in the governance of air quality is any individual or group who "can affect" or "is affected by" (following Freeman 1984) air quality or the governance of air quality. Although the concept of stakeholder engagement used here is rooted in the need to bridge the gap between scientific evidence and policymaking and policy implementation, that gap itself is perceived differently depending on the stake and the stakeholder's perspective.

The chapter also introduces the policy cycle as a heuristic device and makes the point that as the policy-cycle stages move from strategy through translation of that strategy into policy measure and on to implementation, the cycle ideally spirals inward and downward from international agreement through international protocol through national policy measures to local implementation, and as it does so, the stakeholders and their roles change. Those who can have an effect on the process obviously change as the level changes; those who are affected remain the same (although their number is proportionate to the scale of the process).

Further, it is argued that at every opportunity for stakeholder engagement, decisions should be transparent and should be based on the best possible evidence. All sectors need to "understand the environmental ramifications of their strategies and activities" (Nilsson 2007, 167), and stakeholders must be aware of the full consequences of their choices, but the role of science and scientists as a stake in the policy process equally needs to "expand (or at least clarify) the scope of choice for decision-making in way that allows for the decision-maker to reduce choice based on his or her own preferences and values" (Pielke 2007, 2–3). Thus, engagement between those who produce evidence (the research community) and those who make decisions—whether those decisions are on policy (policy community) or on courses of action (end users)—must be transparent.

From Science to Policy: The Role of Evidence and the Role of the Citizen

In order for policy to have a desirable impact, it is necessary that producers of "sustainability knowledge" (knowledge that supports sustainable development) engage with those who take decisions (i.e., not only policymakers and implementers, but also end users such as potential polluters) and communicate with them. They all can be claimed as "stakeholders," and they all have their own knowledge(s) to bring to the table. To ensure that the best-available knowledge on required environmental conditions, such as good air quality, is reflected in all policy and decision making, a continual process of environmental policy integration (EPI) has to be initiated. EPI has been described as "the metamorphosis of environmental protection policy frames to sustainability thinking. . . . This is a process whose face extends horizontally across government agencies and the private sector, but also vertically into societal values and institutions" (Kasperson 2007, xv). For the European Union, this integration was emphasized by Article 6 of the European Community Treaty (European Union 2006), which states that "environmental protection requirements must be integrated into the definition and implementation of the Community policies and activities." The United Nations similarly says that scientific research programs should be supported, "including their socio-economic and human aspects *at the community level, national, subregional, regional, and global levels*, to complement and encourage synergies between traditional and conventional scientific knowledge and practices" (United Nations 1992, chap. 35, para. 9, emphasis added). This policy integration process generally demands innovation in political, organizational, and procedural activities.

But should the involvement of scientific advisors and scientific knowledge holders as stakeholders in the policy process be any different from the involvement of those with other stakes and with different knowledge(s) to bring to the process, and if so, on what grounds? The rationale for stakeholder involvement is certainly often different when looked at from the perspective of the scientific knowledge holder. From this perspective, the rationale is often purpose driven—what Andy Stirling (2005) calls "instrumental" (we come back to this topic later). However, the rationale changes from the perspective of the policy actor so that the scientist becomes a stakeholder in the policymakers' policy process. Here scientific knowledge must be weighed alongside other "evidence" (Elliot Major 2000). Nonetheless, if we look at the scientific research cycle, we can see that it bears more than a passing resemblance to the policy cycle in the identification and assessment of causes, identification, and evaluation of control of prevention options (i.e., "measures") leading to policy advice. All these stages involve social processes, even though technical experts may treat the identification of options as an "objective" process. However, the starting point of air quality governance can be deemed to be some actor or actors' observation of a challenge (pollutant state, exposure, health and/or environmental effects) that is supposed to be caused by air pollution. Science is then usually requested or incited or takes the lead itself to investigate whether there is a causal association between air pollution and the observed effects—the process is objective at this stage. The public as end beneficiaries of policy and others as end implementers are affected by the process at higher levels, and thus they too become "stake" holders—indirectly and directly. This higher-level process then becomes the wider evaluation of control and prevention options, which—in order to be rigorous—needs to involve less-objective processes, such as economic valuation.

Thus, for the purposes of this chapter there are in fact three important stakeholder communities: first, research science and scientists; second, policy and policymakers; and, finally, "other stake" holders such as industry and polluters, including the private sector and citizens or the general public. There is no real rationale for distinguishing among the latter group of stakeholders, even though their interests may differ; from the perspective of creating "usable" policy knowledge (cf. Haas and Stevens, chap. 5 in this volume), the rationales for their inclusion are similar, and both commercial and business interests as well as grounded-knowledge stakeholders (including those who have a vested interest) have a democratic right to involvement alongside more general citizens who may be impacted by policy decisions. In this chapter, we use a "trichotomization." Like all models, this division is not pure and clean. Technical experts sometimes work for private companies and for interest groups. Nonetheless, all three groups of stakeholders should come together to determine what the appropriate actions are to change the challenge posed by air pollution.

Greater coherence and coordination between different levels of government will support the effective delivery of increasingly interdependent environmental and sustainable development objectives (Duit and Galaz

2008). New structures and better coordination mechanisms within organizations, greater resources and capacity, as well as improved information, decision-support, and public-participation mechanisms can help delivery (Folke, Hahn, Olsson, et al. 2005; Plummer and Armitage 2007). The need for integration of governance at different scales is increasingly important, and in attempts to identify opportunities to support policy integration it is valuable to consider how competencies are distributed between governance levels.

There is a growing tendency for international regulations (United Nations treaties, conventions, and protocols), supranational regulations (e.g., European Union directives), and agreements (e.g., the Malé Declaration) to be transposed into national laws. Thus, translation from strategic policy through measures to be implemented is critical and requires careful consideration of stakeholders. It is important to note that as the "action arena" changes from international to local, so the policy cycle spirals downward and inward to involve more local actors and more local-knowledge stakeholders. The success of EPI will also be affected by the nature of the sector (e.g., transport, agriculture, health, and so on) and the extent to which environmental impacts are inherent to the sector's activities: the relationship is often complex. Further, the perceptions held by society at large and by specific stakeholders will also be important, as will their ability to influence decision making and implementation. And, overall, efforts to support policy integration need to be matched to the particular sector and organizations involved.

If engagement of appropriate stakeholders is to be effective, it needs to take place at all stages of the policy cycle. At any stage of the cycle, there may be barriers to decision making, to effective stakeholder engagement, and, as a result, to implementation of policy outputs and therefore to any resultant outcomes or impact of decisions made.

The case studies presented here show how these theoretical principles are rarely followed, and, indeed, they may not always even appear relevant to the actors involved.

Science to Policy: The Role of Democracy and the Role of Reliable Evidence

One way of ensuring the reliability of an outcome is through the use of what is called "evidence-informed policymaking." One way to ensure an evidence base is for knowledge producers to generate "usable knowledge." Peter Haas and Casey Stevens state that, "in short, usable knowledge encompasses a substantive core that makes it usable for policymakers and a procedural dimension that provides a mechanism for transmitting knowledge from the scientific community to the policy world" (chap. 5 in this volume). This definition is useful to the practitioners precisely because it takes its own internal rationale from the policy arena into which the knowledge is being put. Some science and technology studies (STS) writers distinguish between knowledge that has its epistemological basis in scientific and technical knowledge and knowledge that is legitimated by social, political, or other norms besides those of science (cf. Lidskog and Sundqvist, chap. 1 in this volume). Usable knowledge is both. It is not just grounded in the social. As Andrew Webster (2007) notes, "technically warranted" knowledge comes from science, but for social relations to be ordered so that that scientific knowledge is more socially useful, technically warranted knowledge must also be "democratically warranted." Similarly, for science to develop knowledge that is "more socially accountable and reflexive" and that "includes a wider, more temporary and heterogeneous set of practitioners, collaborating on a problem defined in a specific localised context" (Gibbons, Limoges, Nowotny, et al. 1994, 3), it must continue to produce "reliable knowledge" (Ziman 1978, 2000), but do so in a way that is "transparent and participative" so as to make scientific knowledge "socially robust" (Gibbons 1999). As Haas and Stevens also note, there are many ways in which the normative usefulness of the knowledge can also be increased, such as by including a wider knowledge base through stakeholder engagement (see also Yearley, Cinderby, Forrester, et al. 2003; Stringer, Dougill, Fraser, et al. 2006), but stakeholder engagement is also a way to create democratic warranty, social robustness, and, ultimately, usability.

These arguments within a broadly conceived STS field are not new. Over the decades, science has engaged first with the policy community, then with "the public." In parallel, there has been a drive to (re)connect publics with policy decision making. Angela Liberatore and Silvio Funtowicz sum up this process well:

> At this point the issue of participation must be dealt with. If participation is limited to a few (most resourceful) actors, behind closed doors and without procedures for determining who is invited to the table and what is the nature of the involvement (ranging from consultation to co-regulation), then representative institutions are pre-empted from control as well as legislative functions. If, on the other hand, efforts are made to establish balance, transparency and procedure for accountability, such participation can usefully complement the role of representative institutions. (2003, 148)

Silvio Funtowicz and Jerome Ravetz (1991) have also made the case for "extended peer review," where either the decision stakes or the levels of uncertainty are high. This chapter attempts to make the link between Liberatore and Funtowicz's democratic-inclusion rationale and the technical-knowledge rationale of extended peer review. The rationale of extended peer review has been extended through the concepts of sustainability science (Kates, Clark, Corell, et al. 2001) and, more recently, resilience (Folke, Hahn, Olsson, et al. 2005). Each new iteration of the underlying idea has reinforced the concepts that a wider knowledge base is necessary to solve complex environmental problems (e.g., see Carpenter, Folke, Scheffer, et al. 2009).

The Four Case Studies

The examples we look at here are diverse, but each one—looked at from the information provider's point of view—is based on the shifting of policy and decision toward a stronger (scientific) evidence base. The cases span levels of governance from strategic policy at the international level to local policy at the city level. The stakeholders involved are not the same. Thus, these examples also cross the levels of air pollution as a transboundary or local issue, with the concomitant change in "opportunity structure for political action" (Lidskog and Sundqvist, chap. 1 in this volume). Further, the driving actors have perceived the engagement of stakeholders to be different in each case due to differing perceptions not only of who should be engaged, but also of *why* stakeholders should be engaged. Looked at individually, the cases may not be particularly helpful, but looked at in the round they show a move toward a rationale for stakeholder inclusion that over time has been inspired by ideas of postnormal science (Funtowicz and Ravetz 1991), sustainability science (Kates, Clark, Corell, et al. 2001), and now resilience (Folke, Hahn, Olsson, et al. 2005). Science practitioners named the following stakeholders in these cases (note the terminology used here is that of the practitioners):

- The setting up of the Convention on Long-Range Transboundary Air Pollution (CLRTAP) focused on engagement among air pollution experts, elected representatives, government officials, industry, nongovernmental organizations (NGOs), and the media.
- The process for the Malé Declaration on Control and Prevention of Air Pollution and Its Likely Transboundary Effects for South Asia

was designed to engage expert organizations, development organizations, government officials, and government (environment) ministries.

• The Air Pollution Information Network for Africa (APINA) sought to engage expert organizations, development organizations, polluting industries, and later the public, but it is now also successfully engaging policy actors.

• A more locally focused project on stakeholder engagement in the setting up of local air quality management areas (AQMAs) in the United Kingdom sought to engage national agencies, neighboring local authorities and public authorities, NGOs, business interests, and "such other bodies or person as the authority considers appropriate" (UK Govt. 1995, Schedule 11).

Thus, it can be contended that CLRTAP sought initially to engage science experts with policy community and industry and to use the media (itself a private entity) to engage the public. Malé initially engaged "expert" organizations, by which it means experts from national and transnational organizations such as the United Nations, but of course there are also political actors within these organizations and technical experts within government ministries. Now that APINA has started to address engaging the public, it arguably seeks to engage the most holistic stakeholder group, but the same claim can now be made for CLRTAP. The AQMA case in the United Kingdom was local and sought to engage the most limited group of stakeholders. Thus, each case can be characterized as a technopolitical process with little role for "lay citizen" participation.

Regional and International Dimensions of Air Quality Governance

CLRTAP

The CLRTAP case has been exhaustively covered in the literature and well covered in other chapters in this book. It provided a flexible framework within which national data, information, and expertise were shared to create a basis for negotiations on emissions limits to be agreed by the different countries. The stakeholders involved in this international collaborative work ranged from elected government and agency representatives and officials to research scientists, officials of regulatory bodies and inspectorate services, commercial and industrial interests, NGO representatives, and from time to time media representatives. Among these stakeholders, a structure was developed that enabled scientific information and results to pass "up" to governments (acting together through the Executive Body) and that provided the framework within which different scientific and technical stakeholders worked and through which scientific assessments were presented in a form usable for negotiation on policy formulation.

The context within which this process took place was dependent on political will and interest; however, once the process started, it was driven largely by scientific knowledge. The flow of communication went two ways and with many iterative loops providing learning opportunities: scientists provided advice on matters about which policymakers had asked, and then policymakers deepened their demand for more detailed information—all in order to provide technically warranted evidence to sustain the case for concrete action. Indeed, the development of formal and informal international science communities across Europe—scientists who understand policy demands around the issue of transboundary air pollution—can be considered one of CLRTAP's great successes.

There are two features of interest in the realm of evidence-based policymaking and stakeholder engagement in relation to the CLRTAP process. One is that despite the involvement of a larger number of scientists and policymakers in the debate at the international and national scales, a relatively small number of individuals in policy and science appeared to provide the "engine" for the whole process. The second feature is the way in which involved scientists proved to be able to learn about the political basis of engagement in the assessments (Siebenhüner 2002). It is clear that although the science and policy communities were heavily involved and successfully interacting, and big industry was engaged politically in this process, the other part of our third stakeholder community—that is, wider society—was not actually heavily engaged in the process. Certain NGOs were heavily involved and were important for the progress made (especially the Swedish NGO Secretariat on Acid Rain). However, the convention itself has recognized that there is little engagement with the public, and it has attempted to address this deficit by holding a series of workshops to understand how such a regional, international convention can usefully engage with citizens (Grennfelt, Lidskog, Lindau, et al. 2006). Nonetheless, despite the lack of "citizen" involvement, CLRTAP has been a large contributor to the successful reduction in some of the most damaging pollutants, so it has had its instrumental effect.

The Malé Declaration

Certain aspects of the evolution of the Regional Air Pollution in Developing Countries (RAPIDC) program used the CLTRAP approach as a springboard and template for the development of air policy in different regions. However, several difficulties and constraints made it necessary to amend the process and adapt the extent of stakeholder engagement to suit local circumstance. Not least of these issues was the fact that most of the countries in these regions are very poor, lack capacity and resources, and are at an earlier stage in the development of their state administrations than the European countries that engaged in CLRTAP.

In 1997, the two focal regions in the RAPIDC program—South Asia and southern Africa—were in a situation where political and scientific circles' main concern about air pollution issues was the prevention of urban air pollution and in particular its impact on human health. However, smoke from the forest fires in Indonesia in 1997, which reached as far as South Asia, woke many countries up to the potential for long-range transport of air pollutants and the risks of the development of an increasingly regional problem about which they knew little. This sudden awareness is in contrast to the situation in Europe prior to CLRTAP, where the impact of acidic deposition on freshwater bodies was already known to be a big problem.

In 1998, the United Nations Environment Programme Regional Resource Center for Asia and the Pacific initiated a "policy dialog" in collaboration with the Stockholm Environment Institute, involving senior government officials of the South Asian region and science experts on air pollution. The presentations at the policy dialog highlighted the growing problem of regional air pollution in the South Asian countries due to expansion of the energy, transportation, and industrial sectors of their economies. The projections and modeling exercises demonstrated by the experts showed that further expansion, without preventive strategies to control the effects of air pollution, would potentially cause long-term damage to the crops, soils, human health, and architectural heritage of South Asia. The meeting also took note of the possibility that localized and national air pollution problems would become transboundary, as had happened in Europe, and of the need to study the phenomena carefully to facilitate prompt subregional cooperative action.

In light of these presentations and discussions, the meeting agreed on a draft declaration to be presented to the governing council of the South Asia Cooperative Environment Programme, which was to have its seventh meeting at Malé in the Maldives in April 1998. The timing was good for some consensus on tackling the likely effects of transboundary air pollution in South Asia because the countries there were perturbed about the growing threat of air pollution problems in their own territories and the risks of a growing regional problem of the type experienced in Europe, which they wished to avoid. Despite the lack of concrete evidence of a serious regional dimension to their air pollution problem, the environment ministers present at the meeting gave their seal of approval to what became the Malé Declaration on Control and Prevention of Air Pollution and Its Likely Transboundary Effects for South Asia, which is the only international agreement of its kind between the eight countries of the region—Bangladesh, Bhutan, India, Iran, the Maldives, Nepal, Pakistan, and Sri Lanka. It is concerned with both the transboundary and shared problems of urban air pollution. These countries also have a lower capacity to undertake the assessments required, in contrast to the European nations' technical ability at the initiation of CLRTAP. There is thus a strong element of capacity building in the South Asian process.

The focus of the Malé Declaration is not to provide scientific results alone or to "impose" technically warranted knowledge from "outside," but to ingrain the process of producing the data and knowledge in the governance systems of the countries themselves so that the likelihood of sustainability is increased. In many cases, the policymaker and scientific communities in these countries were not used to engaging together in concrete, long-lived programs.

In partly due to the lack of general awareness of air pollution, especially of regional and nonurban components, the main actors driving the Malé Declaration always saw wider engagement of industry and citizen stakeholders as part of the process, as a way to raise awareness of air pollution issues, and as important for the declaration's success in general. The declaration is one of the few intergovernmental networks to have a formal mechanism for stakeholder engagement in its development. This engagement occurs through a meeting of key regional stakeholders from the science and NGO communities each year, which is held back to back with the intergovernmental meeting. In addition, there are national stakeholder meetings to engage more deeply at the national level and to discuss issues and results of relevance to each country and the Malé Declaration goals. Practitioners realize that further stakeholders need to be brought in to embed the intergovernmental process fully in the region—for example, through more involvement of the polluting sectors—and activities are increasing the breadth and organization of stakeholder engagement (e.g., through the South Asian Youth Environment Network).

APINA

In southern Africa, APINA—formed during activities held under the RAPIDC program in 1997—has also developed a regional science–policy process to address its growing problem of air pollution. In comparison to Europe and Asia, southern Africa has a less widespread and more localized air pollution problem, and in many ways this regional approach is precautionary in nature and designed to enhance the region's capacity to assess the air pollution problem. However, the rate of urbanization in Africa is the highest in the world, and, associated with this trend, there is a high annual rate of increase in transport and industrial output. These factors will lead to a future of higher emissions and greater problems unless measures are taken early. APINA has assessed the air pollution situation and developed a number of policy dialogs to share its data and interact with policymakers. These events have resulted in the Harare Resolution (1998), the Draft "Maputo Declaration" (2003), and the Lusaka Agreement (2008), but no formal intergovernmental "convention" as yet. However, APINA's and others' activities are now increasing awareness of air pollution issues at ministerial level in sub-Saharan Africa, as demonstrated by the successful phase-out of leaded gasoline there and the ministerial recommendations from the 2006 Better Air Quality for the Cities of Sub-Saharan Africa Conference. Thus, the APINA context is moving toward one of implemented "paper" agreements (cf. the discussion of Oran Young's work in chap. 1 in this volume).

The crucial iterative and discursive link between the information generators (scientists) and end users (policymakers) was initially weak. However, by 2008 APINA had held three policy dialogs, attended by many of the key stakeholders, including scientists/academics, NGOs, policymakers (with some at ministerial level) and, crucially, representatives from polluting sectors. The ministers at the second APINA policy dialog in Maputo, Mozambique, suggested that APINA create national focal points seated at the ministries of environment in the affiliated Southern African Development Community (SADC) countries to increase the science–policy discourse. At the third policy dialog, held in Lusaka, Zambia, there was more ministerial representation than previously. APINA is now strongly linked with SADC; SADC officers have recommended that the Lusaka Agreement be incorporated into national policy development at the highest political levels. APINA's development has been driven mainly by the scientific community, and the network has successfully engaged different parts of that community across the region, mirroring the linkages achieved in the science community engaged in the CLRTAP process. This community has then developed a permanent link to the policymaking apparatus, up to ministerial and governmental level. Other stakeholders have also been involved in the process at various points, and the private sector has been a particularly visible stakeholder group at key meetings. There is thus good stakeholder "buy-in."

The linkages between the scientific stakeholders and policymakers that proved so vital in the CLRTAP process are features of the development of air policy across both South Asia and southern Africa. In both Malé and APINA, further involvement of other stakeholders has been encouraged from the start, perhaps to a greater degree than initially in Europe. The extent to which this greater pool of stakeholders will affect the success of these initiatives in comparison to CLRTAP is too early to say because there are so many other differences between the circumstances of these regions and those of Europe in the 1970s and 1980s. However, for CLRTAP the public was largely disengaged once the issue was on the political agenda, and public stakeholder engagement has only recently reemerged as an issue. In Asia and Africa, the low public awareness of existing problems and the need for preventive measures mean that the public's role will be more vital throughout the process and may need to be continuous rather than iterative.

Local Dimensions of Air Quality Governance: AQMA

This case was overtly an empirical experiment to create "more socially accountable and reflexive" knowledge, after Michael Gibbons, Camille Limoges, Helga Nowotny, et al. (1994). The aim was the cocreation of citizen and scientific knowledge that was both socially robust and scientifically reliable (Forrester, Potts, Cinderby, et al. 2002). However, the process of setting national air quality targets in the United Kingdom also happened as part of a larger policy process. This case focuses on the most grounded stage of the policy cycle, the process of setting up local AQMAs. In the United Kingdom, there was early concern about air pollution in cities, leading to a range of control measures. By the 1990s and as a result of the European Union imperative to improve the air quality in its cities (Directive 96/62/EC), the central government initiated a program asking every local authority in the country to perform a technical assessment of current air quality and then to model that assessment to come to a decision regarding areas that might exceed national guidelines on air quality in the future (UK Govt. 1995, Part IV, section 83). As a result of that technical process, areas that were modeled to exceed those target limits were considered AQMAs. Although statutorily not required to make those areas meet targets, local authorities were statutorily obliged to institute realistic "Action Plans" that might allow those areas to meet the targets. What was implemented in the setting up of AQMAs was primarily policy output—that is, the declaration of an area and then a plan. These two levels in this case involved successful stakeholder engagement.

At the AQMA declaration stage, the widely accepted guidelines suggested that stakeholders should include government agencies, neighboring local authorities, businesses, and any stakeholders that the air quality officers "thought appropriate" (UK Govt. 1995, Schedule 11; NSCA 1999, 2). This determination produced mixed results, with some local authorities perceiving engagement as a communications exercise and others perceiving it as a more participatory process (Dorfman, Beattie, Burnet, et al. 2006). It did, however, provide a vehicle for the local technical community to engage with the local policymaker community; interestingly, experience in York suggested that air quality officers were unconvinced that the technically warranted knowledge alone would sway policy decisions makers (i.e., would produce "usable knowledge"), so they sought to involve citizen stakeholders as a means of communicating to policymakers (Cinderby and Forrester 2005). Yet even where local authority officers have been successful in engaging citizen stakeholders in the cocreation of policy, such as in the decision over the boundaries of York's AQMA (as described in Cinderby and Forrester 2005), the literature tells us that that engagement is apparently not replicated elsewhere even within the United Kingdom (Petts and Brooks 2006). It remains that "although local authorities were supportive of the theoretical underpinning of greater stakeholder inclusion in local air quality management processes, the actual process of consulting was generally perceived to be more complex, uncertain and problematic" (Air Quality Management Resource Centre 2006, 3).

These qualifications do not mean that there was no useful engagement. Paul Dorfman, Nurul Leksmono, Jim Longhurst, and their colleagues (2006) talk of how "embedded engagement" in Sheffield led to integrated consultation—where the citizen consultation process happened as part of an ongoing relationship with citizen stakeholders rather than being an "add-on." However, this engagement revolved around countering a perceived lack of trust of official information (Yearley, Cinderby, Forrester, et al. 2003). The very successful partnerships fostered by Sheffield City Council, where local people took part in scientific monitoring of local air pollution by performing tasks such as changing diffusion tubes (an apparatus to monitor levels of nitrogen dioxide), may arguably bring the issue of air pollution down to the local level, but they are only the first few rungs on any "ladder" of participation in policy-making or governance. They may put people in charge of handling information, but not its formulation. Citizen stakeholders are allowed to see information, use it, but not influence or alter its meaning. Steven Yearley (chap. 9 in this volume) makes the point that "safe" levels were not on the agenda for discussion and that local "participation" tended toward consultation over the location of monitoring equipment. Nonetheless, Sheffield's work has been cited as an example of good practice in community involvement, but we argue here that although it exhibits good science communication and good stakeholder engagement, it does not show meaningful stakeholder participation in the wider governance of air quality above the local level. It is important to emphasize here that the central government designed the local authority AQMA process to complement national air quality standards: national targets were imposed, and thus local authorities and local stakeholders were unable to question the values set in order to determine whether there was a need for an AQMA or not in their locality. This imposition was particularly problematic when international scientific evidence was pointing to either the absence of safe levels (in the case of particulate matter) or lower safe levels than those set by national government.

Discussion: Lessons and Concepts for Stakeholder Engagement

In this chapter, we have defined the "stakeholder" broadly as any possessor of an interest in scientific research around air quality (i.e., can influence it or is influenced by it) and "engagement" as making the outputs of that research usable to the policy and end-user community. However, these definitions encompass two different rationales for engaging stakeholders: first, to create better, more inclusive knowledge; and, second, to engage the end user in order to influence her or his action. The latter demands a discussion of the proper role of science and scientific advice in the policy process. Roger Pielke cites Lord Robert May as saying, "The role of the scientist is not to determine which risks are worth taking, or deciding what choices we should take, but the scientist must be involved in indicating what the possible choices, constraints and possibilities are. . . . The role of the scientist is not to decide between the possibilities but to determine what the possibilities are." Pielke continues: "Must scientists [thus] be at the mercy of politics? The answer is 'no,' but empowerment [of the role of science] depends on understanding the difference options available for relating to policy and politics" (2007, v). The practical application of this science role can build on insights presented elsewhere in this volume (e.g., in chaps. 1, 5, and 12).

Stakeholders change as we move from the international to the local level, and the stakeholder's role also changes—as does the rationale according to the actor's evolving perspective and the different stages of the policy cycle. Thus, we argue here that at the metalevel—looked at from the perspective of the analysis, meaning the holistic level of the whole policy cycle, not just its component parts—we need not just dialog but "trialogue" between the primary conceptual stakeholder communities of policy (politician, bureaucrats, policymakers, and enforcers); science (experts); and "public" (polluters, including the private sector and citizens, as well as implementers of policy). These "three" communities are, of course, not mutually exclusive. They are, it can be argued, searching for different types of knowledge or knowledges that can be held to be "true" for different reasons. These three communities have their own, self- and externally recognized grounding for knowledge. Politicians search for "usable knowledge"—that is, knowledge that has a technical pedigree but that is also clearly implementable; researchers search for technically warranted, reliable knowledge; and "the public" (including interest groups) and the private sector (including industry) often have rationales that reflect the complexity of the issues in a locally grounded or economically pragmatic manner. Thus, this description can be simplified to the contention that the three "types" of stakeholder look for different types of knowledge; given the fact that there are multiple policy cycles at different levels of governance, however, the direction of the flow of knowledge between these different "grounded-knowledge communities" may change from one policy-cycle stage to another and from one policy-cycle level to another. The process of stakeholder engagement and communication between stakeholders thus has to adapt to suit not only the actors, but also the policy-cycle stages.

The cases show how the variation panned out on the ground and different concepts of who or what a stakeholder is (or should be). They also demonstrate that the absence of social scientists in these decision-making spheres can lead to missed opportunities. Guidance to practitioners failed to suggest *why* stakeholders should be engaged; many practitioners were rushed into performing a tick-box exercise that resulted in little meaningful engagement, but rather mostly consultation or communication and thus not full, cooperative engagement or participation (see Forrester, Swartling, and Lonsdale 2008). Bridging science and policy would certainly gain in efficiency if participants in the policy process started by thinking about why to do it as well as how to do it. Nonetheless, the cases learned—in a pragmatic way if not in an academic manner—from one another: in particular, Malé and APINA learned from CLRTAP. In one way, the UK urban case study can also be seen as being nested within wider European concerns of which CLRTAP is a part. There is as yet no local-level African case to parallel the UK one, although there are Asian examples of engagement at the city level (see SEI, KEI, UNEP, et al. 2004; Schwela, Haq, Huizenga, et al. 2006). For now, it is important to note that the cases, as they are described here, took place at different times and at different points in the policy cycle and in entirely different circumstances.

Science to Policy—Theory to Practice

Many air quality scientists are familiar with theoretical definitions of who the major stakeholders are, but when it comes to engaging with different stakeholder groups, practicing the ideal of participatory engagement (see Forrester, Swartling, and Lonsdale 2008) is not always followed as closely as might be possible or desirable. Although it has been shown that engaging stakeholders can result in better policy output (see, for example, the local AQMA case where the policy output was much stronger as a result of citizen engagement, and the APINA case where the uptake of the policy advice is higher as a result of stakeholder engagement), and stakeholder engagement is one way of focusing knowledge on relevant user communities, where systemic uncertainty exists—that is, where scientific knowledge producers are unsure of which course of action to take to contribute to "usable" knowledge—then the engagement of stakeholders can also contribute to the good governance of science and help to direct research priorities. Such engagement also enables scientists to make sure that their research findings have more direct policy impact.

However, from the science and policy perspective, those who can influence the process most significantly, such as industry, are often favored in stakeholder engagement processes because technopolitical processes are usually outcome led. Thus, the social science theory needs to take into account the practicalities of who is able to influence the process, and the policy dialogs need to take into account the democratic (and the democratization of science [Forrester, 2006]) argument. Processes such as the RAPIDC program and APINA have been carried out, until now, largely divorced from higher-level theory such as that characterized by the work of, Helga Nowotny, Jerome (Jerry) Ravetz, and others, reflecting the gap between social science researchers and natural scientists who aim at having an impact on policy. They rarely meet. Practitioners are also finding that old models of engagement and communications are less useful in practice, and experience from participatory research at the local level (e.g., the UK case) may also not be sufficient when dealing with regional or global issues.

The past decade of RAPIDC project work has been spent generating technical knowledge of the emission, deposition, and impact of air pollutants, and although this work needs to continue to develop the important building blocks upon which policy formulation can develop, the focus is now changing toward looking for implementable solutions and "usable knowledge." Importantly, both the Malé and APINA processes have recognized the necessity to foster science engagement with policy in order to make scientific knowledge useful for policymaking, but both have also recognized the need to bring broader stakeholder engagement into the science–policy process.

The demand for inclusion requires academics and scientists (and policymakers) to move out of their disciplinary framework: the level of cross-disciplinary and cross-grounded knowledge-community interpretation required for good interstakeholder communication is not to be underestimated (Folke, Hahn, Olsson, et al. 2005; Tress, Tress, and Fry 2005). With the acknowledgment that scientists are experts in their own field of competence, it is worth asking whether they should also be expected to be experts in synthesizing and communicating scientific results and whether this task might better involve professional communicators to form intermediary links between scientists and other stakeholders. The policy community can also be gainfully involved in this role of mediation because it would help their search for usable knowledge. In the worst cases, stakeholder engagement addresses just one-way transmission of information; in the best cases, it produces cooperative inquiry into the definition and understanding of the problem and potential solutions.

One particular strength of the RAPIDC "policy dialog" approach—as used in both Malé and APINA—is that scientists can engage with the policy community's needs in situations where such dialog is not the norm. Awareness of issues has been increased significantly by the RAPIDC work: that is, awareness among the scientific communities in Malé Declaration countries and awareness among the policy communities in those countries. APINA has also engaged with African industrial interests, which are significant emitters of pollutants and have been involved in most meetings. Therefore, policy dialogs held by means of the funding provided through RAPIDC have shown themselves a useful tool in developing communication between scientists and government officials at all levels and in helping all parties reach "convergence" (cf. Lidskog and Sundqvist, chap. 1 in this volume), resulting in governance that is potentially as successful as CLRTAP. An important message from these experiences is that the policy cycle should not be separate from the science-challenge cycle—that is, the cycle of producing scientific knowledge—and, indeed, stakeholder engagement in both public policy and science policy (also known as "the governance of science") can be used to draw these two cycles together to provide useful and usable output as well as mutually beneficial outcomes.

Citizen Engagement in the Governance–Policy Cycle

The citizen has a democratic right to be a stakeholder in *all* of the phases of the policy cycle. This is not to say that the citizen's role is the same in all phases. In some phases, they may be an observer and stand farther down the "ladder of participation," yet in other phases they may be a contributor of knowledge, and in yet others they may be an implementer. Specific stakeholders such as industry, polluters, and local residents have local knowledge of their own, and this knowledge needs to be engaged. This right to be a stakeholder is above and beyond any right the citizen has to be involved as a citizen. Certain rationales for citizen stakeholder inclusion need to be emphasized, such as the democratic right (Pateman 1970; Forrester 1999; Wilsdon and Willis 2004).

Stirling (2005) notes three "meta" rationales for stakeholder engagement: the "normative democratic," where the participation is a democratic right; the "instrumental," where engagement is purpose driven toward specific predetermined ends; and the "substantive," where stakeholder engagement can lead toward better ends. John Forrester (1999) has also cited three rationales for stakeholder engagement: democracy (akin to Stirling's normative democratic); economics (leading to more efficient public services); and improvements to scientific knowledge. However, the assumption that stakeholder engagement always leads to *more efficient* governance is probably erroneous.

Nonetheless, it is contended and supported in the literature that stakeholder engagement can lead to *better* governance—what Stirling calls the substantive rationale, which in the case of the governance of air quality can also produce improvements to scientific knowledge. A top-down, expert perception is that citizens should be engaged only at the bottom stage of the policy cycle (the UK AQMA case study). This perception is problematic. The citizen should have a democratic *and* an instrumental *and* a substantive input to all the stages: this combination brings the citizen back into stakeholder.

However, it is more difficult to bring wider stakeholders into an already running process over which they feel they have little or no control (e.g., the AQMA discussions). The UK case also shows how it can be problematic when (good) citizen stakeholder engagement takes place at the local level but is not clearly linked to any stakeholder engagement that may have taken place at "higher" levels of the policy cycle: there is a disjointedness between strategic policy setting and local stakeholder engagement (Air Quality Management Resource Centre 2006; cf. Cinderby and Potts 2007). The lesson here from the local and regional levels is that all three grounded-knowledge communities—science, policy, and public—really need to be brought together at the start and continue to "trialogue" until the end of the process if the end result is to be judged a success by everyone involved, with buy-in from all the different stakeholder communities. Thus, the stakeholder engagement process really needs to engage the end user effectively and start thinking about appropriate solutions that bridge the differently grounded knowledge communities right from the start. In addition, the UK AQMA case shows that technical officers from local authorities were not engaged in the development of the remit within which they worked and could not question the targets or other aspects of the project that had an influence on local policy; thus, one can argue that local technical expertise was also disenfranchised. The opportunity for cocreation of scientific knowledge (cf. the discussion of the STS approach in Lidskog and Sundqvist, chap. 1 in this volume) with any stakeholders was simply not present, although cocreation of policy knowledge is possible.

There is a loose consensus among policy analysts that the policy process and therefore policy change are often incremental in nature and that patience is a virtue (see Lindblom 1959). John Hudson and Stuart Lowe (2004) more recently offered a useful summary of the impact of policy networks on policy change, but this impact, too, is incremental. European and UK examples demonstrate that the translation of strategic policy into policy measures (and on "down" to implementation) is not an easy or straightforward task. The incremental nature and complexity must be recognized, alongside the barriers that make it incremental: time, money, politics, priorities, and windows of opportunity. The approach advocated in this chapter is pragmatic, appreciates the nature of policy, and attempts to assess, analyze, and overcome some of the barriers that need not be there (e.g., *perceptions* of cost) as well as some of the ethical issues. These barriers and issues may be addressed by better communication and understanding as a result of greater stakeholder engagement (see Plummer and Armitage 2007; Plummer 2010). Thus, although this chapter advocates a holistic overview approach to stakeholder engagement with the "whole" policy cycle, it also recognizes that stakeholder engagement will continue separately in each section of the cycle: integration, translation, implementation, and evaluation. This incrementalization brings us back to the necessity for integration of stakeholders: Stephen Dovers distinguishes between "policy integration" and "integration in research": for the development of an integrative capacity you need a "sophisticated understanding of the interactions between highly complex non linear and often closely dependent human and natural systems" (2005, 3). Thus, stakeholder engagement, itself necessary for integration, is a nonlinear and complex process.

Conclusion

As the air quality governance policy cycle spirals down from a high-level "strategic" dialog cycle to the regional, national, and then local levels, it enters the place where pragmatic "scheme-level" dialog must take place. This process dovetails with the change from high-level policymakers as stakeholders to end users as stakeholders and parallels a change in the focus of the communication within the stakeholder engagement from one primarily of transmission of a message to one that includes reception of local needs and capability as well as a cocreation of knowledge. The successful understanding of local effects can also help the successful transmission of the important message to policymakers. Knowledge is an important driver. Stakeholder engagement is and should be an iterative and complex process. For instance, in the local air quality management case in York, the independent corroboration of the technical message from the air quality officers allowed local officials to push for a stronger policy than they at first thought might be politically acceptable. However, we have also seen in this same example the failure of this good stakeholder engagement at the local level to be picked up and replicated at the national or international level. Thus, a mechanism must be found to allow multiple knowledge(s) to feed back "up" to higher governance levels. An iterative, continuous process of stakeholder engagement would arguably do this, but greater guidance on the rationales for stakeholder engagement would also help practitioners.

The best science in the world without communication, stakeholder engagement, and engagement with the policy process is not going to make much difference. Further, the impacts of air pollution as well as solutions for its amelioration need to be packaged in a socioeconomic context for decision makers, but practitioners should not lose sight of the potential offered by the social science literature. For example, Yearley (chap. 9 in this volume) notes that the arguments used in technopolitical stakeholder engagement and the form that negotiations take marginalize "those factors that are usually used to justify public participation in scientific processes." In all of the cases cited in this chapter, the engagement of science with policymakers is recognized as important and has been attempted with reasonable success. However, the process must better understand the social as well as the economic and the technical; stakeholder engagement is one way to locate that process socially, but such engagement is rarely attempted in a manner that is systematically influenced by the academic literature. Bringing together the democratic right thesis for the definition of stakeholder with the extended peer-review rationale is one way to frame this issue, but considering Yearley's thesis that there are "other" factors to justify public participation in scientific processes is also worthwhile (see also Forrester 1999 and Stirling 2005).

Multistakeholder engagement, which we suggested at the outset is necessary for producing usable knowledge, itself necessitates good communication, and expert discourses must be translated. Although for successful outcomes we argue that the process must be located in technically warranted knowledge, we also believe that it must also be democratically warranted (as well as economically viable). It must be open and meaningful to all stakeholders, not just the science stakeholders and the policy stakeholders. However, making regional-scale concerns and policymaking meaningful to "local" citizens is more difficult than when dealing with local issues, but it is, we argue, necessary. An important point is that all three of our stakeholder communities—expert, policy, and citizen—and their different knowledges should be included at the problem identification stage: that is, as early in the policy cycle and the scientific research cycle as possible. This point has important implications for the "epistemic communities approach" advocated by Rolf Lidskog and Göran Sundqvist, (chaps. 1 and 12 in this volume).

One thing that is certain is that as the list of stakeholders grows, so the norms of science—which suffice for disciplinary scientists to interact socially together—will no longer suffice in the new milieu created by successful EPI: scientists should certainly adapt their message to make it interpretable and "usable," but, conversely, policymakers and other stakeholders may need to adapt the way they use science compared to other forms of knowledge and other forms of evidence, even when they are acting in the policy sphere. What was once a relatively simple message-transmission process is now a much more complex information-sharing and colearning process. Further, more effort needs to be expended not only on bringing the stakeholder communities together but on facilitating communication between them and ensuring (social) learning across and between the groups. Finally, social scientists need to note that descriptions and models of effective stakeholder involvement are often ideals that are difficult to obtain in reality. We must ask ourselves how we can help practitioners overcome this barrier.

Acknowledgments

Although John Forrester worked closely with the air pollution scientists who are coauthors—and two of whom were interviewed for this paper—he is not an air pollution scientist himself. The interpretation and analysis of the data in the case studies are his and do not necessarily represent the views of the scientists concerned. Thanks are also due to Dr. Paul Rosen and Professor Mike Ashmore, who read and commented on previous drafts of this paper. Thanks are also due to Dr. Julia Garritt of Open Eye Research for the term *trialogue*.

The Stockholm Environment Institute work on the developing country cases, Malé and APINA, was initiated under the auspices of the Swedish international development agency–funded RAPIDC program (http://www.sei-international.org/rapidc/apina.htm). The UK work was funded by the UK Economic and Social Research Council's Science in Society Programme (www.sci-soc.net/).

References

Air Quality Management Resource Centre. 2006. Consultation as Science Communication? The Case of Local Air Quality Management. Economic and Social Research Council Science in Society project report. Bristol: University of the West of England. Available at http://www.uwe.ac.uk/aqm/esrc.

Carpenter, Stephen, Carl Folke, Marten Scheffer, and Frances Westley. 2009. Resilience: Accounting for the Noncomputables. *Ecology and Society* 14 (1): article 13. Available at http://www.ecologyandsociety.org/vol14/iss1/art13.

Cinderby, Steve, and John Forrester. 2005. Facilitating the Local Governance of Air Pollution Using GIS for Participation. *Applied Geography* 25: 143–158.

Cinderby, Steve, and Laura Potts. 2007. Suspicious Cartographers: Some Realities of Research into Stakeholder Understanding of the Causes and Possible Prevention of Breast Cancer. *Science & Public Policy* 34 (5): 345–354.

Dorfman, Paul, Clare Beattie, Frank Burnet, Dave Gibbs, Jim Longhurst, Emma Weitkamp, and Nurul Leksmono. 2006. A Conceptual Model of the Role of Complex Science in Local Authority Consultations about Air Quality Management. *Local Environment* 11 (4): 399–419.

Dorfman, Paul, Nurul Leksmono, Jim Longhurst, Dave Gibbs, Frank Burnet, Emma Weitkamp, and Clare Beattie. 2006. *Steps to Better Practice: Guidance for Local Authorities on LAQM Consultation.* Bristol: University of the West of England.

Dovers, Stephen. 2005. Clarifying the Imperative of Integration Research for Sustainable Environmental Management. *Journal of Research Practice* 1 (2): manuscript1. Available at http://jrp.icaap.org/index.php/jrp/issue/view/2.

Duit, Andreas, and Victor Galaz. 2008. Governance and Complexity—Emerging Issues for Governance Theory. *Governance: An International Journal of Policy, Administration, and Institutions* 21 (3): 311–335.

Elliot Major, Lee. 2000. What Is Evidence Based Policy? *The Edge* (UK Economic and Social Research Council) 3: 20–21.

European Union. 2006. Consolidated Version of the Treaty Establishing the European Community. *Official Journal of the European Union* C 321: E1–E331.

Folke, Carl, Thomas Hahn, Per Olsson, and Jon Norberg. 2005. Adaptive Governance of Social-Ecological Systems. *Annual Review of Environment and Resources* 30: 441–473.

Forrester, John. 1999. The Logistics of Public Participation in Environmental Assessment. *International Journal of Environment and Pollution* 11: 316–330.

Forrester, John. 2006. Science and Public Participation: The Democratisation of Science. In *Blackwell Encyclopedia of Sociology*, ed. George Ritzer, 4088–4091. Oxford, UK: Wiley-Blackwell.

Forrester, John, Laura Potts, Steve Cinderby, and Paul Rosen. 2002. Creating "Science's New Social Contract"? Paper presented at the European Association for the Study of Science and Technology annual conference, York, UK, 31 July–3 August.

Forrester, John, Åsa Swartling, and Kate Lonsdale. 2008. *Stakeholder Engagement and the Work of SEI: An Empirical Study*. Stockholm: Stockholm Environment Institute.

Freeman, R. Edward. 1984. *Strategic Management: A Stakeholder Approach*. Boston: Pitman.

Funtowicz, Silvio, and Jerome Ravetz. 1991. A New Scientific Methodology for Global Environmental Issues. In *Ecological Economics: The Science and Management of Sustainability*, ed. Robert Costanza, 137–152. New York: Columbia University Press.

Gibbons, Michael. 1999. Science's New Social Contract with Society. *Nature* 402 (supp.): c81–c84.

Gibbons, Michael, Camille Limoges, Helga Nowotny, Simon Schwartzmann, Peter Scott, and Martin Trow. 1994. *The New Production of Knowledge: The Dynamics of Science and Research in Contemporary Society*. Thousand Oaks, CA: Sage.

Grennfelt, Peringe, Rolf Lidskog, Lars Lindau, Rob Mass, Frank Raes, Göran Sundqvist, and Jenny Arnell. 2006. *Towards Robust European Air Pollution Policies: Constraints and Prospects for a Wider Dialogue between Scientists, Experts, Decision-makers, and Citizens. A Workshop Report*. IVL Report no. B1671. Stockholm: Swedish Environmental Research Institute.

Hudson, John, and Stuart Lowe. 2004. *Understanding the Policy Process: Analysing Welfare Policy and Practice*. Bristol: Polity Press.

Kasperson, Roger. 2007. Foreword. In *Environmental Policy Integration in Practice: Shaping Institutions for Learning*, ed. Måns Nilsson and Katerina Eckerberg, xv–xvi. London: Earthscan.

Kates, Robert, William Clark, Robert Corell, J. Michael Hall, Carlo Jaeger, Ian Lowe, James McCarthy, et al. 2001. Sustainability Science. *Science* 292 (5517): 641–642.

Liberatore, Angela, and Silvio Funtowicz. 2003. "Democratising" Expertise, "Expertising" Democracy: What Does This Mean, and Why Bother? *Science & Public Policy* 30 (3): 146–150.

Lindblom, Charles. 1959. The Science of Muddling Through. *Public Administration Review* 19: 78–88.

Nilsson, Måns. 2007. Shaping Institutions for Learning. In *Environmental Policy Integration in Practice: Shaping Institutions for Learning*, ed. Måns Nilsson and Katerina Eckerberg, 163–168. London: Earthscan.

NSCA (National Society for Clean Air and Environmental Protection). 1999. *Consultation for Local Air Quality Management: The How to Guide*. Brighton, UK: NSCA.

Pateman, Carole. 1970. *Participation and Democratic Theory*. Cambridge, UK: Cambridge University Press.

Petts, Judith, and Catherine Brooks. 2006. Expert Conceptualisations of the Role of Lay Knowledge in Environmental Decisionmaking: Challenges for Deliberative Democracy. *Environment & Planning A* 38 (6): 1045–1059.

Pielke, Roger. 2007. *The Honest Broker: Making Sense of Science in Policy and Politics*. Cambridge, UK: Cambridge University Press.

Plummer, Ryan. 2010. Looking forward from the Nexus of Research and Practice: Learning from Issues Emerging with Adaptive Co-Management. In *Overcoming the Challenges of "Doing Participation" in Environment and Development*, ed. John Forrester and Åsa Gerger Swartling, 21–27. Stockholm: Stockholm Environment Institute.

Plummer, Ryan, and Derek Armitage. 2007. Crossing Boundaries, Crossing Scales: The Evolution of Environment and Resource Co-Management. *Geography Compass* 1 (4): 838–849.

Schwela, Dieter, Gary Haq, Cornie Huizenga, Han Wha-jin, Herbert Fabian, and May Ajero. 2006. *Urban Air Pollution in Asian Cities*. London: Earthscan.

SEI, KEI, UNEP, WHO, and CAI–Asia (Stockholm Environment Institute, Korea Environment Institute, United Nations Environment Program, World Health Organization, and Clean Air Initiative for Asian Cities–Asian Development Bank). 2004. *A Strategic Framework for Air Quality Management in Asia*. Stockholm: SEI, KEI, UNEP, WHO, and CAI–Asia.

Siebenhüner, Bernd. 2002. *How Do Scientific Assessments Learn? A Comparative Study of the IPCC and LRTAP*. Belfer Center for Science and International Affairs Discussion Paper 2002-05. Cambridge, MA: Kennedy School of Government, Harvard University.

Stirling, Andy. 2005. Opening Up or Closing Down? Analysis, Participation, and Power in the Social Appraisal of Technology. In *Science and Citizens: Globalization and the Challenges of Engagement*, ed. Melissa Leach, Ian Scoones, and Brian Wynne, 218–231. London: Zed Books.

Stringer, Lindsay, Andy Dougill, Evan Fraser, Klaus Hubacek, Christina Prell, and Mark Reed. 2006. Unpacking "Participation" in the Adaptive Management of Social–Ecological Systems: A Critical Review. *Ecology and Society* 11(2): article 39. Available at http://www.ecologyandsociety.org/vol11/iss2/art39.

Tress, Bärbel, Gunther Tress, and Gary Fry. 2005. Integrative Studies on Rural Landscapes: Policy Expectations and Research Practice. *Landscape and Urban Planning* 70: 177–191.

UK Govt. (United Kingdom Government). 1995. *The Environment Act 1995*. Available at http://www.legislation.gov.uk/ukpga/1995/25/contents.

United Nations. 1992. *Agenda 21: The United Nations' Programme of Action from Rio*. United Nations, New York. Available at http://www.un.org/esa/dsd/agenda21/.

Webster, Andrew. 2007. Crossing Boundaries: Social Science in the Policy Room. *Science, Technology, & Human Values* 32 (4): 458–478.

Wilsdon, James, and Rebecca Willis. 2004. *See-Through Science: Why Public Engagement Needs to Move Upstream*. London: Demos.

Yearley, Steven, Steve Cinderby, John Forrester, Peter Bailey, and Paul Rosen. 2003. Participatory Modelling and the Local Governance of the Politics of UK Air Pollution: A Three-City Case Study. *Environmental Values* 12: 247–262.

Ziman, John. 1978. *Reliable Knowledge: An Exploration of the Grounds for Belief in Science*. Cambridge, UK: Cambridge University Press.

Ziman, John. 2000. *Real Science: What It Is, and What It Means*. Cambridge, UK: Cambridge University Press.

IV

Environmental Governance and Research

12

Science–Policy–Citizen Dynamics in International Environmental Governance

Rolf Lidskog and Göran Sundqvist

This book focuses on the regulatory regime for measuring, sorting, and controlling airborne substances harmful to the environment and public health. In this regime, actors, instruments, practices, and ideas are drawn together to create clean air. By developing norms, rules, and knowledge, actors are expected to reduce emissions of harmful substances. What is seen as a "simple" international agreement—such as the Convention on Long-Range Transboundary Air Pollution (CLRTAP) and its protocols or the European Union (EU) air pollution policy—is only the tip of an iceberg. What seems to be a static and solid regulatory regime, guided by formal agreements, is the result of dynamic and unpredictable processes in which actors struggle to disseminate their understandings of the main causes of pollution and its relevant remedies throughout society.

As emphasized in this book, European air pollution regulation is in transition. The relatively stable regulation of long-range air pollution is currently being challenged by organizational change, scientific advances, and new political priorities. The regulation is currently in a formative phase, and governmental bodies, nation-states, and expert communities are seeking new ways to establish further measures and more robust regulations. In these efforts, the relationships among science, policy, and citizens are pivotal.

In this concluding chapter, we return to the book's overall objectives: to evaluate the regulation of transboundary air pollution and to develop a theoretical understanding of the science–policy–citizen dynamic in international environmental governance. These two objectives are closely related. Without a theoretical understanding of the relationships among expertise, policymaking, and citizen involvement, it is impossible to analyze the trajectory of international air pollution–abatement efforts or their current status and future prospects. Likewise, improving our understanding of the dynamic relationship between science, policy, and citizens requires detailed empirical studies to reveal what in these processes leads to the regulation—or nonregulation—of transboundary issues.

This book answers the urgent need for a better understanding of how regulation is shaped. The point of departure is the realization that regulation is dynamic and that various actors strive to influence regulatory processes. The empirical focus is on European regulation of transboundary air pollution, but the findings are also highly relevant to other areas of international environmental governance.

Unlike several previous studies of transboundary regulation in general and of air pollution regulation in particular, this book delves more deeply into processes that render air pollution governable. It claims that the dynamics of this kind of regulation are underresearched, not least in terms of how various actors and processes influence each other. By focusing on processes rather than products, strategies rather than interests, learning rather than knowledge, this book provides a nuanced view of how air pollution is rendered governable. In addition, because this book does not view the field of international regulation as the object solely of nation-state policymaking (supported by scientific communities) but also includes the role of citizens, it contributes to the ongoing discussion of how to construct socially robust and trusted regulation.

The introductory chapter presented three approaches to the study of international environmental governance: the institutional approach, the epistemic-community approach, and the science and technology studies (STS) approach. Each approach has its pros and cons, and although convergence between them is neither possible nor desirable, we believe that cross-fertilization is necessary to improve our theoretical understanding of the science–policy–citizen dynamic in international environmental governance. By bringing together contributions from the academic fields of international relations (IR) and STS as well as from other areas of social science, this book contributes to a broader and deeper understanding of the issue.

This concluding chapter proposes new prospects for IR, not least in achieving a more dynamic conception of the science–policy–citizen interplay, which is highly relevant to deepening our understanding of European clean air policy efforts. It also aims to enrich the STS approach by relating it to IR findings and perspectives, thereby making the STS approach more suitable for the study of international environmental governance.

In the second section, we turn our attention to the regulation of European air pollution and, based on findings from previous chapters, evaluate its trajectory, current situation, and future challenges. After this empirically oriented section, we consider how to improve our theoretical understanding of international regulation. The third section starts with a broad overview in which we situate the current discussion of how to create more efficient and trustworthy air pollution policies in a wider sociopolitical context. In the fourth section, we discuss the possibility of convergence between the three analytical perspectives mentioned earlier. The fifth and sixth sections are devoted to discussing the three building blocks of this book—namely, policy, science, and citizens. These spheres are obviously not independent and autonomous, but before discussing their relationships we describe them individually using results from the previous chapters. Finally, we examine the relationships and dynamic between policy, science, and citizens.

All chapters in this book implicitly or explicitly ask what role social science should play in the development of international environmental governance; we accordingly conclude by discussing this normative question. We argue that the prescriptive role of social science should not be ignored. This is not to say that social scientists should be political advisers in any restricted sense; rather, their responsibility should not be confined to the nation-state but should encompass the wider society in their critical evaluations of regulatory strategies. We conclude that the role of social science is more concerned with identifying possible alternatives than with narrowing the field and defending specific choices.

Air Pollution Policy: Experiences and Prospects

European air pollution policies are in transition. Many commentators and policy analysts have cited CLRTAP as a successful example of international environmental governance, and the resulting cuts in the emissions of many airborne pollutants are definitely impressive. Nevertheless, these evaluators question the extent to which CLRTAP can continue to be a central instrument for ongoing efforts to reduce emissions. Is it possible to make further emissions cuts, and if so, how should they be made? Growing political complexity, new scientific findings, lack of public support, and competition with other problems—both environmental and nonenvironmental—are seen as major obstacles to advancing policy development in this area.

As demonstrated in the preceding chapters, CLRTAP's success can be questioned or at least problematized. Emissions of several airborne substances, not least sulfur, have successfully been reduced, but these reductions are only in part explained by CLRTAP's policy development. Other factors, such as economic restructuring and technological development (e.g., low-sulfur fossil fuels and desulfurization equipment), have greatly contributed to the reduced emissions. In addition, emissions of other airborne pollutants—not least ammonia and to some extent nitrogen oxides and particulate matter—are still far from sufficiently reduced.

CLRTAP has obviously been an important political tool. Perhaps its chief contributions have been to focus attention on and to create and disseminate knowledge about the problem of air pollution; in terms of actually dealing with the problem, it has been only moderately successful. These successes should not, however, be overlooked. Creating and disseminating knowledge about an environmental problem is important, irrespective of whether it is solved through political decisions or through technological development and economic restructuring that take place far from the arena of international negotiations. At the same time, most air pollutants are far from being successfully dealt with, either by political means or by economic forces and technological development. The current situation is complicated, and politicians and experts engaged in air pollution–abatement efforts are pondering how to create relevant, effective, and trusted policies to combat problems caused by transboundary air pollution.

Alongside a greening of society, we are also witnessing graying processes. Whereas emissions of acidifying substances have decreased in Europe, North America, and Japan, they are increasing in countries that are experiencing rapid industrial development (Moldan 2007, 100). China, Brazil, and India face problems of acidification. Furthermore, air pollution today concerns much more than lake and forest acidification and terrestrial eutrophication. Buildings, crops, biodiversity, and human health are threatened by air pollution. Increasing attention is being paid to the intercontinental transport of pollutants, which means that regional conventions—such as CLRTAP and the EU air pollution policy—can be only part of a solution, which must be globally organized. Some commentators have suggested that transboundary air pollution problems should now be incorporated into the climate-change convention (Pleijel and Grennfelt 2007; Pleijel 2009; cf. Somerville 2008). This suggestion has been made in part because global warming is increasingly considered the main air pollution issue (not to mention the number one environmental problem) and in part because substantial linkages exist between global warming and certain transboundary problems—for example, tropospheric ozone, which is both an air pollutant and a greenhouse gas.

The energy and transport sectors—not least shipping with its high emissions—are other major sources of both air pollution and greenhouse gases (EEA 2006; Pleijel and Grennfelt 2007).

Complicating the situation are doubts as to whether the current approach to air pollution is relevant, feasible, and desirable. Policy development has hitherto involved a close relationship between science and policy, in which national representatives and scientific experts have co-constructed protocols and abatement strategies. Policymakers and experts are increasingly realizing that public support is needed if we are to develop abatement strategies for further reducing emissions. There is, however, great uncertainty about how to configure relationships with citizens so as to facilitate policy development. In existing policy documents, citizens are often understood as passive receptors whose function is simply to support proposed policies, not substantially influence them. An important question is to what extent this view of citizens actually benefits regulatory work. Several chapters of this volume ask to what degree CLRTAP and the EU's air pollution policies can serve as mechanisms for learning not only for the national representatives and experts involved, but also for other actors. The multipollutant/multieffect protocol signed in Gothenburg in 1999 illustrates this problem. The protocol is complicated and difficult to translate into a language comprehensible to actors not involved in the negotiations. Socially robust—relevant, efficient, and trusted—regulation of transboundary air pollution is being sought; however, to understand the conditions and feasibility of this search, we must locate the issue in a broader sociopolitical and theoretical context.

A Changed Context for Transboundary Regulation

Politics concerns the power to influence society—to exert control by restricting certain activities and enabling others. The nation-state has traditionally been the locus of politics, having the administrative capacity to develop policy, the legitimacy to make binding decisions, the economic resources to carry them out, and the coercive power to enforce them (Giddens 1985). Regulation has been understood as a unilateral state activity of formulating and implementing rules backed by various types of sanctions (Black 2002).

Due to globalization processes, the power of the territorially bounded nation-state is in question. Other actors—such as transnational corporations, interest organizations, and social movements—also influence policymaking. The "container model of society," in which the nation-state is seen as providing durable borders for social processes, is becoming obsolete (Beck 2006, chap. 1).

Globalization processes not only challenge the nation-state's sovereignty but offer new ways of exerting power and influence. Globalization does not mean the end of state power and state regulation, but it does force the nation-state to use other strategies to exert influence and manage activities (Rose 1999; Lidskog, Soneryd, and Uggla 2009).

This shift is visible in current trends in regulation, such as the recent emphasis on legitimacy and trustworthiness. Attention was earlier directed at problems *within* risk analysis—that is, how risks were defined and approached by technical and calculative means. Since the 1990s, however, the legitimacy of government and regulation has increasingly been emphasized (Power 2007). Organizations must ponder not only how to deal with risks, but also how to deal with actors who may question both the legitimacy of current methods for regulating risk and the trustworthiness of organizations responsible for this regulation. Prompted by several failures, regulation has started to account for and deal with public opinion and public perceptions of risks, not only in order to handle criticism, but to forestall it (Löfstedt 2005). Programs for risk communication, public relations, stakeholder dialog, and public involvement are today integral to both public and corporative governance (Gouldson, Lidskog, and Wester-Herber 2007). Calls for more inclusive and transparent processes, public dialog, and democratic engagement are widespread in society (Irwin 2006). A heightened concern for stakeholder involvement and public inclusion can be seen as a strategy to influence perception, shape understandings, and produce legitimacy.

Knowledge, many scholars argue, is the emerging measure of social and political power in an increasingly dynamic and diffuse political landscape consisting of multiple stakeholders (Rosenau 1998). The opinion is often voiced that more science is needed for better-informed action plans in which the consequences of various options are more clearly articulated. It has even been claimed that scientific consensus is a key mechanism for fostering trust among nation-states with regard to international environmental agreements (Imber 1996).

At the same time, scientific and societal developments make it more difficult than ever to reach scientific consensus. The contingency of all knowledge claims is frequently articulated not only by scientists, but also by other stakeholders and the general public (Yearley 2005). In cases in which science itself is fraught with uncertainty and is slower to reach consensus than demanded by the political decision-making process, it is questionable whether science can provide a sound basis for political action (Lidskog and Sundqvist 2004; Collins and Evans 2007). In addition, growing public distrust in science forms a further set of problems for regulation.

In the social sciences and science studies, several proposals have been made concerning the need to configure new relationships between science and society. In this situation, characterized by the spatial diffusion of knowledge sites, plural rationalities, scientific controversies, and public distrust, there has been a call to "rethink science," to make it more socially robust and context sensitive (Nowotny, Scott, and Gibbons 2001; Irwin and Michael 2003; Leach, Scoones, and Wynne 2005; Callon, Lascoumes, and Barthe 2009). Although the proposals differ, they all seem to agree on the need for inclusive strategies of knowledge production (Lidskog 2008). Citizens should not be seen as mere passive recipients of scientific knowledge, but as active evaluators and partners in producing it. Science should become more reflexive and transparent and make its own often implicit assumptions and values visible to others.

Because of widespread public distrust of science, political institutions are pondering how to restore its credibility (Kasperson and Kasperson 2005; Bouder, Slavin, and Löfstedt 2009). The CLRTAP body and the EU are well aware of this situation. The hitherto closed-room exercise of expertise and the tendency toward technocratic decision making have resulted in decreased trust and interest on the part of the public, politicians, environmental groups, and the media. The regulation of transboundary air pollution has gone from being characterized by visibility and public awareness to being dominated by complex science-based abatement strategies, unknown and largely incomprehensible to the general public (Lidskog and Sundqvist 2007).

According to the Commission of the European Communities (European Commission), the solution to the problem of public distrust or at least indifference is increased transparency of expert work and policy regulation—that is, making visible how expertise is selected, used, and diffused in public policymaking (European Commission 2001). These efforts to "democratize expertise" aim to restore the credibility of expertise. However, from a social science perspective, more elaborate and far-reaching strategies are needed to bring scientific experts, policymakers, and citizens closer to each other or at least to foster communication and interaction among them so that they can learn from each other. As stated earlier, those involved in CLRTAP are aware of the need to connect regional and local issues as well as to gain public support for policy. Both CLRTAP and EU air pollution policy are poorly designed for the purpose of inviting and including public input; furthermore, they seem better designed to elicit public support for predetermined policies than to let citizens take part in the construction of these policies. We must reflect on how the relationships among experts, policymakers, and citizens are configured. We start this discussion by investigating how IR researchers have come to terms with transboundary environmental problems and what they have found to be essential for developing efficient and social robust regulation.

Assessing International Relations Theory

Over the past three decades, social scientists have devoted increasing attention to the field of international environmental agreements. This research was based initially on a *realist perspective* on international cooperation, which occupied a hegemonic position from the early 1950s to the early 1980s. This perspective strongly assumed that international cooperation could be explained by state interests, with cooperation seen as occurring only when strong states wanted it to happen. In this respect, international institutions are seen as epiphenomenal, which can be explained by and largely reduced to state interests. Furthermore, these institutions are anarchic because they have no independent or autonomous guiding principles and are dependent on state initiatives, to which their performance can be reduced (Underdal 2008, 51). According to realists, state interests concern "material ends, specifically, the accumulation of power and wealth relative to other states" (Danish 2007, 207; cf. Mearsheimer 1994–1995). International cooperation is thereby understood to result from the distribution of power between engaged nation-states. States are viewed as rational and autonomous, an understanding that provides a strong foundation for explaining international cooperation. If some states change their interests due to international cooperation, this change can be explained as a response to pressure from other, stronger nation-states, not from international institutions as such.

This realist view of state cooperation is still alive but has been complemented by a view ascribing a more independent status to international institutions. The *institutional approach,* developed in the 1980s, suggests that not only states have power, but some international institutions have it, too. Through the creation of certain institutions to govern states in specific issue areas and often labeled "international regimes" (Rittberger

1993), new incentives were created for state action. International institutions can reduce the transactional costs between states, pool resources (including information), and effectively organize collective rewards and punishments (Danish 2007, 210). Incentives are thereby created for cooperation between nation-states. In this sense, international cooperation offers a way for individual states to pursue their own interests by relinquishing some of their autonomy and delegating it to the international institution. Thus, international institutions have the power to change states' external context and to exploit the states' self-interest to facilitate certain actions.

This institutional approach, according to one of its leading proponents, Oran Young, consists today of three complementary analytical strands (Young 2008, 7–8). The first is the *collective-action perspective*, based on the realist approach, which explains international cooperation with reference to the stable self-interest of involved nation-states. Rational actions may individually lead to negative outcomes on an aggregated level, as described by "the tragedy of the commons" (Hardin 1968) and "the free-rider problem" (Olson 1965). In these cases, international institutions can offer a solution, and nation-states can be expected to support cooperation to prevent these negative outcomes.

The second strand is the *social practices perspective*, which is more sociological in nature. It focuses on the process of actor socialization, by which actors change identities when joining new groups. This means that successfully established international cooperation tends to become inert and path dependent. International cooperation is therefore not reducible to strong states' interests because social structures, norms, and ideas developed via international cooperation become important causal factors.

The third strand is the *knowledge–action perspective*, which is the least developed of the three strands. In this view, international institutions and institutional changes are not explained by either state interests or group loyalty. On the contrary, they are explained by new knowledge, new leadership, and new patterns of governance. This perspective turns the realist approach upside down, arguing that ideas explain state interests and international cooperation, not the other way around.

Some institutionalists are now trying to marry these three strands (Young 2008, 8) by means of a synthesizing drive to establish a broad and encompassing paradigm for the study of international relations (Biermann 2008; Young, King, and Schroeder 2008; cf. Young and Levy 1999). However, integrating these divergent strands is nothing if not problematic. There is a risk that the synthesizing impulse may lack both precision and a distinct explanatory model if it ignores the divergent assumptions that underpin the individual approaches. The institutional approach, impelled by its integrative ambition, tries to integrate the old realist perspective, giving state interests a pivotal role, with more sociological and constructivist perspectives that attach this role to norms, ideas, and knowledge. What is framed as a synthetic approach thus also risks becoming a hegemonic effort to make one approach superior to the others.

A synthetic approach may result in vagueness, eclecticism, and inconsistency in which perspectives and approaches grounded in different interpretive frameworks are brought together under one umbrella without any deeper reflection on whether doing so is possible or desirable and on how crucial difficulties may be masked in the process. There are good reasons to develop distinct approaches that can then be contrasted by emphasizing the factors and causal relationships that shape international governance.

The *epistemic-community approach* is one of the best examples of a distinct approach and is close to what Young calls the "knowledge–action perspective." In the epistemic-community approach, knowledge and science are treated as important explanatory factors of international cooperation, suggesting that ideas can change state interests. Peter Haas (1997, 2007) argues clearly against the view that state interests are the prime driver of international cooperation and suggests that ideas can play this role instead. This approach, considering state interests as the result of norms, ideas, and knowledge, is often called a *constructivist approach* in contrast to the realist institutional approach.

Another important way the constructivist approach differs from institutionalism is its focus on agency, which explains how ideas are transformed into policy that can change state interests. Haas suggests that "epistemic communities are important actors that are responsible for developing and circulating causal ideas and associated normative beliefs and, thus, help to identify state interests and preferences" (2007, 792). This function is needed largely because in the age of globalization state actors have become more uncertain about their own interests; in this situation, ideas from epistemic communities help them to see a way forward—that is, help to shape their interests (Haas 2007, 794). The epistemic-community approach stresses the role of agency: things happen because actors—in particular international organizations and networks of expertise—facilitate the transmission of ideas to nation-states. An epistemic community is a "transmission belt" that helps ideas to travel from research to policy by shaping state interests and is a central factor explaining international cooperation. The epistemic-community approach focuses on learning rather than on knowledge; if no learning occurs, no epistemic community exists (Haas 2007, 801). The existence of an epistemic community makes "speaking truth to power" possible (Haas and Stevens, chap. 5 in this volume). The epistemic-community approach is a distinct model based on assumptions and causal relationships not easily assimilated into the overall framework of the institutional approach, as Young (2008) seems to believe.

Like the epistemic-community approach, the *STS approach* focuses on learning and knowledge, though it is not entirely at ease with the suggestion that strong and almost independent ideas play an important role in international environmental governance. Sheila Jasanoff, for example, argues that the model of "speaking truth to power"—with its idea that "technical input to policy problems has to be developed independently of political influences; meaning that the 'truth' so generated acts as a constraint, perhaps the most important one, on subsequent exercises of political power"—is never actualized (2003b, 225). On the contrary, she claims that politicians and the public must be engaged rather than taught if change is to occur, and this engagement cannot happen if knowledge is developed in isolation outside such a process.

To believe that ready-made knowledge influences policy (and thereby society) from the outside in an almost mechanistic way leaves too much unexplained regarding how influence is exerted. According to STS, scientific knowledge never moves freely; rather, a network is needed for knowledge to travel from one context to another (Latour 1993). This network is made invisible by talk of "speaking truth to power." The issue is not simply "speaking" or "truth," but constructing relationships that make it possible both to produce and to disseminate truth. Obviously, this clarification does not necessarily mean that the epistemic-community approach is wrong, only that the conditions under which knowledge travels from one group to another should be investigated. Detailed empirical analyses are called for, which is what Peter Haas and Casey Stevens (chap. 5 in this volume) do in their explorations of the *conditions* required to speak truth to power. However, the epistemic-community and STS approaches have different foci concerning the extent to which the transmission of knowledge between groups needs to be explained and how to assess the independence of knowledge in relation to the policy process (cf. chaps. 5 and 7 in this volume). They seem, however, to agree on the need for a better understanding of the conditions necessary for science–policy interaction.

The epistemic-community and STS approaches also share a focus on dynamic learning and complex processes of knowledge transmission. They differ, however, in an important way concerning the matter of causality. STS argues that it is wrong to see either interests or ideas as prime drivers shaping international institutions: both are needed but should be assessed as the result of co-production (Jasanoff 2004a). This means that interests and ideas produce each other interdependently, neither being the prime driver.

STS has developed largely parallel to IR research, with little cross-fertilization between the two fields. We believe that much might be gained from a more explicit and critical discussion of how to relate these strands to each other. The co-production thesis lets us take a step further in the old, ongoing discussion of whether state interests or ideas are the prime drivers shaping international institutions. The co-production thesis also makes it more germane to discuss citizens' possible contribution to shaping international environmental governance. Nevertheless, the STS community has thus far displayed only limited interest in developing a more coherent framework for evaluating international environmental governance. In doing so, it might learn a great deal from studying IR research. Deeper analysis of the relationships among science, policy, and citizens is greatly needed—analysis that takes advantage of all three approaches to the study of international environmental governance. In what follows, we contribute to this effort in two steps: in the next section by discussing the findings from the volume chapters and in the subsequent section by making a general examination of how to understand the science–policy–citizen dynamic in a more integrative way.

International Governance: Science, Policy, and Citizens

International governance concerns how to develop constructive relationships between groups and stakeholders as well as among science, policy, and citizens. It concerns not only learning and knowledge, but also power and struggle. As noted in the introductory chapter, the governing of transboundary air pollution is a complex and dynamic activity in which knowledge, norms, institutions, and actors struggle to shape policy. Policy, science, and citizens—the building blocks structuring the chapters of this volume—are therefore not separate but rather overlapping and interrelated spheres.

Policy and Institutions

Part I of this volume concerns policy and institutions. As stated in chapter 1 in the volume, environmental policy and its institutions are currently in a state of transition. In recent decades, state-centered, government-led policy has been criticized and replaced by more governance-oriented policymaking. But why? Does this change mean that governments are no longer important? And how should the change be assessed?

The critique of a state-centered analysis of policy and politics is presented from two partly overlapping perspectives. First, studies of globalization suggest that nation-states are challenged by globalizing processes such as corporations acting on a global market and nongovernmental organizations (NGOs) acting across national borders (Rosenau 1998; Benner, Reinicke, and Witte 2004). Governments are more reactive than active and must adapt to the effects of globalization. Second, studies of regulation conclude that state-driven regulation has evolved from a command-and-control model to a decentered regulation, the state serving as a coordinator in a governance process in which corporations, NGOs, and even citizens substantially influence the regulative process (Black 2002; Power 2007; Lidskog, Soneryd, and Uggla 2009). Taken together, these two trends suggest that a state-centered policy process is neither possible nor desirable. The nation-state is no longer the privileged concern of politics, but more of an object for other actors to act upon.

The shift "from government to governance" neatly illustrates these changes (Kooiman 1993; Borraz 2007). The shift implies that distinctions between the local and the global, on one hand, and the private and the public, on the other, have become more difficult to maintain. Policy is not developed only in the formal political sphere, but also in what Ulrich Beck (1992) calls the "subpolitical sphere." Policy develops through the actions of transnational corporations, research laboratories, media agencies, and interest organizations. This shift does not mean that nation-states are no longer significant. The nation-state is still an important regulatory actor, but it rarely exercises its power through command and control. In advanced liberal societies, the state rules at a distance by inducing actors to conduct themselves in line with state interests (Rose 1999). What role the national government plays in a regulatory process is an empirical question: in a governance-oriented process, the government can still play a key role as a decision-making coordinator.

Social scientists have not restricted themselves only to describing the shift from government to governance. They have also evaluated it, paying attention to negative consequences such as increased political complexity, opaque decision-making processes, and blurred political accountability (Elander 2002; Benner, Reinicke, and Witte 2004; Winter 2006).

The three chapters constituting part I offer key insights into central policy trends in international environmental governance. Jørgen Wettestad (chap. 2) starts from the institutional perspective and delves into questions of regime effectiveness, which is a hallmark of this approach. He finds that nation-states have not driven CLRTAP efforts. On the contrary, a significant part of what he considers the problem-solving capacity of the CLRTAP institution is based on a flexible organization of the science–policy relationship, implying that policy adapts to improvements of knowledge. This organization clearly exemplifies how other actors—in this case, knowledge brokers—act upon nation-states. These actors are in fact substantial CLRTAP decision makers.

Bernd Siebenhüner (chap. 4) follows up on this approach when arguing that CLRTAP activities should be understood as science-driven policy processes, whereas the EU clean air program should be seen as more policy driven. From this distinction, we can draw the interesting conclusion that the air can be governed in various ways. One reason for this difference is that these two international institutions are adapting to each other in a way that makes them complementary. Henrik Selin and Stacy VanDeever (chap. 3) examine the institutional linkages between CLRTAP and the EU air policy program in detail. They argue strongly against the idea of government-centered decision making and instead propose a more complex understanding of decision making. Decision makers are dependent on other institutions and actors on various levels, and this dependency can be constructively and strategically used to produce synergetic effects and regulatory authority.

Traditional top-down ways of managing environmental problems are questioned because state sovereignty is in question due to globalization processes. Governance—policymaking that involves a wide range of public and private actors on various policy levels—is presented as a more feasible and efficient way to handle transboundary environmental problems (Winter 2006). Clean air policy efforts are accordingly not driven by independent nation-states but are instead complex activities, heavily dependent on input from a range of actors and facilitated by the existence and development of international institutions. In this process, as all three chapters in part I emphasize, knowledge is crucial, as are shared knowledge production, flexible organization for science–policy interaction, and an improved interface between scientific knowledge and policymaking.

However, none of these chapters discusses in any detail how this interaction works in practice, which is the topic of part II.

Expertise and Learning

Part II concerns expertise and learning. Commentators argue that international clean air policy efforts in Europe have been dependent largely on expert knowledge and that this dependency has been the case since the early 1970s. But how has knowledge been important, what kinds of knowledge have exerted influence, and why? Moreover, should a strong dependency on knowledge be seen as good?

According to institutionalists, scientific input is important for designing effective international environmental institutions. When analyzing whether this connection is necessary and, if so, how significant a factor it is, knowledge is assumed to represent "a 'neutral' input on behalf of the resource or environment in question" (Andresen and Skjærseth 2007, 186), which as such can be used in policy negotiations. Science is thus not part of the policy process but is something neutral and outside that provides important input into the policy process, not least when political conflicts over questions of values are small (Andresen and Skjærseth 2007, 200–201). This view of the role of science in international institutions is a legacy of the realist approach, which focuses on negotiations between interest-driven nation-states.

As discussed earlier, epistemic-community and STS scholars have criticized this view. According to the epistemic-community approach, ideas (including knowledge and science) are much more important as drivers than institutionalists recognize. In addition, STS researchers argue that expert knowledge is neither a neutral input nor developed in isolation from policymaking; rather, knowledge and policy are always codependent (see chaps. 7 and 8 in this volume).

The four chapters constituting the second part of the book scrutinize the role of expert knowledge in European air pollution regulation. Haas and Stevens (chap. 5) argue that for knowledge to influence policy, it must be developed independently of the policy process; before any attempt to influence policymaking, scientific consensus should be firmly established. This view is close to the realist argument that knowledge enters the policy process from the outside, as an independent and neutral input. However, the hallmark of the epistemic-community approach is the view that under certain conditions science can influence state interests and decision making. The combination of the view that expert knowledge is developed at a distance from political decision making and the view that this distanced development benefits the creation of more robust regulation is in fact also traceable in the STS approach.

Some STS scholars suggest that science and policy should be kept apart and that issues to be handled by scientific experts can be separated from issues that are politicians' responsibility. For example, Harry Collins and Robert Evans (2002, 2007) argue that a prerequisite for assessing an issue as knowledge based is that it must be possible to identify a "core set" or group of experts who are the most knowledgeable in the field; in addition, there must be consensus in the knowledge community that this core set exists and is recognized as the center of expertise. Without a core set, it is impossible to regard the issue as experts' responsibility. If there is such a group, technical issues may legitimately be delegated to these experts. This perspective is close to Haas and Stevens's (chap. 5) view of expert groups as independent from the policy process and thereby granted the authority to influence policymaking.

Other STS scholars, not least Sheila Jasanoff (2003a, 2005) and Brian Wynne (2003, 2005), have criticized this view. They argue that no issues are independent of political considerations and that other groups can always contribute to the development of expert knowledge, especially in framing processes, when issues are shaped and broad questions—such as "Why is this important to know more about?" and "How important are the results presented by the scientific experts?"—are settled. This idea also underlies the concept of co-production: although epistemic communities or core sets of independent scientists may be working in isolation, their activities and produced knowledge are always connected with the policy process. The implication is that Haas and Stevens as well as Collins and Evans may be right in a certain sense: science is often produced in settings independent of the formal policymaking sphere and can act with social authority gained outside the political sphere. On another level, however, science can be seen as co-produced with policy: the development of scientific knowledge and the shaping of expertise always occur in a political setting in which various actors consider how to construct policy-relevant science and, through knowledge production, to influence policymaking.

Rolf Lidskog and Håkan Pleijel (chap.8) find in their study of the regulation of ground-level ozone that although scientific advice was *formally* separated from policy negotiations, it was in fact *substantially* integrated with them. The object of study—ozone exposure and its effects—was not constructed in isolation from political considerations. On the contrary, it was developed and shaped in a close relationship with policy as scientists tried to make their knowledge policy relevant in a way that matched policymakers' search for science-based policies.

In a historical study of the role of expertise in air pollution abatement efforts, Göran Sundqvist (chap. 7) goes one step further, stating that even scientific consensus is not particularly important in giving science influence on policymaking; on the contrary, Sundqvist claims it could may even be a hindrance. He argues that both the epistemic-community and the STS approaches oversell their findings, at least regarding the regulation of transboundary air pollution. Scientists do not work in isolation, and they do not start to influence policy only after scientific consensus is achieved. Neither do they try to hide uncertainties as a strategy to increase scientific authority when presenting results to politicians and the general public, as STS scholars argue. Sundqvist contends that in the regulation of transboundary air pollution, "soft" science, in which uncertainties and conflicting views are explicitly acknowledged, has been productive and that "soft experts" have quite successfully managed to interact with policymakers and thereby influence regulation.

What seems important is not that scientific experts deliver packages of definite knowledge, but that they create learning processes in which actors gain new perspectives and find new ways to approach an issue. Such processes are a crucial factor in successful regimes. Many studies of international environmental governance ascribe a key role to such learning processes, not least when it comes to explaining policy change, which concerns both the understanding of an issue and how to manage it (cf. Clark, Jäger, and van Eijndhoven 2001, 14; Social Learning Group 2001). This role is the focus of Atsushi Ishii's contribution (chap. 6), in which he finds that learning results can fundamentally change the understanding of the policy process, including its goals, in what he calls "paradigmatic learning." However, in the cases he analyzes, learning results in both changed policy and changed science. Viewing science and policy as separate processes conceals how they influence each other. Scientists and politicians rarely develop their understanding of an issue in completely parallel processes. This does not mean that no boundaries separate them, only that they develop in close—but not always visible—connection with each other.

However, Haas and Stevens argue that science is better equipped to influence policymaking if it develops concepts, mobilizes support, and produces results independently of the policy process. Science does not develop in a social and political vacuum, though, or without taking political considerations into account. To exert influence, science must always be connected to the policy process. A science "unplugged" from the policy process has no possibility of influencing policymaking, so experts must connect, attach, and interact. What remains to be done in further research is to deepen our understanding of when and how these connections are most successfully established.

All chapters in part II emphasize that knowledge is crucial for policymaking, so we must understand better how it works and under what conditions and through what practices it exerts power. The conclusion these chapters come to is that knowledge speaks to power only under certain conditions and that social scientists have an obligation to investigate these conditions more deeply. Despite current calls to involve citizens and stakeholders in regulation and knowledge production, none of the chapters in this section pays any particular attention to these other actors; the final part of this volume addresses this element.

Citizens and Involvement

Part III focuses on citizen involvement, which we argued in the introductory chapter is the missing link in studies of international environmental governance. But is it really? And if so, why is it important to involve citizens in policymaking? What roles might citizens and other stakeholders play in environmental governance, and what contributions might be expected from this involvement?

Over the past decade, there has been a strong focus on citizen involvement in international environmental governance. The United Nations Convention on Access to Information, Public Participation in Decision-making, and Access to Justice in Environmental Matters (Århus Convention), adopted in 1998, is a clear example, as are the EU's many initiatives to strengthen public involvement in relation to new technology as well as regulation concerning health and the environment (UNECE 1998; European Commission 2001; Wynne and Felt 2007). An important underlying reason for many of these initiatives is increasing public unease with science and technology and the many crises in state-centered regulation, exemplified by the controversial issue of bovine spongiform encephalopathy, or "mad cow disease," and continuing discussion of how to deal with genetically modified organisms and climate change (Löfstedt 2005; Lezaun and Soneryd 2007; Boyce and Lewis 2009; Ockwell, Whitmarsh, and O'Neill 2009; Ricci, Bellaby, and Flynn 2010).

The increased focus on citizen involvement challenges the classical-liberal understanding of representative democracy, which assumes that elected politicians represent citizens (Jasanoff 2005, chap. 11; Lidskog and Elander 2007; Beck 2010). The liberal perspective views politicians as legitimately acting on the behalf of citizens because they know and represent the citizens' wishes. From this perspective, state interests are the sum of constituent preferences, so it is unnecessary to involve citizens beyond the elections for political representatives. Once elected, politicians act on the behalf of the citizens.

However, many analysts have identified cracks between the state and its constituents, expressed in the form of mistrust and criticism of government decisions. A broader and more critical focus on the relationships between the state, citizens, and other stakeholders is a strong current trend in social science; for example, Beck (1992, 2006) identifies the growing power of actors outside the formal political sphere, where an increasing number of important issues are dealt with in nongovernmental fora, and subpolitical actors—such as research laboratories, corporations, and international media agencies—make decisions that strongly influence society without taking the governmental route. The governance approach emphasizes that the important question is not what actors have the formal power, but how power is amassed and exerted through networks, coalitions, and partnerships of public and private actors (Cheema and Rondinelli 2007; Okereke, Bulkeley, and Schroeder 2009; Stern and Seligmann 2004). The neo-Foucauldian approach states that less government does not mean less governance. Instead, governing takes place not from above, but through less visible technologies, some of which are strongly coordinated by the state, that configure organizations and citizens in relation to interests and power (Burchell, Gordon, and Miller 1991; Barry, Osborne, and Rose 1996; Dean 1999; Rose 1999).

On a more practical level, the solution to a tense relationship between government representatives (including government experts) and citizens requires not only an expansion of the space for public involvement, but also a focus on how citizens are enrolled. In general, this focus entails awareness of the asymmetrical power relationships between involved actors. Creating socially robust regulation implies merging science with democracy, and citizens and scientists should be mobilized in more symmetrical ways when policy is shaped (Elam and Bertilsson 2003; Lidskog 2008).

Citizen involvement always risks being framed in a technocratic way—that is, when the public is enrolled in scientific discourses and disciplined with the aim of winning legitimacy for a predetermined government policy. In this case, we find an instrumental justification for citizen engagement. According to Wynne, who has published extensively on citizen engagement with science and technology, this manipulation is all too often perceptible to the public and leads to alienation instead of to engagement (see, e.g., Wynne 2005). Citizens are taught and not listened to; they are assumed not to know enough to be allowed to influence the issue at stake. This approach has been called *the deficit model* of the public understanding of science, implying that citizens have a knowledge deficit in relation to experts and therefore should be educated. STS scholars have heavily criticized this approach (Irwin and Wynne 1996; Leach, Scoones, and Wynne 2005). Part of the problem is that framing an issue as a risk issue dissolves its political aspects; public knowledge is not demanded, and citizens are given few reasons to engage in regulatory work (Jasanoff 2005). The solution is to repoliticize the involvement process, putting policy experts, government officials, and scientific experts on the same level as citizens and inviting them all to develop, discuss, and negotiate how an issue should be framed.

The three chapters in part III explore the role and significance of citizen involvement in environmental regulation. Steven Yearley starts chapter 9 by asking whether citizen engagement is needed in air pollution regulation. Are citizens really a missing link in international environmental governance, which concerns complex issues remote from everyday life concerns? Yearley raises a number of important questions and sheds light on why citizens have difficulties influencing the regulation of air pollution.

However, as emphasized in chapter 1, the spatial identity of an environmental problem is not given from above, but rather is constructed through actors' strategies and negotiations. In this way, the framing of air pollution problems is always important. This conclusion serves as the point of departure for Arthur Petersen, Leendert van Bree, and Willemijn Tuinstra in chapter 10. They demonstrate that framing is crucial if stakeholders and citizens are to be involved in clean air policy efforts. If an issue is framed in a technocratic way, citizens and stakeholders have limited opportunities to exert influence and therefore lack incentives for involvement. This argument is in line with a point we made earlier: if an environmental problem is framed in terms of risk, the democratic and political aspects of the issue are dissolved, and the public will have few reasons to be involved. To make an issue more conducive to citizen involvement, frames and framing processes should be made visible to and negotiable for involved and invited parties, not least citizens—which is our concluding message in this chapter.

In chapter 11, John Forrester and his colleagues bring together Yearley's emphasis on the difficulty citizens have in engaging in translocal issues and Petersen, van Bree, and Tuinstra's suggestion regarding the importance of framing. The chapter articulates an international outlook on stakeholder participation. Like Yearley, its authors stress that it is more difficult for citizens to engage with regional problems than with local problems. At the same time, though, they emphasize that environmental problems, including air pollution, have both local and regional dimensions, so more is involved than just making the problems more visible. Like Petersen, van Bree, and Tuinstra, the chapter 11 authors single out framing. As emphasized by STS scholars, framing touches on deeply political aspects, so a focus on framing will facilitate citizen engagement. Instead of seeing certain environmental problems as nonlocal in character, which makes them ill-suited for citizen involvement, social science studies should ask why certain issues are framed in a way that makes them less attractive for citizen engagement.

The chapters in this part raise not only questions about the nature of environmental problems—their framings and associated spatial identities—but also fundamental questions about the configuring of citizens. Discussions of citizen involvement always risk treating citizens as locally situated, possessing limited understanding of both the nature of specific environmental problems and the institutional structures that govern them. As various social theorists have suggested, this treatment conveys a wrong view of citizens, whom we should instead understand as globally connected and capable of critically evaluating both scientific claims and political decisions (e.g., Nowotny, Scott, and Gibbons 2001; Irwin and Michael 2003). Citizens obviously always have local connections, which is true also for scientists and policymakers, but such connections do not mean that they are unable to deliberate on complex issues or critically evaluate standpoints and knowledge. Environmental issues, such as air pollution, have local and global connections, but so do citizens. We still need to determine the implications of this understanding for citizen engagement and the possibility that citizens can influence international environmental governance.

Based on the three chapters of part III, citizen involvement can be said to be important for international environmental governance. Nevertheless, better definitions of the forms this involvement should take are still needed. As Yearley notes, existing institutions governing transboundary air pollution are poorly designed to exploit citizen contributions. New methods are needed to facilitate citizen involvement in environmental governance, and new institutional structures are needed to facilitate—but also delimit—this involvement. Replacing science-centered governance, in which citizens are seated ringside, with a public-centered process, in which expertise and political representatives are subordinated to public opinion, does not seem a suitable solution (Irwin 1995). Instead, more nuanced views of how scientists, policymakers, and citizens interact, negotiate, and shape environmental governance should be developed. The next section is devoted to this crucial issue.

The Science–Policy–Citizen Dynamic

According to IR theory, an institution is constituted by systems of norms, rules, and decision-making procedures. An institution defines the roles played by its actors and guides the interactions between them. International regimes are institutions agreed upon by states, created with the aim of governing their behavior concerning specific issues.

IR scholars agree that science can play a key role in shaping and implementing a regime. However, in mainstream IR research (based on a realist perspective) science is located outside the regime and is seen as an external and neutral input that can be used in policy negotiations and is thereby important for regime development. From this perspective, science is not directly involved in negotiations between state interests but can nevertheless play a role when a regime is designed. Science and policy are understood as independent spheres—policy being the essential constituent of a regime, but science sometimes serving as an important ally.

International environmental governance concerns how to develop constructive relationships not only with science and expertise, but also with stakeholders and citizens. However, IR research has not considered the role of citizens in international regimes in any deeper sense. One reason for this neglect may be that it sees citizens as represented by nation-states.

We argue that regimes should have clear mechanisms for involving both science and citizens. Even when citizens are viewed—by policy analysts and by political representatives and bodies—as actors who must be taken into account when designing and implementing regimes, they are often treated as actors external to the regime. Mainstream IR research sees both science and citizens as part of the surroundings of a regime, approaching it from the outside.

In contrast, we argue that environmental governance cannot be properly understood without taking both expert knowledge and citizen involvement into account and perceiving them as internal to institutions or at least without clarifying how they are connected with and how they influence regime development.

Science as Staged Activity

As presented earlier in this chapter, Oran Young has recently formulated an ambitious and broad research program for IR studies of international environmental regimes. He makes strenuous efforts to diagnose mismatches between environmental regimes and biogeophysical systems (Young, King, and Schroeder 2008; cf. Young 2002). Diagnosing misfits and creating a better match between institutions and environmental problems calls for knowledge—primarily scientific knowledge, but sometimes also what is called "informal knowledge" (Young 2002, 66). Environmental institutions should be designed to promote science's ability to influence policy (cf. Andresen, Skodvin, Underdal, et al. 2000; Mitchell 2008, 96). However, this strong focus on knowledge does not mean that science is an integral part of institutions; as a corrective factor determining the fit or misfit between institutions and natural systems, it *must* be external to institutions. Nature cannot be properly understood by political representatives, so science is invoked from the outside to contribute to a more adequate and valid view of natural systems and how they work. According to many IR scholars, science is external, though often crucial, to the design of institutions for managing environmental matters.

The epistemic-community approach claims that scientific knowledge is decisive in shaping environmental regimes. It also considers knowledge to be external to institutions—although in cases in which consensual knowledge is established and conveyed by an epistemic community, it becomes integral to the institution and a main characteristic of it. According to the epistemic-community approach, expert knowledge is both external and internal, depending on the particular knowledge-development phase (Haas 2007; Haas and Stevens, chap. 5 in this volume).

In contrast, the STS approach states that the cognitive dimension always constitutes part of institutions. The STS standpoint is similar to that of cultural anthropology, which claims that an institution is a convention, but that for such "a convention to turn into a legitimate social institution it needs a parallel cognitive convention to sustain it" (Douglas 1987, 46). This cognitive dimension, crucial for all institutions, consists of shared knowledge. From this perspective, viewing knowledge as an external evaluator of the proper match between institutions and nature is problematic.

STS forcefully argues that separating knowledge and science from policy creates a false dichotomy. Science always needs social enforcement because scientific propositions have no power in themselves but are made credible through social interests and strategies (Latour 1987). These strategies are part of what has been called "stage management," in which science and policy are balanced to present convincing regulatory science that tells what should be done and why doing it is important (Hilgartner 2000). Stage-management activities always rest on cognitive conventions that are part of institutions. In the analysis of experts' work in the public arena, a distinction must be made between front-stage and backstage activities and among the strategies used by experts to manage the boundaries between them (Hilgartner 2000, 7). Backstage management concerns the process of knowledge production, which is uncertain, controversial, risky, and involving (Latour 1998). In front-stage activities, science becomes public, most often portraying its results as certain and as having been produced independently of political considerations.

When expert knowledge is communicated outside the expert group, a cultural distance must be bridged (Shapin 1995; Sundqvist 2003). When elaborating their strategies, scientists need to consider various stakeholders' views and expectations of science. What should be kept backstage and what should be put front stage differ depending on whom the experts are trying to persuade and on the issue under consideration.

In front-stage presentations, science and policy usually appear to be separate activities, belonging to different spheres of society, what Bruno Latour (1993) calls "purified presentations." Underlying such presentations, however, are processes of hybridization in which science and policy are explicitly mixed and co-produced.

Whereas experts consciously employ stage management to meet other actors' expectations (e.g., negotiators, industrial corporations, environmental movements, and the public), thereby creating better opportunities to influence them, more fundamental and deeply embedded processes also exist, processes not necessarily recognized by involved actors. The hybrid processes of science and policy in the domain of public decision making rest within a broader frame of what Jasanoff calls "civic epistemology" (2005, chap. 10). This frame is close to what Mary Douglas calls "shared beliefs" (1987, 128). Shared beliefs constitute all institutions; they are what make institutions socially legitimate. Civic epistemologies are crucial parts of political cultures, whose role is to confer meaning on issues and to shape the identities of the members of a specific culture. They can, for example, confer meaning on technological projects or regulatory efforts and ascribe certain capabilities to actors.

In turning to the concept of civic epistemology, we move away from a priori assumptions about what the public should know or understand about science. Instead, we pose the question, salient in any democracy, of how knowledge comes to be perceived as reliable in political settings. Put differently, a civic epistemology conceptualizes the credibility of science in contemporary political life as a phenomenon to be explained, not taken for granted (Jasanoff 2005, 250).

Civic epistemologies do not deal with technical details but answer broad questions such as "what is the purpose; who will be hurt; who benefits; and how can we know?" (Jasanoff 2003b, 240). Although these epistemologies are fundamental constructs embedded in particular political cultures, they are at the same time influenced by strategies developed by actors and by the various science–policy hybrids the actors construct. Our point is that whereas stage management can be strategically and flexibly used by actors, civic epistemologies are more fixed and stable, which means that actors must adapt to them while not necessarily being conscious of them. Stage management concerns how science is strategically presented, whereas civic epistemologies concern how the relationships among policy, science, and citizens are configured. If science and policy are seen as parts of institutions, then we should pay more attention to processes of stage management and the role of civic epistemologies. This focus would let us understand and explain specific science–policy–citizen relationships, assessing them as parts of the same project of developing civic epistemologies—for example, when regulating environmental issues.

Though the concepts of stage management and civic epistemologies are suitable analytical tools with which to investigate the relationships among science, policy, and citizens, they are not sufficient in themselves. In the next section, we present a third analytical concept useful in deepening our understanding of international institutions: framing.

The Importance of Framing

To make an issue governable, it must be simplified and packaged in a way that makes it suitable for political management (Fischer 2003; Lidskog, Soneryd, and Uggla 2009). This task is done through processes of framing, in which unstable and complex phenomena are stabilized and a space for political action is carved out.

In framing processes, boundaries and hierarchies are established, and the relationships among science, policy, and citizens are configured and legitimated. A great many STS studies critically scrutinize framing processes in public decision making, and a key finding in them is that frames concerning technical issues are usually dominated and influenced by experts in a technocratic way (Wynne 2001, 2005). Technocratic frames entail a strong focus on technical details at the expense of broader questions such as "Why is this important?" "Who has the relevant knowledge?" and "What is the driving force behind this development?" By narrowing the issue, scientific experts exaggerate the scope, power, and importance of scientific knowledge in the public domain, thus neglecting cultural factors and ignoring citizen competence (Wynne 2001, 460; Jasanoff 2010; cf. Wynne 1992, 115). In this way, public policy creates more alienation than involvement on the part of citizens. However, issues can be framed more democratically, which implies focusing on their connections to public concerns (Marres 2007).

Framing processes must therefore be analyzed; otherwise, it is difficult to understand either conflicts or communication between groups. How a problem is framed influences what is seen as a proper remedy for it—that is, what solution is seen as relevant, effective, and legitimate. A basic assumption in STS is that citizens can always contribute to the discussion of how issues are and should be framed, although they may not necessarily be able to do so in relation to technical details. However, because technical details are produced and presented within broader frames, laypeople can also contribute independently and competently to public decision making regarding all technical issues (Wynne 2003).

An important reason for encouraging citizen involvement in environmental governance is so that citizens can evaluate the frames proposed by experts. In this way, citizen involvement is "an instrument for holding expertise to cultural standards for establishing reliable public knowledge, standards that constitute a culture's distinctive civic epistemology" (Jasanoff 2003a, 397–398). If public involvement is lacking, the science–policy relationship risks offering advice that society sees as irrelevant because the answers delivered are to wrong or misguided questions.

Nevertheless, science is often accorded the most prominent role in public discussion of risks, paradoxically even when public dialog is striven for. One reason for this assignment, according to Wynne, is that many environmental issues are technically framed, with questions of risk, safety, and effectiveness placed at the center (2005, 66; 2010, 291). Scientific expertise is needed to answer such questions, the thinking goes, because experts have the resources and competence to know the "true" nature of the issue at stake. Such frames, however, are based on an unexamined and preordained meaning in which alternatives are made invisible. However, a technocratic framing reduces citizens' role to one of trusting or distrusting experts; alternative framings are considered irrelevant, and no space is made available for a substantial public input.

IR studies often accept this kind of narrow framing. Science is seen as separate from political considerations, the public is excluded as a significant actor, and critical analyses of the policymaking frames are lacking. Focusing on the effectiveness of environmental regimes in terms of fits/misfits in relation to a predefined environmental issue conveys the impression that discrete and unambiguous environmental problems exist "out there," ready to be acted upon, and that scientists are assigned the role of finding the optimal match between a problem in nature and a solution in society (cf. Young 2008, 27–31; Schroeder, King, and Tay 2008, 266). This view is close to what Jasanoff (2003b) calls the "technology of hubris," which aims to facilitate management and control. It implies the preemption of political discussion and limits the capacity to discover, understand, and internalize challenges that arise outside a particular frame. The solution to this limitation is, according to Jasanoff, to develop a "technology of humility," which aims to facilitate deliberation on the substance of decision making as well as reflection on framing processes and issue formation. In this way, the focus shifts from teaching citizens about technical details to communication and mutual social learning.

All regulatory work obviously requires expertise to design and implement rules, and transboundary environmental issues are no exception to this rule. We conclude, however, that if established frames are hidden and taken for granted, limited space is left for citizen involvement and broader discussion of the aims and scope of environmental regulation. Frames not only diagnose environmental problems and propose appropriate remedies but also imply proper relationships among science, policy, and citizens.

Co-produced Configurations

STS contributes to international environmental governance by facilitating research into the science–policy–citizen dynamic. The co-production thesis states that the social and cognitive orders are interdependent because they are part of the same institution. Knowledge is not explained by the social (e.g., social order, social institutions, and social groups), or vice versa. Instead, neither of them should be taken for granted, and both must be critically studied. How nature is understood also implies a proposed solution to social problems and a stable political order because solving environmental problems means defining and controlling not just nature, but people as well (Jasanoff 2004b, 30–31). Being a scientific expert—for example, by working as an adviser within an international environmental regime—means possessing knowledge about nature and about society. Environmental experts exert influence on both nature and society. The expert's identity and credibility become part of a project in which nature, science, and social institutions are stabilized. The implication is that processes of separating science from society and politics, which are frequently undertaken in modern societies, always start from co-produced hybrids. Below the tip of the iceberg where separation is publicly presented, we always find a multiplicity of collaborations and connections among actors, activities, and spheres, which superficially present themselves as pure, independent, and separated.

Studies of co-production processes do not a priori take a stand on what the relationships among science, policy, and citizens are or should be. Instead, they examine these processes with an open mind, starting from the assumption that there can be many configurations between them. A specific configuration should always be critically studied and normatively evaluated because other configurations are always possible.

From Instrumental Relevance to Shaping Visions

All chapters in this book have explicitly or implicitly raised a key question: What role should social science play in the development of international environmental governance? Whereas natural scientists and technical experts have filled the ranks of those working to combat transboundary air pollution, social scientists have remained largely aloof, studying the efforts of other groups without influencing ongoing work. Social scientists can thus be criticized for withdrawing to an ivory tower, but they can also be criticized for developing overly close relationships with political leaders, for taking an instrumental and strategic view of political decision making, and for being blind to the ideological and sociopolitical contexts in which their expert work is embedded.

Though the role of social science in environmental governance is contested, it is naive to claim that science (whether natural or social) should take no responsibility for societal development. Irrespective of the role it plays, science is part of society and influences society in one way or another. The questions to be asked instead are what role science should play and whether it can find a position that is neither in the ivory tower nor at the prince's knee.

How the relationships between society and nature as well as between science and policy are framed has consequences for how we view the environment and environmental issues (Irwin 2001). Furthermore, products of science (irrespective of disciplinary affiliation) embody beliefs about not only how the world is, but also how it ought to be (Taylor 1989; Urry 2000; Wynne 2010). In efforts to combat air pollution, experts invoke the future when predicting the consequences of various courses of action. These expectations about future development are performative. They help us to manage uncertainties, set priorities, amass resources, coordinate activities, and facilitate actions, thereby increasing the likelihood that we will bring a particular future into being (Brown and Michael 2003; Borup, Brown, Konrad, et al. 2006; Kitzinger 2008). In this sense, experts play a normative role, whether they are conscious of it or not, in constructing and stabilizing the world.

As demonstrated in this book, the regulatory objects of transboundary air pollution are not stable entities. Scientific advances, political considerations, and social practices are involved in constructing what to regulate. In the 1980s, acidifying substances were put at the center, not least because of their effects on certain ecosystems. Fine particles and ground-level ozone—with their consequences for human health—have currently emerged as important regulatory objects (Pleijel 2007). In addition, science plays a key role in negotiating and constructing the spatial identity of environmental issues. In this way, a geographical location at risk is constructed and becomes an object for action.

Thus, science—including social science—constructs and dissolves regulatory objects, creating some opportunities for political action and eliminating others. It thereby becomes an intrinsic part of environmental governance. Social scientists should reflect on their own role in constructing expectations and visions of society. What ideas about society and about nature are social scientists fostering? What meanings of policy, science, and citizenship are they configuring? What paths of development do they regard as necessary, desirable, or realistic, and what paths as unnecessary, undesirable, or unrealistic? What forms of regulation do they regard as relevant, efficient, and robust, and what forms do they not? The prescriptive function of social science should not be ignored, but explicitly discussed.

Science obviously plays the important role of providing input to shape the technical design of regulatory frameworks, CLRTAP being an example of how science has made substantial contributions to regime development. However, being part of this effort necessitates even greater awareness of one's role as an expert and of the visions of society produced in the effort. As discussed earlier in this chapter, the question of how to make science matter always risks being instrumentally framed, thereby restricting the discussion to how science can increase the efficiency of international regulation and national policymaking. Within such frames, scientists are indeed called upon, though not to democratize policymaking or to make different visions of society feasible. They are instead invited to help decision makers convince citizens and other stakeholders that a current or proposed policy is a necessity. In this way, social science is accorded the function of helping policymakers when they have problems implementing predetermined policies or of helping them make political processes smoother and more conflict free.

This responsibility must not reduce social scientists to being political advisers in a narrow sense. This point is crucial because science has responsibilities not only toward the state, but also toward the wider society. Science needs to reflect on its relationships with policymakers as well as with citizens and other stakeholders. As other studies have demonstrated, science has a long tradition of cooperating with the nation-state and industry but neglecting its connections with citizens (Irwin 1995; Jasanoff 2005). There is a pressing need to create new and constructive ways for science to interact with citizens. In the attempt to realize this need, social science has the important task of analyzing various possible trajectories of environmental issues and public responses to them, while investigating and constructing alternative framings and visions of society.

This book has sought to develop our understanding of how international environmental governance is shaped. We have emphasized that regulation develops through dynamic—and often conflictual—interactions among nation-states, transnational corporations, interest organizations, and research communities. In most cases, the established frames and processes of issue formation are not facilitating public involvement. As a consequence, we have a paradoxical situation in which international environmental governance is developed at a distance from citizens, but at the same time that citizen involvement and public inclusion are sought by policymakers, regulators, and scientists.

No simple and universally applicable formula exists for bringing decision makers and citizens closer to each other. As emphasized in this concluding chapter, an important task is to facilitate wider-reaching interaction and cooperation between actors. This goal can be accomplished by developing spaces for cooperation as well as contestation, spaces in which actors are welcome to deconstruct established frames and construct new ones—that is, to discuss the formation of the issues at stake. No frame—whether technical or democratic—should be excluded from discussion but should be the object of reflection, deliberation, and negotiation because frames tell not only what something is, but also what it ought to be. Bringing questions of framing and issue formation to the table can serve to foster citizen involvement and learning processes for all actors. These framings, with their implicit or explicit visions, can then be discussed and acted upon in broader discussions among policymakers, scientists, and citizens. They might be seen as utopian visions, but many ongoing activities in European air policy should take advantage of ideas for framing issues to attract the interest of various groups, including citizens, instead of marginalizing—or even alienating—them. Doing so will permit the construction of more socially robust regulation that is politically viable, environmentally effective, and publicly trusted.

References

Andresen, Steinar, and Jon Birger Skjærseth. 2007. Science and Technology: From Agenda Setting to Implementation. In *The Oxford Handbook of International Environmental Law*, ed. Daniel Bodanski, Jutta Brunnée, and Ellen Hey, 182–202. Oxford, UK: Oxford University Press.

Andresen, Steinar, Tora Skodvin, Arild Underdal, and Jørgen Wettestad. 2000. *Science and Politics in International Environmental Regimes*. Manchester, UK: Manchester University Press.

Barry, Andrew, Thomas Osborne, and Nikolas Rose, eds. 1996. *Foucault and Political Reason: Liberalism, Neo-Liberalism, and Rationalities of Government*. Chicago: University of Chicago Press.

Beck, Ulrich. 1992. *Risk Society: Towards a New Modernity*. London: Sage.

Beck, Ulrich. 2006. *The Cosmopolitan Vision*. Cambridge, UK: Polity Press.

Beck, Ulrich. 2010. Climate for Change, or How to Create a Green Modernity? *Theory, Culture & Society* 27 (2–3): 254–266.

Benner, Torsten, Wolfgang H. Reinicke, and Jan M. Witte. 2004. Multisectoral Networks in Global Governance: Towards a Pluralistic System of Accountability. *Government and Opposition* 39 (2): 191–210.

Biermann, Frank. 2008. Earth System Governance: A Research Agenda. In *Institutions and Environmental Change: Principal Findings, Applications, and Research Frontiers*, eds. Oran R. Young, Leslie A. King, and Heike Schroeder, 277–301. Cambridge, MA: MIT Press.

Black, Judith. 2002. *Critical Reflections on Regulation*. Center for Analysis of Risk and Regulation Discussion Paper no. 4. London: London School of Economics and Political Science.

Borraz, Olivier. 2007. Risk and Public Problems. *Journal of Risk Research* 10 (7): 941–957.

Borup, Mads, Nik Brown, Kornelia Konrad, and Harro van Lente. 2006. The Sociology of Expectations in Science and Technology. *Technology Analysis and Strategic Management* 18 (3): 285–298.

Bouder, Fredric, David Slavin, and Ragnar Löfstedt. 2009. *The Tolerability of Risk*. London: Earthscan.

Boyce, Tammy, and Justin Lewis, eds. 2009. *Climate Change and the Media*. New York: Peter Lang.

Brown, Nik, and Mike Michael. 2003. Sociology of Expectations: Retrospecting Prospects and Prospecting Retrospects. *Technology Analysis and Strategic Management* 15 (1): 3–18.

Burchell, Graham, Colin Gordon, and Peter Miller, eds. 1991. *The Foucault Effect: Studies in Governmentality*. Hemel Hempstead, UK: Harvester Wheatsheaf.

Callon, Michel, Pierre Lascoumes, and Yannick Barthe. 2009. *Acting in an Uncertain World: An Essay on Technical Democracy*. Cambridge, MA: MIT Press.

Cheema, G. Shabbir, and Dennis A. Rondinelli, eds. 2007. *Decentralizing Governance: Emerging Concepts and Practices*. Washington, DC: Brookings Institution Press.

Clark, William C., Jill Jäger, and Josee van Eijndhoven. 2001. Managing Global Environmental Change: An Introduction to the Volume. In *A Comparative History of the Social Responses to Climate Change, Ozone Depletion, and Acid Rain*, vol. 1 of *Learning to Manage Global Environmental Risks*, by the Social Learning Group, 3–20. Cambridge, MA: MIT Press.

Collins, Harry M., and Robert Evans. 2002. The Third Wave of Science Studies: Studies of Expertise and Experience. *Social Studies of Science* 32 (2): 235–296.

Collins, Harry, and Robert Evans. 2007. *Rethinking Expertise*. Chicago: University of Chicago Press.

Danish, Kyle W. 2007. International Relations Theory. In *The Oxford Handbook of International Environmental Law*, ed. Daniel Bodanski, Jutta Brunnée, and Ellen Hey, 205–230. Oxford, UK: Oxford University Press.

Dean, Mitchell. 1999. *Governmentality*. London: Sage.

Douglas, Mary. 1987. *How Institutions Think*. London: Routledge & Kegan Paul.

EEA (European Environment Agency). 2006. *Air Quality and Ancillary Benefits of Climate Change Policies*. EEA Technical Report no. 4/2006. Copenhagen: EEA.

Elam, Mark, and Margareta Bertilsson. 2003. Consuming, Engaging, and Confronting Science: The Merging Dimensions of Scientific Citizenship. *European Journal of Social Theory* 6 (2): 233–251.

Elander, Ingemar. 2002. Partnership and Urban Governance. *International Social Science Journal* 54 (172): 191–204.

European Commission (Commission of the European Communities). 2001. *White Paper on Governance: Report of the Working Group "Democratising Expertise and Establishing Scientific Reference Systems."* Final version of 7 February 2001. Brussels: Commission of the European Communities, May. Available at http://ec.europa.eu/governance/areas/group2/report_en.pdf.

Fischer, Frank. 2003. *Reframing Public Policy: Discursive Politics and Deliberative Practices*. Oxford, UK: Oxford University Press.

Giddens, Anthony. 1985. *A Contemporary Critique of Historical Materialism*. Vol. 2 of *The Nation-State and Violence*. Cambridge, UK: Polity Press.

Gouldson, Andrew, Rolf Lidskog, and Misse Wester-Herber. 2007. The Battle for Hearts and Minds: Evolutions in Organisational Approaches to Environmental Risk Communication. *Environment and Planning* C 25 (1): 56–72.

Haas, Peter M. 1997. Scientific Communities and Multiple Paths to Environmental Management. In *Saving the Seas: Values, Scientists, and International Governance*, ed. L. Anathea Brooks and Stacy D. VanDeveer, 193–228.College Park: Maryland Sea Grant College.

Haas, Peter M. 2007. Epistemic Communities. In *The Oxford Handbook of International Environmental Law*, ed. Daniel Bodanski, Jutta Brunnée, and Ellen Hey, 791–806. Oxford, UK: Oxford University Press.

Hardin, Garrett. 1968. The Tragedy of the Commons. *Science* 162 (859): 1243–1248.

Hilgartner, Stephen. 2000. *Science on Stage: Expert Advice as Public Drama*. Stanford, CA: Stanford University Press.

Imber, Mark F. 1996. The Environment and the United Nations. In *The Environment and International Relations*, ed. John Vogler and Mark F. Imber, 138–154. London: Routledge.

Irwin, Alan. 1995. *Citizen Science: A Study of People, Expertise, and Sustainable Development*. London: Routledge.

Irwin, Alan. 2001. *Sociology and the Environment: A Critical Introduction to Society, Nature, and Knowledge*. Cambridge, UK: Polity Press.

Irwin, Alan. 2006. The Politics of Talk: Coming to Terms with the "New" Scientific Governance. *Social Studies of Science* 36 (2): 299–320.

Irwin, Alan, and Mike Michael. 2003. *Science, Social Theory, and Public Knowledge*. Maidenhead, UK: Open University Press.

Irwin, Alan, and Brian Wynne, eds. 1996. *Misunderstanding Science? The Public Reconstruction of Science and Technology*. Cambridge, UK: Cambridge University Press.

Jasanoff, Sheila. 2003a. Breaking the Waves in Science Studies. *Social Studies of Science* 33 (3): 389–400.

Jasanoff, Sheila. 2003b. Technologies of Humilities: Citizen Participation in Governing Science. *Minerva* 41: 223–244.

Jasanoff, Sheila. 2004a. The Idiom of Co-Production. In *States of Knowledge: The Co-Production of Science and Social Order*, ed. Sheila Jasanoff, 1–12. London: Routledge.

Jasanoff, Sheila. 2004b. Ordering Knowledge, Ordering Society. In *States of Knowledge: The Co-Production of Science and Social Order*, ed. Sheila Jasanoff, 13–45. London: Routledge.

Jasanoff, Sheila. 2005. *Designs of Nature: Science and Democracy in Europe and the United States*. Princeton, NJ: Princeton University Press.

Jasanoff, Sheila. 2010. A New Climate for Society. *Theory, Culture & Society* 27 (2–3): 233–253.

Kasperson, Jeanne X., and Roger E. Kasperson. 2005. *Publics, Risk Communication, and the Social Amplification of Risk*. Vol. 1. London: Earthscan.

Kitzinger, Jenny. 2008. Questioning Hype, Rescuing Hope? The Hwang Stem Cell Scandal and the Reassertion of Hopeful Horizons. *Science as Culture* 17 (4): 417–434.

Kooiman, Jan. 1993. *Modern Governance: New Government–Society Interactions*. London: Sage.

Latour, Bruno. 1987. *Science in Action: How to Follow Scientists and Engineers through Society*. Cambridge, MA: Harvard University Press.

Latour, Bruno. 1993. *We Have Never Been Modern*. Cambridge, MA: Harvard University Press.

Latour, Bruno. 1998. From the World of Science to the World of Research? *Science* 280 (5361): 208–209.

Leach, Melissa, Ian Scoones, and Brian Wynne, eds. 2005. *Science and Citizens: Globalization and the Challenge of Engagement*. London: Zed Books.

Lezaun, Javier, and Linda Soneryd. 2007. Consulting Citizens: Technologies of Elicitation and the Mobility of Publics. *Public Understanding of Science* (Bristol, England) 16 (3): 279–297.

Lidskog, Rolf. 2008. Scientised Citizens and Democratised Science: Re-Assessing the Expert–Lay Divide. *Journal of Risk Research* 11 (1–2): 69–86.

Lidskog, Rolf, and Ingemar Elander. 2007. Representation, Participation, or Deliberation? Democratic Responses to the Environmental Challenge. *Space and Polity* 11 (1): 75–94.

Lidskog, Rolf, Linda Soneryd, and Ylva Uggla. 2009. *Transboundary Risk Governance*. London: Earthscan.

Lidskog, Rolf, and Göran Sundqvist. 2004. From Consensus to Credibility: New Challenges for Policy-Relevant Science. *Innovation* 17 (3): 205–226.

Lidskog, Rolf, and Göran Sundqvist. 2007. Regulating European Air: The Co-Production of Science and Policy. In *Transboundary Air Pollution: Scientific Understanding and Environmental Policy in Europe*, ed. Håkan Pleijel, 177–202. Lund, Sweden: Studentlitteratur.

Löfstedt, Ragnar E. 2005. *Risk Management in Post-Trust Societies*. New York: Palgrave Macmillan.

Marres, Noortje. 2007. The Issues Deserve More Credit: Pragmatist Contributions to the Study of Public Involvement in Controversy. *Social Studies of Science* 37 (5): 759–778.

Mearsheimer, John J. 1994–1995. The False Promise of International Institutions. *International Security* 19 (3): 5–49.

Mitchell, Ronald B. 2008. Evaluating the Performance of Environmental Institutions: What to Evaluate and How to Evaluate It? In *Institutions and Environmental Change: Principal Findings, Applications, and Research Frontiers*, ed. Oran R. Young, Leslie A. King, and Heike Schroeder, 79–114. Cambridge, MA: MIT Press.

Moldan, Filip. 2007. From Fish Death to Dynamic Modelling: Tracking Trends in Acidification. In *Transboundary Air Pollution: Scientific Understanding and Environmental Policy in Europe*, ed. Håkan Pleijel, 81–101. Lund, Sweden: Studentlitteratur.

Nowotny, Helga, Peter Scott, and Michael Gibbons. 2001. *Re-Thinking Science: Knowledge and the Public in an Age of Uncertainty*. Cambridge, UK: Polity Press.

Ockwell, David, Lorraine Whitmarsh, and Saffron O'Neill. 2009. Reorienting Climate Change Communication for Effective Mitigation: Forcing People to Be Green or Fostering Grass-Roots Engagement? *Science Communication* 30 (3): 305–327.

Okereke, Chukwumerije, Harriet Bulkeley, and Heike Schroeder. 2009. Conceptualizing Climate Governance beyond the International Regime. *Global Environmental Politics* 9 (1): 58–78.

Olson, Mancur. 1965. *The Logic of Collective Action: Public Goods and the Theory of Groups*. Cambridge, MA: Harvard University Press.

Pleijel, Håkan, ed. 2007. *Transboundary Air Pollution: Scientific Understanding and Environmental Policy in Europe*. Lund, Sweden: Studentlitteratur.

Pleijel, Håkan, ed. 2009. *Air Pollution and Climate Change: Two Sides of the Same Coin?* Stockholm: Swedish Environmental Protection Agency.

Pleijel, Håkan, and Peringe Grennfelt. 2007. Air Quality Globalisation: Prospects for the Future. In *Transboundary Air Pollution: Scientific Understanding and Environmental Policy in Europe*, ed. Håkan Pleijel, 203–218. Lund, Sweden: Studentlitteratur.

Power, Michael. 2007. *Organized Uncertainty: Designing a World of Risk Management.* Oxford, UK: Oxford University Press.

Ricci, Miriam, Paul Bellaby, and Robert Flynn. 2010. Engaging the Public on Paths to Sustainable Energy: Who Has to Trust Whom? *Energy Policy* 38 (6): 2633–2640.

Rittberger, Volker, ed. 1993. *Regime Theory and International Relations.* Oxford, UK: Clarendon Press.

Rose, Nikolas. 1999. *Powers of Freedom: Reframing Political Thought.* Cambridge, UK: Cambridge University Press.

Rosenau, James N. 1998. Governance and Democracy in a Globalizing World. In *Cosmopolitan Democracy: An Agenda for a New World Order*, ed. Daniele Archibugi and David Held, 28–57. Cambridge, UK: Polity Press.

Schroeder, Heike, Leslie A. King, and Simon Tay. 2008. Contributing to the Science–Policy Interface: Policy Relevance of Findings on the Institutional Dimensions of Global Environmental Change. In *Institutions and Environmental Change: Principal Findings, Applications, and Research Frontiers*, ed. Oran R. Young, Leslie A. King, and Heike Schroeder, 261–275. Cambridge, MA: MIT Press.

Shapin, Steven. 1995. Cordelia's Love: Credibility and the Social Studies of Science. *Perspectives on Science* 3 (3): 255–275.

Social Learning Group. 2001. *A Comparative History of the Social Responses to Climate Change, Ozone Depletion, and Acid Rain.* Vol. 1 of *Learning to Manage Global Environmental Risks.* Cambridge, MA: MIT Press.

Somerville, Richard C. J. 2008. *The Forgiving Air: Understanding Environmental Change.* Boston: American Meteorological Society.

Stern, Susan, and Elisabeth Seligmann, eds. 2004. *The Partnership Principle: New Forms of Governance in the 21st Century.* Berlin: Alfred Herrhausen Society.

Sundqvist, Göran. 2003. Recovery in the Acid Rain Story: Transparency and Credibility in Science-Based Environmental Regulation. *Journal of Environmental Policy and Planning* 5 (1): 57–79.

Taylor, Charles. 1989. The Self in Moral Space. In *Sources of the Self. The Making of the Modern Identity*, 25–52. Cambridge: Cambridge University Press.

Underdal, Arild. 2008. Determining the Causal Significance of Institutions: Accomplishments and Challenges. In *Institutions and Environmental Change: Principal Findings, Applications, and Research Frontiers*, ed. Oran R. Young, Leslie A. King, and Heike Schroeder, 49–78. Cambridge, MA: MIT Press.

UNECE (United Nations Economic Commission for Europe). 1998. *Convention on Access to Information, Public Participation in Decisionmaking, and Access to Justice in Environmental Matters.* Signed in Aarhus, June 1998. Geneva: UNECE.

Urry, John. 2000. *Sociology Beyond Societies. Mobilities for the Twenty-First Century.* London: Routledge.

Winter, Gerd, ed. 2006. *Multilevel Governance of Global Environmental Change: Perspectives from Science, Sociology, and the Law*. Cambridge, UK: Cambridge University Press.

Wynne, Brian. 1992. Uncertainty and Environmental Learning: Reconceiving Science and Policy in the Preventive Paradigm. *Global Environmental Change* 2 (2): 111–127.

Wynne, Brian. 2001. Creating Public Alienation: Expert Cultures of Risk and Ethics on GMOs. *Science as Culture* 10: 445–481.

Wynne, Brian. 2003. Seasick on the Third Wave? Subverting the Hegemony of Propositionalism. Response to Collins and Evans (2002). *Social Studies of Science* 33 (3): 401–417.

Wynne, Brian. 2005. Risk as Globalizing "Democratic" Discourse? Framing Subjects and Citizens. In *Science and Citizens: Globalization and the Challenge of Engagement*, ed. Melissa Leach, Ian Scoones, and Brian Wynne, 66–82. London: Zed Books.

Wynne, Brian. 2010. Strange Weather, Again. Climate Science as Political Art. *Theory, Culture & Society* 27 (2–3): 289–305.

Wynne, Brian, and Ulrike Felt. 2007. *Taking European Knowledge Society Seriously*. Expert Group on Science and Governance, EUR 22700. Brussels: European Commission DG Research; Science, Economy, and Society Directorate.

Yearley, Steven. 2005. *Making Sense of Science: Understanding the Social Study of Science*. London: Sage.

Young, Oran R. 2002. *The Institutional Dimensions of Environmental Change: Fit, Interplay, and Scale*. Cambridge, MA: MIT Press.

Young, Oran R. 2008. Institutions and Environmental Change: The Scientific Legacy of a Decade of IDGEC Research. In *Institutions and Environmental Change: Principal Findings, Applications, and Research Frontiers*, ed. Oran R. Young, Leslie A. King, and Heike Schroeder, 3–45. Cambridge, MA: MIT Press.

Young, Oran R., Leslie A. King, and Heike Schroeder, eds. 2008. *Institutions and Environmental Change: Principal Findings, Applications, and Research Frontiers*. Cambridge, MA: MIT Press.

Young, Oran R., and Marc A. Levy. 1999. The Effectiveness of Environmental Regimes. In *The Effectiveness of International Environmental Regimes: Causal Connections and Behavioral Mechanisms*, ed. Oran R. Young, 1–32. Cambridge, MA: MIT Press.

Index

Politics, Science, and the Environment

Peter M. Haas and Sheila Jasanoff, editors

Peter Dauvergne, *Shadows in the Forest: Japan and the Politics of Timber in Southeast Asia*

Peter Cebon, Urs Dahinden, Huw Davies, Dieter M. Imboden, and Carlo C. Jaeger, eds., *Views from the Alps: Regional Perspectives on Climate Change*

Clark C. Gibson, Margaret A. McKean, and Elinor Ostrom, eds., *People and Forests: Communities, Institutions, and Governance*

The Social Learning Group, *Learning to Manage Global Environmental Risks.* Volume 1: *A Comparative History of Social Responses to Climate Change, Ozone Depletion, and Acid Rain.* Volume 2: *A Functional Analysis of Social Responses to Climate Change, Ozone Depletion, and Acid Rain*

Clark Miller and Paul N. Edwards, eds., *Changing the Atmosphere: Expert Knowledge and Environmental Governance*

Craig W. Thomas, *Bureaucratic Landscapes: Interagency Cooperation and the Preservation of Biodiversity*

Nives Dolsak and Elinor Ostrom, eds., *The Commons in the New Millennium: Challenges and Adaptation*

Kenneth E. Wilkening, *Acid Rain Science and Politics in Japan: A History of Knowledge and Action toward Sustainability*

Virginia M. Walsh, *Global Institutions and Social Knowledge: Generating Research at the Scripps Institution and the Inter-American Tropical Tuna Commission, 1900s–1990s*

Sheila Jasanoff and Marybeth Long Martello, eds., *Earthly Politics: Local and Global in Environmental Governance*

Christopher Ansell and David Vogel, eds., *What's the Beef? The Contested Governance of European Food Safety*

Charlotte Epstein, *The Power of Words in International Relations: Birth of an Anti-Whaling Discourse*

Ann Campbell Keller, *Science in Environmental Politics: The Politics of Objective Advice*

Henrik Selin, *Global Governance of Hazardous Chemicals: Challenges of Multilevel Management*

Rolf Lidskog and Göran Sundqvist, eds., *Governing the Air: The Dynamics of Science, Policy, and Citizen Interaction*